AF752555

Current Topics in Microbiology and Immunology

Volume 427

Series Editors

Rafi Ahmed
School of Medicine, Rollins Research Center, Emory University, Atlanta, GA, USA

Shizuo Akira
Immunology Frontier Research Center, Osaka University, Suita, Osaka, Japan

Klaus Aktories
Faculty of Medicine, Institute of Experimental and Clinical Pharmacology and Toxicology, University of Freiburg, Freiburg, Baden-Württemberg, Germany

Arturo Casadevall
W. Harry Feinstone Department of Molecular Microbiology & Immunology, Johns Hopkins Bloomberg School of Public Health, Baltimore, MD, USA

Richard W. Compans
Department of Microbiology and Immunology, Emory University, Atlanta, GA, USA

Jorge E. Galan
Boyer Ctr. for Molecular Medicine, School of Medicine, Yale University, New Haven, CT, USA

Adolfo Garcia-Sastre
Department of Microbiology, Icahn School of Medicine at Mount Sinai, New York, NY, USA

Bernard Malissen
Parc Scientifique de Luminy, Centre d'Immunologie de Marseille-Luminy, Marseille, France

Rino Rappuoli
GSK Vaccines, Siena, Italy

The review series *Current Topics in Microbiology and Immunology* provides a synthesis of the latest research findings in the areas of molecular immunology, bacteriology and virology. Each timely volume contains a wealth of information on the featured subject. This review series is designed to provide access to up-to-date, often previously unpublished information.

2019 Impact Factor: 3.095., 5-Year Impact Factor: 3.895

2019 Eigenfaktor Score: 0.00081, Article Influence Score: 1.363

2019 Cite Score: 6.0, SNIP: 1.023, h5-Index: 43

More information about this series at http://www.springer.com/series/82

Samuel Wagner · Jorge E. Galan
Editors

Bacterial Type III Protein Secretion Systems

Responsible Series Editor: Jorge E. Galan

Springer

Editors
Samuel Wagner
Institute of Microbiology
and Infection Medicine
University of Tübingen
Tübingen, Germany

Jorge E. Galan
Boyer Center for Molecular Medicine
Yale University School of Medicine
New Haven, CT, USA

ISSN 0070-217X ISSN 2196-9965 (electronic)
Current Topics in Microbiology and Immunology
ISBN 978-3-030-52122-6 ISBN 978-3-030-52123-3 (eBook)
https://doi.org/10.1007/978-3-030-52123-3

© Springer Nature Switzerland AG 2020
This work is subject to copyright. All rights are reserved by the Publisher, whether the whole or part of the material is concerned, specifically the rights of translation, reprinting, reuse of illustrations, recitation, broadcasting, reproduction on microfilms or in any other physical way, and transmission or information storage and retrieval, electronic adaptation, computer software, or by similar or dissimilar methodology now known or hereafter developed.
The use of general descriptive names, registered names, trademarks, service marks, etc. in this publication does not imply, even in the absence of a specific statement, that such names are exempt from the relevant protective laws and regulations and therefore free for general use.
The publisher, the authors and the editors are safe to assume that the advice and information in this book are believed to be true and accurate at the date of publication. Neither the publisher nor the authors or the editors give a warranty, expressed or implied, with respect to the material contained herein or for any errors or omissions that may have been made. The publisher remains neutral with regard to jurisdictional claims in published maps and institutional affiliations.

This Springer imprint is published by the registered company Springer Nature Switzerland AG
The registered company address is: Gewerbestrasse 11, 6330 Cham, Switzerland

Preface

Many bacteria that engage in pathogenic or symbiotic relationship with eukaryotes utilize protein secretion machines to deliver bacterially encoded effector proteins into their target cells. A prominent example among these protein injection machines is the so-called type III secretion system (T3SS). These are evolutionary related to bacterial flagella and, consequently, they share several structural and functional features.

Early discoveries on flagellar and T3SSs were largely made utilizing genetic approaches. Much of the knowledge gained in these early years, however, continues to provide a treasure of information for current investigations. Subsequently, biochemistry and structural biology started to reveal molecular details of individual components of these systems and provided a coarse picture of the entire macromolecular complex. In more recent years, technological progress, in particular, in the areas of super-resolution fluorescence microscopy, single particle cryo-electron microscopy, as well as cryo-electron tomography, coupled to an increased ability to biochemically handle large membrane-embedded protein complexes has substantially pushed the boundaries of our knowledge of T3SS. This volume comprehensively reviews our current understanding of T3SS regulation, assembly, structure, and function, and it provides an overview of the evolution of type III secreted effector proteins.

This book starts with a chapter on the unification of the nomenclature of type T3SSs. The independent discovery of these systems in various bacterial pathogens and symbionts has resulted in a confusing diversity of names for highly conserved proteins, which has made it difficult to seamlessly translate discoveries in related systems. The unification of the T3SS nomenclature as originally proposed by Hueck in 1998 was intensively discussed at the international T3SS meeting held in Tübingen in 2016 and a consensus was achieved for a broad framework to unify the nomenclature. The first chapter of this book provides an overview of the initial discoveries, introduces the extented unified nomenclature for T3SS, and suggests rules for its future use. The second chapter reviews the transcriptional and post-transcriptional regulatory mechanisms that control the expression of virulence-associated type III secretion systems in various pathogens. Chapters three through seven present and discuss our current understanding of the assembly and structure of the flagellar and T3SS injection machines, and relate their structures to secretion functions. The actual export mechanisms and the energy transduction in type III secretion machines are reviewed in chapter eight. Chapter nine focuses on the mechanism of hook and needle length control as well as on the mechanism of substrate switching to reprogram the secretion machines from the export of early to later substrates. Chapter ten reviews the information on the needle tip complex of the injectisomes and the process of host cell sensing and translocon formation at the host target cell membrane. Finally, in the last chapter, the diversity and evolution of type III secreted effector proteins is presented by reviewing three different families of effector proteins.

It should become obvious to the reader that although T3SSs are arguably the most studied bacterial secretion machines, much remains to be discovered and many aspects are still poorly understood. It is our hope that the comprehensive overview provided by this book sparks new ideas to uncover the remaining mysteries hidden by these fascinating protein secretion machines.

Last but not least, we want to thank our colleagues, who have contributed the different chapters and made this book possible.

Tübingen, Germany — Samuel Wagner
New Haven, USA — Jorge E. Galan

Contents

A Unified Nomenclature for Injectisome-Type Type III Secretion Systems

Samuel Wagner and Andreas Diepold

Contents

Abstract The independent naming of components of injectisome-type type III secretion systems in different bacterial species has resulted in considerable confusion, impeding accessibility of the literature and hindering communication between scientists of the same field. A unified nomenclature had been proposed by Hueck more than 20 years ago. It found little attention for many years, but usage was sparked again by recent reviews and an international type III secretion meeting in 2016. Here, we propose that the field consistently switches to an extended version of this nomenclature to be no longer lost in translation.

S. Wagner
Section of Cellular and Molecular Microbiology, Eberhard Karls University Tübingen, Interfaculty Institute of Microbiology and Infection Medicine, Elfriede-Aulhorn-Str. 6, 72076 Tübingen, Germany
e-mail: samuel.wagner@med.uni-tuebingen.de

Excellence Cluster "Controlling Microbes to Fight Infections" (CMFI), Elfriede-Aulhorn-Str. 6, 72076 Tübingen, Germany

German Center for Infection Research, Partner-Site Tübingen, Elfriede-Aulhorn-Str. 6, 72076 Tübingen, Germany

A. Diepold (✉)
Department of Ecophysiology, Max Planck Institute for Terrestrial Microbiology, Karl-von-Frisch-Str. 10, 35043 Marburg, Germany
e-mail: andreas.diepold@mpi-marburg.mpg.de

Current Topics in Microbiology and Immunology (2020) 427: 1–10
https://doi.org/10.1007/82_2020_210
© Springer Nature Switzerland AG 2020
Published Online: 16 May 2020

The discovery of injectisome-type type III secretion systems (T3SS) was a gradual process. The original observation of a growth defect of *Yersinia* under low Ca^{2+} conditions was linked to virulence and secretion of effector proteins into the supernatant and host cells over a period of more than thirty years (Kupferberg and Higuchi 1958; Higuchi et al. 1959; Gemski et al. 1980; Zink et al. 1980; Ben-Gurion and Shafferman 1981; Straley and Brubaker 1981; Heesemann et al. 1984; Goguen et al. 1984; Cornelis et al. 1989). In the following years, T3SS [first named as such in 1993 (Salmond and Reeves 1993)] were found and characterized in many, mainly pathogenic species.[1] The proteins associated with the system were named as they were discovered or sequenced, according to species-specific conventions. In *Yersinia*, genes previously identified for their role in the low-calcium response were named "Lcr", whereas genes sequenced afterward were named "Ysc" A-U (for *Yersinia* secretion) according to their synteny. Yet more multifariously, *Salmonella* pathogenicity island 1 (*S*PI-1) components were named "Inv" (for invasion) or Spa (for surface presentation of antigens), or kept their previous names "Prg" (PhoP repressed gene) or "Org" (for oxygen repressed gene). In *Shigella*, some T3SS components were named for their molecular weight ("Spa", e.g., Spa33, for surface presentation of Ipa[2]), and some based on synteny ("Mxi" for membrane expression of Ipa), with letter identifiers that largely differ from the *Yersinia* and/or *S*PI-1 homologs. In plant pathogens, the proteins of the two main subfamilies were called "Hrp" (hypersensitive plant response) or "Hrc" (for Hrp conserved), with partially, but not fully matching letter identifiers between the two families. T3SS that were identified later, such as the ones in pathogenic *Escherichia coli*, received unique protein names (e.g., "Esc" for *E. coli* secretion components), but were given the identifier of homologous proteins, often the *Yersinia* homologs.

In contrast to other fields where similar problems with different protein names were overcome relatively early, such as for the flagellum (Iino et al. 1988), the incompatible nomenclature persisted for the T3SS and perceptibly hampered (and still hampers) mutual understanding and exchange between researchers working on often highly conserved T3SS components in different bacteria. As an example, few of even the most widely read researchers would know offhand that YscL, MxiN, OrgB, SsaK, and HrpE denote the same protein (the ATPase regulator in *Yersinia*, *Shigella*, *Salmonella* *S*PI-1, *S*PI-2, and *Pseudomonas syringae*, respectively).

Unifying the nomenclature of T3SS components is therefore an investment that both current and future researchers in the field, as well as related research areas (for whom the current nomenclature is even less understandable), would greatly benefit from. During the Type III Secretion Meeting in Tübingen in April 2016, overwhelming support for such a unification of nomenclature across species was expressed by the participating researchers (Fig. 1).

---

[1]It also became clear that core components of the newly discovered secretion systems are homologous to the bacterial flagellum (Galán et al. 1992), which harbors a T3SS for the export of its distal components. In this note, "T3SS" will refer to the non-flagellar injectisome type III secretion systems (see (Desvaux et al, 2006) for a more thorough discussion).

[2]Which itself stands for invasion plasmid antigens.

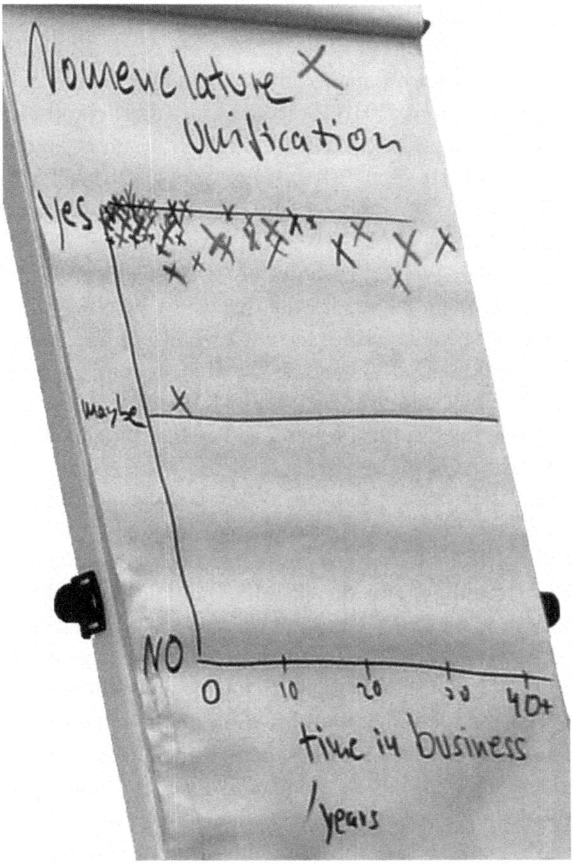

Fig. 1 Whiteboard showing the votes concerning the unification of the T3SS nomenclature. The voting was done during the T3SS Meeting 2016 in Tübingen, Germany

In fact, a first attempt for the unification of the T3SS nomenclature had already been brought up much earlier, in a review by Christoph Hueck from 1998, who named the T3SS proteins Sct (for secretion and cellular translocation), with identifiers that largely followed the *Yersinia* convention (Hueck 1998). This proposal was almost ignored for more than a decade; however in the last years, the Sct nomenclature was increasingly used, first in reviews studying the evolution and phylogeny of the T3SS (Gophna et al. 2003; Pallen and Gophna 2007; Abby and Rocha 2012; Gazi et al. 2012; Hu et al. 2017), then also in general reviews on the T3SS (Diepold and Wagner 2014; Abrusci et al. 2014; Galán et al. 2014; Tampakaki 2014; Diepold and Armitage 2015; Portaliou et al. 2016; Notti et al. 2016; Gold and Kudryashev 2016; Deng et al. 2017; Nazir et al. 2017; Erhardt 2017; Cascales 2017; Wagner et al. 2018; Dey et al. 2019; Diepold 2019; Lara-Tejero 2019; Lyons and Strynadka 2019; Renault et al. 2019; Miletic et al.

2019; Senthilkumar et al. 2019; Singh and Wagner 2019; Lara-Tejero and Galán 2019; Habenstein et al. 2019; Gitsels et al. 2020; dos Santos et al. 2020), and in original research papers, which include (Diepold et al. 2015; Zilkenat et al. 2016; McDowell et al. 2016; Soto et al. 2017; Diepold et al. 2017; Morgan et al. 2017; Kuhlen et al. 2018; Yu et al. 2018; Morgan et al. 2018; Butan et al. 2019; Bernal et al. 2019; Hu et al. 2019; Hausner et al. 2019; Tseytin et al. 2019; Drehkopf et al. 2020) or exclusively apply the common nomenclature (Rocha et al. 2018; Torres-Vargas et al. 2019; Milne-Davies et al. 2019; Wimmi et al. 2019; Johnson et al. 2019; Westerhausen et al. 2020; Kuhlen et al. 2020; Lindner et al. 2020).

As an alternative, it was discussed whether the flagellar names or identifiers could not be adapted for the eleven conserved components (e.g., FliP or SctP for the FliP homolog SctR) to stress the close relation of both secretion systems. However, this would require a parallel, completely new set of identifiers for the injectisome-specific proteins and is thus not likely to be broadly accepted by the injectisome field.

For this reason, and given the increasing utilization of the Sct nomenclature in the recent literature, we propose to fully adopt the Sct nomenclature from now on. Table 1 lists the Sct protein names, as well as the species-specific names for major T3SS families, and the flagellar homologs. In this table, we have included four protein names that were not part of the original Sct nomenclature (Hueck 1998): SctA, SctB, and SctE for the translocators, as proposed by Portaliou and colleagues (Portaliou et al. 2016), and SctG for the pilotin proteins.

For the identification of T3SS components in a specific species, we propose to denote the species in full at the first mentioning (e.g., *Shigella flexneri* SctD) and to add species initials in front of the Sct protein name (e.g., *Sf*SctD) when characteristics of a specific T3SS of one species are discussed. In case of more than one T3SS per species, subscript numbers should be used to distinguish the system: *St*$SctD_1$ for *Salmonella* typhimurium *S*PI-1 and *St*$SctD_2$ for *S*PI-2. For all proteins that are not widely conserved across different T3SS (such as YscX/Y in *Yersinia*, OrgC in *S*PI-1, EspA in EPEC), as well as the T3SS chaperones and effectors, we suggest to continue using the original name.

To facilitate the use of the unified Sct nomenclature, as well as providing a resource for identifying proteins in already published reports, we are in the process of adding the common protein names to databases such as UniProt, NCBI Protein, KEGG, BioCyc. We also set up a T3SS wiki to gather and organize all available information on T3SS in one common source (t3sswiki.science).

We invite the community to use the unified Sct nomenclature in their future research communication and hope that this will lead to a better understanding across species, the emergence and rigorous scrutiny of general ideas and concepts, and easier access for researchers from other fields, in order to advance research on the type III secretion system.

Table 1 Translation table for T3SS components

Functional name	Flagellar homolog		Ysc	Inv-Mxi-Spa		Ssa-Esc		Hrp-Hrc 1	Hrp-Hrc 2
		Sct name	*Yersinia*	*Shigella*	*Salmonella SPI-1*	*Salmonella SPI-2*	*Escherichia coli*	*Pseudomonas syringae*	*Ralstonia solanacearum*
Secretin	–	SctC	YscC	MxiD	InvG	SsaC	EscC	HrcC	HrcC
Outer–inner membrane ring protein	–	SctD	YscD	MxiG	PrgH	SsaD	EscD	HrpQ	HrpW
Inner–inner membrane ring protein	FliF	SctJ	YscJ	MxiJ	PrgK	SsaJ	EscJ	HrcJ	HrcJ
Core export apparatus protein R	FliP	SctR	YscR	Spa24 (SpaP)	SpaP	SsaR	EscR	HrcR	HrcR
Core export apparatus protein S	FliQ	SctS	YscS	Spa9 (SpaQ)	SpaQ	SsaS	EscS	HrcS	HrcS
Core export apparatus protein T	FliR	SctT	YscT	Spa29 (SpaR)	SpaR	SsaT	EscT	HrcT	HrcT
Core export apparatus protein U, switch protein	FlhB	SctU	YscU	Spa40 (SpaS)	SpaS	SsaU	EscU	HrcU	HrcU
Major export apparatus protein	FlhA	SctV	YscV	MxiA	InvA	SsaV	EscV	HrcV	HrcV
Base-pod connector	–	SctK	YscK	MxiK	OrgA	SsaX	EscK	HrpD	–
Major pod protein	FliM + FliN	SctQ	YscQ	Spa33 (SpaO)	SpaO	SsaQ	EscQ	$HrcQ_{A+B}$	HrcQ
Stator	FliH	SctL	YscL	MxiN	OrgB	SsaK	EscL	HrpE	HrpF
ATPase	FliI	SctN	YscN	Spa47 (SpaL)	InvC	SsaN	EscN	HrcN	HrcN
Stalk	FliJ	SctO	YscO	Spa13 (SpaM)	InvI	SsaO	EscO	HrpO	HrpD
Needle filament protein	–	SctF	YscF	MxiH	PrgI	SsaG	EscF	HrpA	HrpY
Inner rod protein/needle adapter	FliE	SctI	YscI	MxiI	PrgJ	SsaI	EscI	HrpB	HrpJ

(continued)

Table 1 (continued)

Functional name	Flagellar homolog		Ysc	Inv-Mxi-Spa		Ssa-Esc		Hrp-Hrc 1	Hrp-Hrc 2
		Sct name	*Yersinia*	*Shigella*	*Salmonella* SPI-1	*Salmonella* SPI-2	*Escherichia coli*	*Pseudomonas syringae*	*Ralstonia solanacearum*
Needle length regulator	FliK	SctP	YscP	Spa32 (SpaN)	InvJ	SsaP	EscP	HrpP	HpaP
Hydrophilic translocator, needle tip protein	–	SctA	LcrV	IpaD	SipD	SseB	–	–	–
Hydrophobic translocator, pore protein	–	SctE	YopB	IpaB	SipB	SseC	EspD	HrpK	PopF1, PopF2
Hydrophobic translocator, pore protein	–	SctB	YopD	IpaC	SipC	SseD	EspB	–	–
Pilotin	–	SctG	YscW	MxiM	InvH	–	–	–	–
Gatekeeper	–	SctW	YopN	MxiC	InvE	SsaL	SepL	HrpJ	HpaA

References

Abby SS, Rocha EPC (2012) The non-flagellar type III secretion system evolved from the bacterial flagellum and diversified into host-cell adapted systems. PLoS Genet 8:e1002983

Abrusci P, McDowell MA, Lea SM, Johnson S (2014) Building a secreting nanomachine: a structural overview of the T3SS. Curr Opin Struct Biol 25C:111–117

Ben-Gurion R, Shafferman A (1981) Essential virulence determinants of different Yersinia species are carried on a common plasmid. Plasmid 5:183–187

Bernal I, Börnicke J, Heidemann J, Svergun D, Horstmann JA, Erhardt M, Tuukkanen A, Uetrecht C, Kolbe M (2019) Molecular organization of soluble type III secretion system sorting platform complexes. J Mol Biol 431:3787–3803

Butan C, Lara-Tejero M, Li W, Liu J, Galán JE (2019) High-resolution view of the type III secretion export apparatus in situ reveals membrane remodeling and a secretion pathway. Proc Natl Acad Sci 116:24786–24795

Cascales E (2017) Inside the chamber of secrets of the type III secretion system. Cell 168:949–951

Cornelis GR, Biot T, Lambert de Rouvroit C, Michiels T, Mulder B, Sluiters C, Sory M-P, Bouchaute M, Vanooteghem J, De Rouvroit CL (1989) The Yersinia yop regulon. Mol Microbiol 3:1455–1459

Deng W, Marshall NC, Rowland JL, McCoy JM, Worrall LJ, Santos AS, Strynadka NC, Finlay BB (2017) Assembly, structure, function and regulation of type III secretion systems. Nat Rev Microbiol 15:323–337

Desvaux M, Hébraud M, Henderson IR, Pallen MJ (2006) Type III secretion: what's in a name? Trends Microbiol 14:157–160

Dey S, Chakravarty A, Guha Biswas P, De Guzman RN (2019) The type III secretion system needle, tip, and translocon. Protein Sci 28:1582–1593

Diepold A (2019) Assembly and post-assembly turnover and dynamics in the type III secretion system. In: Current topics in microbiology and immunology

Diepold A, Armitage JP (2015) Type III secretion systems: the bacterial flagellum and the injectisome. Philos Trans R Soc B Biol Sci 370:20150020

Diepold A, Kudryashev M, Delalez NJ, Berry RM, Armitage JP (2015) Composition, formation, and regulation of the cytosolic C-ring, a dynamic component of the type III secretion injectisome. PLoS Biol 13:e1002039

Diepold A, Sezgin E, Huseyin M, Mortimer T, Eggeling C, Armitage JP (2017) A dynamic and adaptive network of cytosolic interactions governs protein export by the T3SS injectisome. Nat Commun 8:15940

Diepold A, Wagner S (2014) Assembly of the bacterial type III secretion machinery. FEMS Microbiol Rev 38:802–822

dos Santos AMP, Ferrari RG, Conte-Junior CA (2020) Type three secretion system in *Salmonella Typhimurium*: the key to infection. Genes Genomics

Drehkopf S, Otten C, Hausner J, Seifert T, Büttner D (2020) HrpB7 from Xanthomonas campestris pv. vesicatoria is an essential component of the type III secretion system and shares features of HrpO/FliJ/YscO family members. Cell Microbiol

Erhardt M (2017) Energy requirements for protein secretion via the flagellar type III secretion system. In: Methods in molecular biology, pp 449–457

Galán JE, Ginocchio C, Costeas P (1992) Molecular and functional characterization of the Salmonella invasion gene invA: homology of InvA to members of a new protein family. J Bacteriol 174:4338–4349

Galán JE, Lara-Tejero M, Marlovits TC, Wagner S (2014) Bacterial type III secretion systems: specialized nanomachines for protein delivery into target cells. Annu Rev Microbiol 68:415–438

Gazi AD, Sarris PF, Fadouloglou VE, Charova SN, Mathioudakis N, Panopoulos NJ, Kokkinidis M (2012) Phylogenetic analysis of a gene cluster encoding an additional, rhizobial-like type III secretion system that is narrowly distributed among Pseudomonas syringae strains. BMC Microbiol 12:188

Gemski P, Lazere JR, Casey T (1980) Plasmid associated with pathogenicity and calcium dependency of *Yersinia enterocolitica*. Infect Immun 27:682–685

Gitsels A, Van Lent S, Sanders N, Vanrompay D (2020) Chlamydia: what is on the outside does matter. Crit Rev Microbiol 0:1–20

Goguen JD, Yother J, Straley S (1984) Genetic analysis of the low calcium response in *Yersinia pestis* mu d1(Ap lac) insertion mutants. J Bacteriol 160:842–848

Gold V, Kudryashev M (2016) Recent progress in structure and dynamics of dual-membrane-spanning bacterial nanomachines. Curr Opin Struct Biol 39:1–7

Gophna U, Ron E, Graur D (2003) Bacterial type III secretion systems are ancient and evolved by multiple horizontal-transfer events. Gene 312:151–163

Habenstein B et al (2019) Structures of type III secretion system needle filaments. In: Current topics in microbiology and immunology. https://doi.org/10.1007/82_2019_192

Hausner J, Jordan M, Otten C, Marillonnet S, Büttner D (2019) Modular cloning of the type III secretion gene cluster from the plant-pathogenic bacterium *Xanthomonas euvesicatoria*. ACS Synth Biol 8:532–547

Heesemann J, Algermissen B, Laufs R (1984) Genetically manipulated virulence of *Yersinia enterocolitica*. Infect Immun 46:105–110

Higuchi K, Kupferberg LL, Smith JL (1959) Studies on the nutrition and physiology of *Pasteurella pestis*. III. Effects of calcium ions on the growth of virulent and avirulent strains of *Pasteurella pestis*. J Bacteriol 77:317–321

Hu J, Worrall LJ, Vuckovic M, Hong C, Deng W, Atkinson CE, Brett Finlay B, Yu Z, Strynadka NCJ (2019) T3S injectisome needle complex structures in four distinct states reveal the basis of membrane coupling and assembly. Nat Microbiol 4:2010–2019

Hu Y, Huang H, Cheng X, Shu X, White AP, Stavrinides J, Köster W, Zhu G, Zhao Z, Wang Y (2017) A global survey of bacterial type III secretion systems and their effectors. Environ Microbiol 19:3879–3895

Hueck CJ (1998) Type III protein secretion systems in bacterial pathogens of animals and plants. Microbiol Mol Biol Rev 62:379–433

Iino T, Komeda Y, Kutsukake K, Macnab RM, Matsumura P, Parkinson JS, Simon MI, Yamaguchi S (1988) New unified nomenclature for the flagellar genes of Escherichia coli and Salmonella typhimurium. Microbiol Rev 52:533–535

Johnson S et al (2019) The structure of an injectisome export gate demonstrates conservation of architecture in the core export gate between flagellar and virulence type III secretion systems. mBio 10(3). https://doi.org/10.1128/mBio.00818-19

Kuhlen L, Abrusci P, Johnson S, Gault J, Deme J, Caesar J, Dietsche T, Mebrhatu MT, Ganief T, Macek B, Wagner S, Robinson CV, Lea SM (2018) Structure of the core of the type III secretion system export apparatus. Nat Struct Mol Biol 25:583–590

Kuhlen L, Johnson S, Zeitler A, Bäurle S, Deme JC, Caesar JJE, Debo R, Fisher J, Wagner S, Lea SM (2020) The substrate specificity switch FlhB assembles onto the export gate to regulate type three secretion. Nat Commun 11:1296

Kupferberg LL, Higuchi K (1958) Role of calcium ions in the stimulation of growth of virulent strains of Pasteurella pestis. J Bacteriol 76:120–121

Lara-Tejero M (2019) The type III secretion system sorting platform. In: Current topics in microbiology and immunology. https://doi.org/10.1007/82_2019_167

Lara-Tejero M, Galán JE (2019) The injectisome, a complex nanomachine for protein injection into mammalian cells. EcoSal Plus 8(2). https://doi.org/10.1128/ecosalplus.ESP-0039-2018

Lindner F, Milne-Davies B, Langenfeld K, Stiewe T, Diepold A (2020) LITESEC-T3SS—Light-controlled protein delivery into eukaryotic cells with high spatial and temporal resolution. Nat Commun (in press) (bioRxiv 807461)

Lyons BJE, Strynadka NC (2019) On the road to structure-based development of anti-virulence therapeutics targeting the type III secretion system injectisome. Medchemcomm 10:1273–1289

McDowell MA, Marcoux J, McVicker G, Johnson S, Fong YH, Stevens R, Bowman LAH, Degiacomi MT, Yan J, Wise A, Friede ME, Benesch JLP, Deane JE, Tang CM, Robinson CV, Lea SM (2016) Characterisation of Shigella Spa33 and Thermotoga FliM/N reveals a new model for C-ring assembly in T3SS. Mol Microbiol 99:749–766

Miletic S, Goessweiner-Mohr N, Marlovits TC (2019) The structure of the type III secretion system needle complex. In: Current topics in microbiology and immunology. https://doi.org/10.1007/82_2019_178

Milne-Davies B, Helbig C, Wimmi S, Cheng DWC, Paczia N, Diepold A (2019) Life after secretion—*Yersinia enterocolitica* rapidly toggles effector secretion and can resume cell division in response to changing external conditions. Front Microbiol 10:2128

Morgan JM, Duncan MC, Johnson KS, Diepold A, Lam H, Dupzyk AJ, Martin LR, Wong WR, Armitage JP, Linington RG, Auerbuch V (2017) Piericidin A1 blocks Yersinia Ysc type III secretion system needle assembly. mSphere 2:e00030-17

Morgan JM, Lam HN, Delgado J, Luu J, Mohammadi S, Isberg RR, Wang H, Auerbuch V (2018) An experimental pipeline for initial characterization of bacterial type III secretion system inhibitor mode of action using enteropathogenic *Yersinia*. Front Cell Infect Microbiol 8:404

Nazir R, Mazurier S, Yang P, Lemanceau P, van Elsas JD (2017) The ecological role of type three secretion systems in the interaction of bacteria with fungi in soil and related habitats is diverse and context-dependent. Front Microbiol 8:38

Notti RQ, Stebbins CE, Erec Stebbins C, Stebbins CE (2016) The structure and function of type III secretion systems. Microbiol Spectr 4:1–30

Pallen MJ, Gophna U (2007) Bacterial flagella and type III secretion: case studies in the evolution of complexity. Genome Dyn 3:30–47

Portaliou AG, Tsolis KC, Loos MS, Zorzini V, Economou A (2016) Type III secretion: building and operating a remarkable nanomachine. Trends Biochem Sci 41:175–189

Renault TT, Guse A, Erhardt M (2019) Export mechanisms and energy transduction in type-III secretion machines. Current topics in microbiology and immunology. Springer, Berlin, Heidelberg, pp 1–17

Rocha JM, Richardson CJ, Zhang M, Darch CM, Cai E, Diepold A, Gahlmann A (2018) Single-molecule tracking in live *Yersinia enterocolitica* reveals distinct cytosolic complexes of injectisome subunits. Integr Biol 10:502–515

Salmond GPC, Reeves PJ (1993) Membrance traffic wardens and protein secretion in Gram-negative bacteria. Trends Biochem Sci 18:7–12

Senthilkumar M, Swarnalakshmi K, Annapurna K (2019) Diversity in type III secreting systems (T3SSs) in legume-rhizobium symbiosis. In: Microbial diversity in ecosystem sustainability and biotechnological applications. Springer Singapore, Singapore, pp 83–107

Singh N, Wagner S (2019) Investigating the assembly of the bacterial type III secretion system injectisome by in vivo photocrosslinking. Int J Med Microbiol 151331

Soto E, Espinosa N, Díaz-Guerrero M, Gaytán MO, Puente JL, González-Pedrajo B (2017) Functional characterization of EscK (Orf4), a sorting platform component of the enteropathogenic *Escherichia coli* injectisome. J Bacteriol 199:JB.00538-16

Straley SC, Brubaker RR (1981) Cytoplasmic and membrane proteins of Yersiniae cultivated under conditions simulating mammalian intracellular environment. Proc Natl Acad Sci USA 78:1224–1228

Tampakaki AP (2014) Commonalities and differences of T3SSs in rhizobia and plant pathogenic bacteria. Front Plant Sci 5:114

Torres-Vargas CE, Kronenberger T, Roos N, Dietsche T, Poso A, Wagner S (2019) The inner rod of virulence-associated type III secretion systems constitutes a needle adapter of one helical turn that is deeply integrated into the system's export apparatus. Mol Microbiol 112:918–931

Tseytin I, Mitrovic B, David N, Langenfeld K, Zarivach R, Diepold A, Sal-Man N (2019) The role of the small export apparatus protein, SctS, in the activity of the type III secretion system. Front Microbiol 10

Wagner S, Grin I, Malmsheimer S, Singh N, Torres-Vargas CE, Westerhausen S (2018) Bacterial type III secretion systems: a complex device for the delivery of bacterial effector proteins into eukaryotic host cells. FEMS Microbiol Lett 365

Westerhausen S, Nowak M, Torres-Vargas CE, Bilitewski U, Bohn E, Grin I, Wagner S (2020) A NanoLuc luciferase-based assay enabling the real-time analysis of protein secretion and injection by bacterial type III secretion systems. Mol Microbiol 745471

Wimmi S, Balinovic A, Jeckel H, Selinger L, Lampaki D, Eisemann E, Meuskens I, Linke D, Drescher K, Endesfelder U, Diepold A (2019) Dynamic relocalization of the cytosolic type III secretion system components ensures specific protein secretion. bioRxiv 869214

Yu X-J, Grabe GJ, Liu M, Mota LJ, Holden DW (2018) SsaV interacts with SsaL to control the translocon-to-effector switch in the Salmonella SPI-2 type three secretion system. MBio 9: e01149-18

Zilkenat S, Franz-Wachtel M, Stierhof Y-D, Galán JE, Macek B, Wagner S (2016) Determination of the stoichiometry of the complete bacterial type III secretion needle complex using a combined quantitative proteomic approach. Mol Cell Proteomics M115.056598

Zink D, Feeley J, Wells J, Vanderzant C, Vickery J, Roof W, O'Donovan G (1980) Plasmid-mediated tissue invasiveness in *Yersinia enterocolitica*. Nature 283:224–226

Transcriptional and Post-transcriptional Regulatory Mechanisms Controlling Type III Secretion

Marcel Volk, Ines Vollmer, Ann Kathrin Heroven and Petra Dersch

Contents

Abstract Type III secretion systems (T3SSs) are utilized by numerous Gram-negative bacteria to efficiently interact with host cells and manipulate their function. Appropriate expression of type III secretion genes is achieved through the

---

Marcel Volk and Ines Vollmer contributed equally

M. Volk · I. Vollmer · A. K. Heroven · P. Dersch (✉)
Department of Molecular Infection Biology, Helmholtz Centre for Infection Research, Brunswick, Germany
e-mail: petra.dersch@uni-muenster.de

P. Dersch
Institute for Infectiology, University Münster, Münster, Germany

Current Topics in Microbiology and Immunology (2020) 427: 11–33
https://doi.org/10.1007/82_2019_168
© Springer Nature Switzerland AG 2019
Published Online: 20 June 2019

integration of multiple control elements and regulatory pathways that ultimately coordinate the activity of a central transcriptional activator usually belonging to the AraC/XylS family. Although several regulatory elements are conserved between different species and families, each pathogen uses a unique set of control factors and mechanisms to adjust and optimize T3SS gene expression to the need and lifestyle of the pathogen. This is reflected by the complex set of sensory systems and diverse transcriptional, post-transcriptional and post-translational control strategies modulating T3SS expression in response to environmental and intrinsic cues. Whereas some pathways regulate solely the T3SS, others coordinately control expression of one or multiple T3SSs together with other virulence factors and fitness traits on a global scale. Over the past years, several common regulatory themes emerged, e.g., environmental control by two-component systems and carbon metabolism regulators or coupling of T3SS induction with host cell contact/translocon-effector secretion. One of the remaining challenges is to resolve the understudied post-transcriptional regulation of T3SS and the dynamics of the control process.

1 Introduction

Type III secretion systems (T3SS) are complex nanomachines of Gram-negative bacterial pathogens of plants, animals and humans that export proteins (called effectors) from the bacterial cytoplasm across the cell envelope and inject them into host cells during infection (Büttner 2012; Portaliou et al. 2016; Galan et al. 2014; Deng et al. 2017; Hueck 1998). Depending on the individual pathogen, the injected effectors act as toxins, adhesins or enzymes and can promote attachment and invasion into host cells, intracellular survival and replication or prevent host defenses. The T3SS is composed of 30–40 proteins that are assembled into a needle-like injectisome structure where they are incorporated as single molecules or in few or multiple copies (Galan et al. 2014; Deng et al. 2017; Diepold and Wagner 2014). Consequently, the synthesis of the T3SS proteins that form or utilize the machinery follows a strict hierarchy (Galan et al. 2014; Diepold and Wagner 2014). Moreover, synthesis, recruitment and unfolding as well as secretion and delivery of the effector proteins into host cells are highly energy consuming processes. This is particular evident by the fact that several pathogens stop growth and replication when the T3SS is induced (Büttner 2012; Portaliou et al. 2016; Deng et al. 2017). Thus, it is not surprising that the production of the T3SS in pathogens is tightly controlled by a plethora of regulatory factors in response to numerous environmental cues and the secretory activity of the injectisome. They act on the transcriptional, post-transcriptional and post-translational level and form a complex feedback-controlled network. In the following overview, we focus on the wealth of T3SS control factors of important human pathogens and discuss current challenges to examine their functions.

2 T3SS Regulation on the Transcriptional Level

2.1 Control by Central Transcriptional Activators

The first central factors identified to regulate the expression of the T3SS components were dimeric, **transcriptional activators of the AraC/XylS family**. This family of activators includes LcrF/VirF in human pathogenic yersiniae, HilD, HilC and RtsA in *Salmonella enterica*, MxiE in *Shigella flexneri* and ExsA in *Pseudomonas aeruginosa* (Francis et al. 2002). A common regulatory principle of these key activators involves the coordinate control of all or of a large subset of components and substrates of the T3SS in the respective pathogen. In several cases, multiple binding sites have been identified upstream the target genes and operons (Schwiesow et al. 2015).

In general, all transcriptional activators of this class are composed of an N-terminal sensing domain for environmental signals and a C-terminal helix-turn-helix (HTH) DNA-binding domain per subunit. DNA binding occurs through the recognition helix in the HTH that binds to specific DNA residues within the major groove (Bustos and Schleif 1993; Schleif 2010). Interestingly, some of these master activators, such as LcrF/VirF of *Yersinia* and ExsA of *P. aeruginosa*, are highly homologous and can complement each other. They were shown to interact with common nucleotide sequence motifs that are often highly variable regarding the distance to the transcriptional start sites, directionality and sequence conservation (Schwiesow et al. 2015; King et al. 2013). Notably, despite the resemblance of the DNA motifs, the oligomeric state of the LcrF/VirF (dimer) and ExsA (monomer) proteins is different when they are recruited to the binding sites. This ultimately results in distinct binding affinities, promoter bending and different kinetics of transcriptional activation (Schwiesow et al. 2015; King et al. 2013). Some of the N-terminal domains were further shown to bind co-factors that induce conformational changes and influence the ability of the protein to regulate transcription. One recent study for example showed that the crucial regulator HilD of *S. enterica* binds long-chain fatty acids such as oleate, which prevents binding of the activator to its target sites (Golubeva et al. 2016) (for more details, see Sect. 4.2).

Additionally, many important bacterial pathogens, including *S. enterica*, possess multiple T3SSs in which the key regulators of the systems are also implicated in the control of the master regulator of the other T3SSs. For instance, HilD coordinates the expression of two other type III secretion machineries of *Salmonella*, the flagellar T3SS and the *Salmonella* pathogenicity island 2 (SPI-2) encoded injectisome required for host defenses (Ellermeier and Slauch 2007) (Fig. 1).

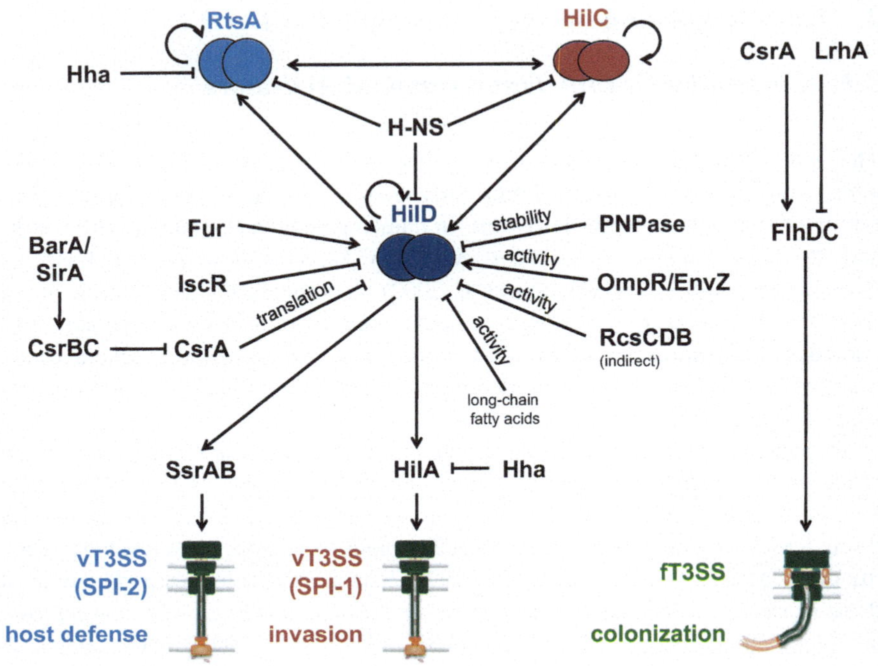

Fig. 1 Regulation of the T3SSs of *Salmonella enteric* serovar Typhimurium. The regulatory network of the flagellar and the two virulence-associated T3SSs (T3SS-1 and T3SS-2) encoded on SPI-1 and SPI-2 is shown. The overview illustrates sensory and regulatory factors, which control the T3SSs on the transcriptional and post-transcriptional level

2.2 Transcriptional Control of the Key T3SS Regulators

Appropriate expression of T3SS components is generally achieved through the integration of several regulatory factors and pathways in the form of a highly structured network that ultimately control the activity of the central AraC/XylS-type transcription activator. This enables the bacteria to tightly control and adjust the production of the T3SS in response to unique environmental cues encountered in the respective host niches. In *S. enterica,* control of the T3SS-1 (encoded on the *Salmonella* pathogenicity island 1 (SPI-1)) is mediated by the AraC/XylS proteins HilC, HilD and RtsA (Erhardt and Dersch 2015) (Fig. 1). They regulate each other and the transcription of their own gene and independently activate the promoter of the OmpR/ToxR family activator HilA. HilA primarily induces transcription of the structural components of the SPI-1 gene cluster, including *prg/org* and *inv/spa* (Ellermeier et al. 2005). Most environmental signals that regulate SPI-1 gene expression are sensed by **two-component systems** and integrated on the level of HilD. The response to (i) osmolarity, desiccation and low temperature occurs through the two-component systems OmpR/EnvZ and the RcsCD/RcsB, and

(ii) presence of certain metabolites is sensed through the BarA/SirA two-component system (Erhardt and Dersch 2015). Hence, HilD constitutes the key activator of *Salmonella* T3SS genes, whereas HilC and RtsA rather act as amplifiers of the activating signal. Control of the key regulator LcrF/VirF of the *Yersinia* T3SS, which is encoded on the *Yersinia* virulence plasmid (pYV), is similarly complex (Fig. 2). Its expression is tightly controlled by nutrients, Mg^{2+} and pH, through the BarA/SirA, PhoP/PhoQ-CsrABC cascade and the cyclic AMP receptor protein (Crp). In addition, diverse cell stresses can be sensed through the two-component systems RcsCD/RcsB and CpxA/CpxR (Schwiesow et al. 2015).

Expression of the T3SS master activators is often further influenced by the action of **specific transcriptional regulators** that promote activator transcription only under certain environmental conditions. Examples of important environmental parameters in this context are the availability of oxygen and ferric/ferrous ions. In this context, Ellermeier and Slauch and Teixido et al. (Ellermeier and Slauch 2008; Teixido et al. 2011) found that the ferric uptake regulator Fur controls HilD expression and more recent studies identified the iron–sulfur cluster coordinating transcriptional regulator IscR as an important transcriptional repressor of the LcrF/VirF and HilD master regulators of *Yersinia* and *Salmonella* (Miller et al. 2014; Vergnes et al. 2017) (Figs. 1 and 2). It is assumed that iron limitation, oxidative

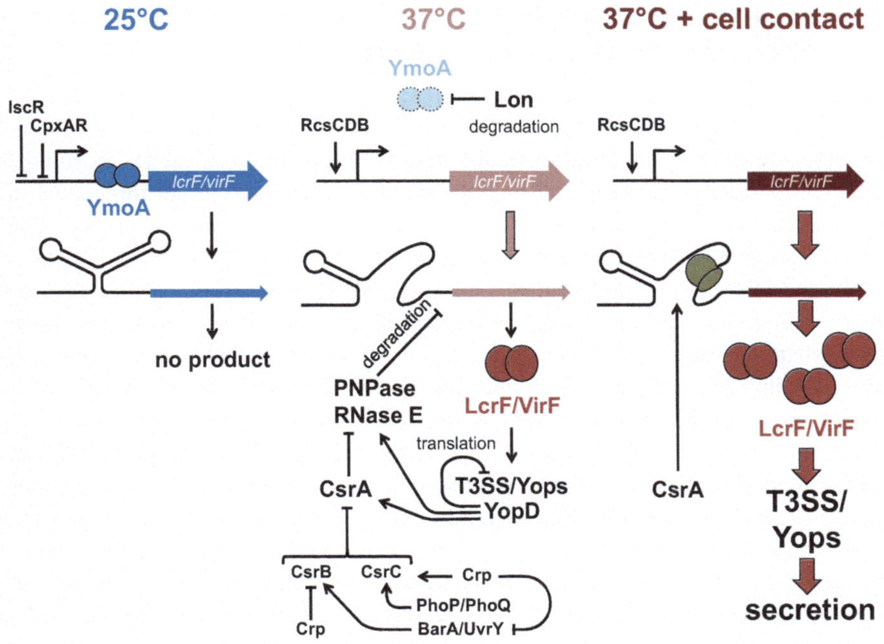

Fig. 2 **Regulation of the Ysc/Yop T3SS of *Yersinia*.** The regulatory network controlling the virulence plasmid-encoded Ysc/Yop T3SS in *Y. pseudotuberculosis* is illustrated. The most important global and *Yersinia*-specific RNA and protein regulators influencing T3SS gene expression at 25 °C (environment), 37 °C (host entry) and 37 °C upon host cell contact are indicated

stress, as well as oxygen limitation as a result of Fe–S cluster damage, affect the activity of IscR. Another common global regulatory protein is the LysR homologue LrhA. While this regulator represses flagella T3SS expression in *Escherichia coli* and *Salmonella*, it, on the contrary, induces the *espA* gene of the pathogenicity island *locus of enterocyte effacement* (LEE) of enterohemorrhagic *E. coli* (EHEC) and the *rtsB* regulator gene of SPI-1 genes in *Salmonella* (Erhardt and Dersch 2015; Shimizu et al. 2015).

Another common feature is that expression of the master T3SS regulators is also governed by global regulators. They are needed for metabolic adaptation and can additionally coordinate expression of the T3SSs with available carbon sources. Among these regulators is Crp. Crp primarily helps the pathogens to manage and optimize their metabolism by checking and ranking uptake and utilization of available and readily digestible carbon sources (Görke and Stülke 2008; Poncet et al. 2009). In this context, Crp is also important to link the nutrient status and carbon metabolism with the regulation of the T3SS either directly or via the control of the post-transcriptional carbon storage regulator system (Csr), which affects the expression of multiple master regulators in many pathogens (for details see Sect. 3.1.2) (Kusmierek and Dersch 2017; Vakulskas et al. 2015).

2.3 Silencing and Activation of T3SS Master Regulator Expression by Modulator Proteins

Under non-inducing conditions, e.g., outside the host, T3SS genes are often subjected to silencing by ancestral **nucleoid-structuring proteins** of the H-NS and the Hha/YmoA family. These global modulators of gene expression are implicated in the xenogeneic repression of many virulence genes with a low GC-content that were acquired by horizontal gene transfer (Navarre et al. 2006; Dorman 2007). H-NS and HhA/YmoA preferentially bind to AT-rich bent promoter regions, and their binding sites often overlap with binding sites of positive transcriptional activators that are able to alleviate transcriptional repression. This silencing and anti-silencing mechanism was found for H-NS and Hha for all three AraC/XylS-type transcriptional activators RtsA, HilC and HilD which control T3SS genes in *S. enterica* (Olekhnovich and Kadner 2007) (Fig. 1). Similarly, Ler (LEE-encoded regulator), a master regulator of the LEE operons of enteropathogenic *E. coli* (EPEC) and EHEC, is able to relieve H-NS mediated silencing of the LEE5 promoter (Laaberki et al. 2006) (Fig. 3). The LEE-encoded T3SS promotes the establishment of intimate attachment structures (intimin-mediated pedestals) of EPEC/EHEC via translocated effectors leading to attaching and effacing (A/E) lesions and severe damage of the intestinal villi (Katsowich et al. 2017; Bhatt et al. 2009). This is enabled by the significantly higher binding affinity (about 40-fold) of the T3SS master activator Ler to the LEE target DNA sequences, in comparison to H-NS (Choi et al. 2016).

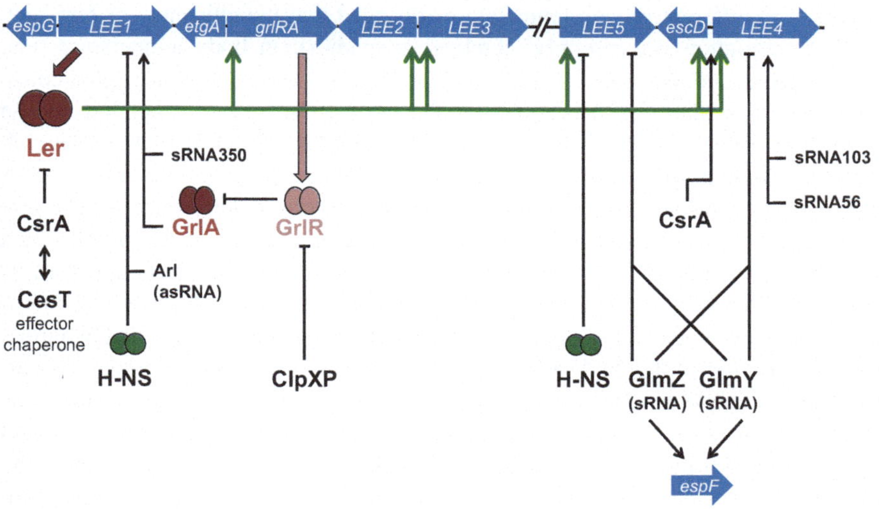

Fig. 3 Regulation of the LEE-encoded T3SS genes of EHEC. The locus of enterocyte effacement (LEE) pathogenicity island of EHEC, including the operons *LEE1-5*, the bicistronic operon *grlRA* and other monocistronic genes, is illustrated. The most important global and EHEC-specific RNA and protein regulators influencing LEE gene expression are indicated

Interestingly, members of the H-NS and Hha/YmoA family are also able to form heterodimers. How this interaction modulates expression of their target genes is not clear yet, as the influence of both family members on gene expression varies significantly between the different pathogens. However, some evidence exists that heterodimer formation influences the binding affinity to their target sites and links control of the affected genes to other environmental control systems (Madrid et al. 2002, 2007). One striking example is the formation of H-NS and YmoA complexes in human pathogenic yersiniae. YmoA is important to block transcription of the T3SS master regulator gene *lcrF/virF* in *Yersinia* to silence sequences downstream of the *lcrF/virF* promoter (Böhme et al. 2012) (Fig. 2). In contrast to its *E. coli* homologue, it is preferentially degraded by the Lon and Clp proteases at body temperature, leading to derepression of *lcrF/virF* transcription upon infection (Jackson et al. 2004).

3 T3SS Regulation on the Post-transcriptional Level

3.1 Control of Translation

3.1.1 Sensory RNAs—RNA Thermometers

One of the first post-transcriptional control mechanisms of T3SS gene expression was discovered in *Yersinia*. A comparison of the amount of the T3SS master regulator at different growth temperatures revealed that the efficiency of *lcrF/virF* mRNA translation was strongly increased at mammalian body temperature (i.e., 37 °C) (Hoe and Goguen 1993). Initial mRNA secondary structure predictions suggested that the *lcrF* ribosome binding site (RBS) is incorporated into a stem-loop. In a later study, the 5′-untranslated region (5′-UTR) of the *lcrF/virF* mRNA was mapped, and structure probing identified a unique *cis*-acting RNA element which forms a two stem-loop structure at moderate temperatures. The first stem-loop stabilizes the second stem-loop which sequesters the *lcrF/virF* RBS by a stretch of four uracils. Opening of this structure is favored at 37 °C and permits ribosome binding at host body temperature (Böhme et al. 2012; Rhigetti et al. 2016) (Fig. 2). The biological relevance of this **RNA thermometer** was verified in animal models with two different *Yersinia pseudotuberculosis* strains expressing a stabilized and a labile variant of the RNA thermometer. The stabilized variant was strongly reduced in its ability to disseminate into the Peyer's patches, liver and spleen and had fully lost its lethality due to the lack of *lcrF* expression, whereas the destabilized version of the thermosensor was attenuated or exhibited a similar, but not a higher mortality (Böhme et al. 2012). This illustrated that the evolved RNA thermometer provides just the appropriate amount of the T3SS master regulator at the appropriate condition and time for an optimal infection efficiency of *Yersinia*.

3.1.2 Translational Control by CsrA/RsmA RNA-Binding Proteins

CsrA/RsmA-type RNA-binding proteins belong to a multicomponent, post-transcriptional regulatory control system that adjusts expression of T3SS genes in several Gram-negative pathogens. A key function of CsrA/RsmA protein family members is the coordinate adaptation of virulence traits, metabolic functions and physiological properties to optimize virulence and fitness of the pathogen during the different stages of the infection (Kusmierek and Dersch 2017; Vakulskas et al. 2015). CsrA/RsmA proteins most commonly prevent the translation of their numerous target mRNAs, including multiple T3SS gene transcripts. They usually bind to GGA-containing sequences that are exposed in single-stranded loop regions of two or more stem-loop structures residing in the 5'-UTRs of the impacted mRNAs, of which one includes the RBS. Simultaneous interaction of the dimeric CsrA/RsmA proteins predominantly hinders ribosome access and translation, which often also reduces the stability of the target RNA (Mercante et al. 2009; Lapouge et al. 2013; Dubey et al. 2005). However, in several cases, CsrA/RsmA protein binding prevents the formation of translation-blocking RNA structures or hinders access of RNases leading to an increase of the total translational efficiency and transcript stability (Ren et al. 2014; Yakhnin et al. 2013). CsrA/RsmA protein activity can be immediately inhibited through sequestration of CsrA by certain non-coding RNAs (e.g., CsrB and CsrC in *Enterobacteriaceae*), which harbor multiple GGA-containing hairpins. Alternatively, CsrA can be inactivated by binding to specific highly abundant GGA-rich mRNAs as well as by interacting proteins (Kusmierek and Dersch 2017).

CsrA/RsmA deficiency impairs synthesis and function of several T3SSs implicated in motility, host colonization and host defense and strongly attenuates virulence. For instance, CsrA induces the expression of the master regulator FlhDC of the flagellar T3SS in *E. coli* (Yakhnin et al. 2013). CsrA activates *flhDC* expression by protecting the *flhDC* mRNA from decay. mRNA degradation studies in *E. coli* support a model in which CsrA binding activates *flhDC* expression by inhibiting the 5′ end-dependent RNase E cleavage pathway (Yakhnin et al. 2013). In addition, the synthesis of the LEE-encoded T3SS for the formation of the intimate attachment structures by EPEC/EHEC is regulated by CsrA on multiple levels (Fig. 3) (Katsowich et al. 2017; Bhatt et al. 2009). First, CsrA directly activates expression of the *escD* and the LEE4 genes and represses expression of several LEE operons through the regulator Ler and GrlA (Bhatt et al. 2009). Moreover, CsrA was found to interact with the T3SS effector chaperone CesT when it is released from its effector under effector-secreting conditions. Sequestration of CsrA by CesT relieves CsrA-mediated repression of *nleA/espI* effector mRNA translation, which enables the bacteria to adjust T3SS gene expression in response to host cell contact (Katsowich et al. 2017).

The CsrA equivalent RsmA protein of *P. aeruginosa* is also implicated in the bacterial response to host cell contact, but the underlying molecular mechanism is quite different. In this pathogen, RsmA controls the expression of the major T3SS transcriptional activator ExsA. The activity of ExsA is controlled by a complex 'partner-switching' cascade involving ExsE, ExsC and ExsD, which is triggered upon host cell contact and secretion of the ExsE effector (Fig. 4, for details see Sect. 4.1) (Intile et al. 2014; Dasgupta et al. 2004; Urbanowski et al. 2005). RsmA activity itself is regulated through the magnesium transporter MgtE and the GacAS two-component system via activation of the Csr/Rsm-type RNAs RsmY and RsmZ (Chakravarty et al. 2017).

CsrA has also a very strong impact on the expression of different virulence-relevant T3SSs in *Yersinia*. Similar to *E. coli*, CsrA activates the FlhDC master regulator of the flagellar T3SS and is thus required for *Yersinia* motility (Heroven et al. 2008). Recent work further demonstrates that CsrA has a major influence on the virulence plasmid-encoded T3SS (Ysc/Yop). Expression of the CsrB and CsrC RNAs in *Y. pseudotuberculosis* is downregulated in the Peyer's patches during acute infection (Nuss et al. 2017), suggesting that more active CsrA protein is required for virulence. In fact, a *csrA*-deficient mutant of *Yersinia enterocolitica* was characterized by reduced secretion of the T3SS effectors YopE and YopH, and a *Y. pseudotuberculosis csrA* mutant was unable to secrete all known Yop effectors, which are usually injected into neutrophils and macrophages to prevent their phagocytic attack (Nuss et al. 2017; Ozturk et al. 2017). In addition to the virulence plasmid-encoded T3SS, *Y. enterocolitica* harbors a second chromosomally encoded T3SS (Ysa). This T3SS is only expressed under very special environmental conditions in vitro and seems to promote intracellular virulence (Bent et al. 2015). This Ysa T3SS system is also influenced by CsrA, but in contrast to the Ysc/Yop T3SS effectors, Ysp effector proteins are over-secreted in the *csrA* mutant (Ozturk et al. 2017).

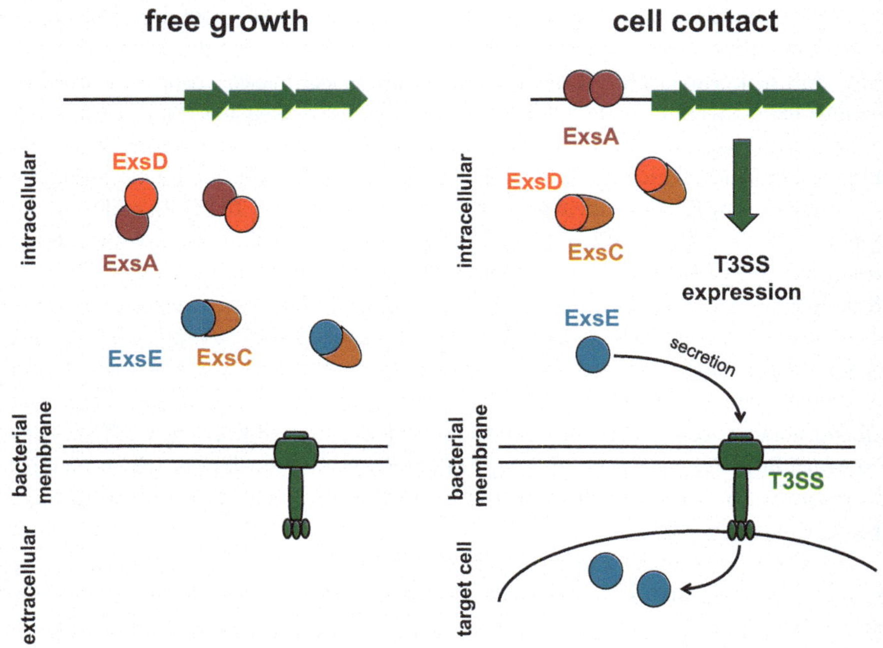

Fig. 4 Control network and partner switching regulating T3SS in *P. aeruginosa*. The regulatory network controlling the T3SS of *Pseudomonas* and the partner switching upon host cell contact is presented. The most important transcriptional and post-transcriptional regulators are illustrated, and the partner-switching mechanisms is presented when the bacteria change from free to the host cell bound stage

Similar to *Yersinia*, CsrA also controls the virulence-associated T3SS of *Salmonella typhimurium*: the cell invasion-promoting T3SS-1, encoded on SPI-1, and the T3SS-2 mediating *Salmonella* replication within macrophages (Fig. 1). CsrA regulates the expression of the common regulator HilD by binding near the ribosome binding site on the *hilD* transcript leading to a decrease of HilD translation and a reduced stability of the *hilD* mRNA (Altier et al. 2000; Martinez et al. 2011).

3.2 *Control of T3SS mRNA Structures and Stability—Influence of Helicases and RNases*

The overall efficiency of mRNA translation depends on the stability and the structure of the transcript. Bacteria generally have several RNA helicases that unfold intrinsic hairpin and other secondary RNA structures. Of special interest here is a certain class of RNA-binding helicases, the **DEAD-box helicases**. They are named after their conserved DEAD amino acid sequence in their catalytic domain and can hydrolyze ATP to dissolve inhibitory duplex RNA structures. One of a set of seven DEAD-box helicases, the DeaD helicase, stimulates translation of the T3SS master regulator ExsA in *P. aeruginosa* to promote expression of the injectisome (Intile et al. 2015). DeaD seems to directly stimulate *exsA* mRNA translation, as DeaD-dependent activation is specific to the *exsA* coding region and the native RBS. The purified protein is able to promote ExsA synthesis using in vitro translation assays (Intile et al. 2015). RNA secondary structure predictions of the 5′-UTR of *exsA* and the proximal coding sequence revealed extensive base pairing with the RBS, indicating that DeaD enhances ribosomal access. DeaD does not alter T3SS expression through RsmA (Intile et al. 2015), but whether DeaD and RsmA function independently or are dependent upon each other is currently unknown.

Both T3SS regulators and structural components are also targeted by different **RNases**. One important RNase implicated in T3SS regulation is the polynucleotide phosphorylase (PNPase). PNPase belongs to the group of 3′-exoribonucleases. It can act alone or it functions as part of the multicomponent degradosome complex together with RNase E, the glycolytic enzyme enolase and helicase RhlB (Mohanty and Kushner 2016). In *Yersinia*, optimal functioning of the virulence plasmid-encoded T3SS was shown to require PNPase. While a *pnp* deletion mutant of *Y. pseudotuberculosis* possessed enhanced levels of the T3SS-encoding transcripts and proteins in contrast to wildtype, secretion of the Yop effectors was strongly reduced (Rosenzweig et al. 2007). However, this is in contrast with initial results of our laboratory in which both T3SS gene transcription and Yop secretion was strongly increased in the absence of PNPase in *Y. pseudotuberculosis* (Kusmierek et al., unpublished). The reason for this discrepancy is unclear, and we are currently testing whether a different genetic background or differences in the growth conditions may be responsible for this discrepancy. It was further found by Rosenzweig et al. that normal T3SS activity was restored when a 70 amino acid peptide (S1 domain) containing one of the RNA-binding domains of PNPase was expressed in the *pnp* mutant. Notably, T3SS expression, but not T3SS activity, was identical in Δ*pnp* strains expressing active or inactive S1 variants, indicating that PNPase influence on T3SS may involve an RNA intermediate (Rosenzweig et al. 2005, 2007; Rosenzweig and Chopra 2013). Likewise, T3SS functioning is impaired in *Y. pseudotuberculosis* with a reduced RNase E activity, and these bacteria further resemble the *pnp* mutant, indicating that they act via a common pathway, e.g., via the degradosome (Yang et al. 2008). In *Yersinia*, the single-strand-specific RNase Y (YbeY) processing 3′-ends of the 16S RNA was also found to control T3SS expression (Leskinen et al. 2015). The precise mechanism is unknown, but the CsrB and CsrC RNAs are among its target, indicating that the influence of YbeY on T3SS occurs through regulation of the Csr system.

S. typhimurium strains harboring a deletion of the 310 nt 3′-untranslated region (3′-UTR) of *hilD* or lacking the RNase E gene *rne* or the PNPase gene *pnp* exhibit increased *hilD* mRNA levels. This resulted in SPI-1 gene overexpression, impaired *Salmonella* growth and uncontrolled invasion of epithelial cells, suggesting that the *hilD* 3′-UTR is a target for degradation by the bacterial degradosome (Fig. 1) (Lopez-Garrido et al. 2014). In contrast, the RNA chaperone Hfq interacts with the

3′-UTR of the *hilD* mRNA and has a positive influence on its stability. Therefore, it has been assumed that this effect occurs through competitive inhibition of the degradosome (Lopez-Garrido et al. 2014; Sittka et al. 2007; Chao et al. 2012).

The degradosome is also implicated in the regulation of T3SS gene expression in pathogenic *E. coli*. In EHEC, the *espADB* translocon genes are controlled by an RNase E-dependent mechanism, in which a small six-codon mini-open reading frame at the 5’-end of the common transcript is recognized and preferentially degraded by RNase E (Fig. 3). However, translation of the mini-open reading frame by ribosomes protects the mRNA and allows a more efficient production of the translocon proteins (Lodato et al. 2012, 2017).

3.3 Regulatory RNAs

Over the past years, a plethora of regulatory RNAs have been identified in many bacterial pathogens that are implicated in virulence regulation. Besides the Csr/Rsm-type RNAs (see also Sect. 3.1.2), several others of them have also been found to control T3SSs. For instance, the small non-coding RNA Spot42 regulates the expression of the chaperone protein VP1682 of components of the T3SS-1 of *Vibrio parahaemolyticus* by basepairing with the RBS and the initial codons of the *vp1682* transcript (Tanabe et al. 2015). Another example constitutes the *Yersinia*-specific non-coding RNA Ysr141. In the absence of Ysr141, a selection of the T3SS components and the effectors YpkA, YscF, YopE, YopK and YopJ and the regulator LcrF were 30–70% decreased (Schiano et al. 2014). Although the molecular mechanism how Ysr141 exerts this effect is largely unknown, initial experiments demonstrated that Ysr141 interferes with *yopJ* mRNA translation. Schiano et al. (Schiano et al. 2014) proposed that all observed changes could be based on the dysregulation of YopJ production as *Yersinia* T3SS was shown to be sensitive to changes in Yop protein levels. However, the observed regulatory influence on YopJ translation was rather small suggesting that other mechanisms contribute to the overall influence of Ysr141 on T3SS regulation. A recent study revealed that the copy number of the *Yersinia* virulence plasmid pYV encoding the Ysc/Yop T3SS increases during infection (Wang et al. 2016). This upregulation is caused by an increased expression of the pYV replicase RepA. It was shown that levels of an antisense RNA, CopA, overlapping with the upstream region of the *repA* gene are strongly decreased under secretion conditions and during infection, leading to a marked decrease of the CopA/*repA* mRNA ratio and an increase of the copy number (Nuss et al. 2017; Wang et al. 2016). The precise regulatory mechanism still needs to be elucidated, but first evidence exists that the secreted translocon protein YopD is involved in this process (Wang et al. 2016). Finally, the unique bacterial translational control system, including the small protein B (SmpB) and the regulatory RNA SsrA/tmRNA, is required for efficient expression of Yop effectors and the flagellar T3SS of *Yersinia* (Okan et al. 2006).

A study by Gruber and Sperandio (2015) identified multiple non-coding EHEC-specific RNAs of which sRNA56, sRNA103 and sRNA350 activate T3SS genes of the LEE pathogenicity island (Fig. 3). sRNA350, encoded in the 3′-UTR of the *cesF* transcript, activates all LEE operons via its influence on the master regulator gene *ler*, whereas sRNA56 and sRNA103 target only the LEE4-encoded *espA* mRNA. Besides these EHEC-specific non-coding RNAs, the well-characterized, paralogous non-coding RNAs GlmZ and GlmY also influence expression of EHEC T3SS genes (Fig. 3). Both RNAs destabilize the LEE4- and LEE5-encoded mRNAs and increase translation of the *espF* transcript (Gruber and Sperandio 2014). GlmZ interacts with the LEE4 mRNA and selectively destabilizes the downstream part of the transcript encoding the translocon *espADB* genes, but not the upstream part encoding *sepL*, which acts as regulatory switch by binding effectors until the T3SS is formed (Gruber and Sperandio 2014). In addition, GlmY also represses LEE4 on the post-transcriptional level. However, this occurs by binding to and sequestering the protein RapZ from GlmZ, which, when released from GlmZ, can destabilize the LEE4 transcript (Gruber and Sperandio 2015). The antagonistic control of LEE4/LEE5 and *espF* by both RNAs is not easy to interpret, as all targeted genes are required for pedestal and A/E lesion formation. However, as suggested by the authors, it is possible that GlmY/GlmZ limits overexpression of the LEE4/LEE5 T3SS components and synchronizes their production with the non-LEE *espF* gene to enable a precise ratio of the structure components and secreted effectors (Gruber and Sperandio 2014). Moreover, a *cis*-encoded antisense RNA (Arl) located downstream of the *ler* gene, covering the last codons and the 3′-UTR, controls T3SS gene expression from the LEE pathogenicity island (Tobe et al. 2014). Interaction of Arl with the *ler* transcript destabilizes the LEE1 mRNA and also hinders final elongation of the Ler protein synthesis.

3.4 Feedback Control Mechanisms Through Secreted Effectors or Anti-sigma Factors

T3SS gene expression in several pathogens is strongly induced upon host cell contact and effector secretion. This includes the T3SS systems of *Yersinia, Salmonella, Shigella* and *Pseudomonas* (Pettersson et al. 1996; Brutinel and Yahr 2008; Zierler and Galan 1995). In *Yersinia*, T3SS gene expression is coupled to secretion via the translocon protein YopD, the YopD chaperone LcrH and the effector LcrQ (YscM1/YscM2 in *Y. enterocolitica*) (Anderson et al. 2002; Anderson and Schneewind 1999). LcrQ and YopD are secreted upon host cell contact reducing their concentration in the bacterial cytoplasm. The role of LcrQ is still unclear, but it is assumed that it assists the YopD–LcrH complex (Cambronne and Schneewind 2002; Li et al. 2014). In addition, to its translocon function, YopD is able to interact with RNA (alone or in complex with the chaperone LcrH) by for example binding to the 5′-UTR of multiple *yop* transcripts and the *lcrF* mRNA

(Fig. 2) (Cambronne and Schneewind 2002; Chen and Anderson 2011). Absence of the YopD protein under secretion conditions strongly decreases their degradation and/or prevents their translation (Chen and Anderson 2011). The RNA-binding mode of the protein is not known, but AU-rich regions in the proximity of the RBS seem important for YopD-mediated repression of the targeted T3SS mRNAs. Another important information shedding more light on the molecular mechanism of cell contact-mediated induction of T3SS genes was obtained by an analysis of Kopaskie et al. (2013). They report that the YopD protein is able to interact with the 30S particles of the bacterial ribosome in an LcrH-dependent manner and could show that YopD in association with YscM1 and LcrH is able to repress YopQ translation. However, how YopD–LcrH interaction with the 30S ribosomal particle is mediated and how this influences the translation of other transcripts of the Ysc/Yop T3SS is still unclear. Yet, it is likely that transient interaction perturbs the formation of the 30S complex before the 50S ribosomal particle associate with the transcript, as no binding of YopD to the assembled 70S ribosome could be detected (Kopaskie et al. 2013).

Another sophisticated feedback control mechanism is promoted by secreted anti-sigma factors and was found for the flagellar T3SS and virulence-associated T3SSs, e.g., from *Bordetella* spp. (Chevance and Hughes 2017; Ahuja et al. 2016). The flagellar master regulator FlhDC activates *fliA*, encoding the alternative sigma factor σ^{28}. This sigma factor promotes expression of the motor force generator and the filament, and its function is inhibited when bound by an anti-σ^{28} factor (FlgM). Upon formation of the hook basal body, the anti-sigma factor is secreted and σ^{28} is released to activate σ^{28} target genes (Chevance and Hughes 2017). Similarly, the secreted antagonist BtrA of the sigma factor BtrS establishes a feedback loop that couples the activity of the T3SS with expression of the T3SS genes. BtrA differentially controls nearly 300 genes, including many T3SS genes, which define six distinct regulatory virulence modules in *Bordetella* (Ahuja et al. 2016).

4 T3SS Regulation on the Protein Level

4.1 Changing Binding Partners

Another mechanism promoting cell contact- and secretion-induced T3SS gene expression is 'partner-switching.' One important control step of this process is the interaction of a T3SS chaperone with its T3SS substrate or a cytoplasmic regulatory factor. Without host cell contact, the secreted T3SS effector remains bound to its chaperone that keeps the substrate in a transport-competent stage and inhibits the activating function of the chaperone (Schulmeyer and Yahr 2017). Upon host cell contact, the chaperone is released from the secreted binding partner (effector or translocon proteins) and interacts with transcriptional activators or suppressors of the transcriptional activators (anti-repressor) to induce T3SS gene transcription.

This changing of partner(s) can implicate multiple co-, anti- or anti-anti-activators that form a complex feedback control cascade linking T3SS gene expression with the secretory activity of the system (Dasgupta et al. 2004; Urbanowski et al. 2005). This partner-switching mechanism was characterized in detail in *P. aeruginosa* by the Yahr group and was shown to include four different proteins (Urbanowski et al. 2005; Schulmeyer and Yahr 2017; Vakulskas et al. 2009). In the absence of host cell contact, the effector protein ExsE interacts with chaperone ExsC. The master T3SS transcriptional activator ExsA is bound to ExsD. Upon cell contact, the released ExsC chaperone, which has a higher affinity to ExsD than ExsA, sequesters the ExsD protein, whereby the ExsA activator is liberated and can activate T3SS gene transcription (Brutinel et al. 2010; Zheng et al. 2007; McCaw et al. 2002; Rietsch et al. 2005; Dasgupta et al. 2006). Interestingly, *Vibrio parahaemolyticus* and *Vibrio cholerae* appear to have functional orthologues of their *Pseudomonas* counterparts to control the T3SS-1 (Zhou et al. 2010). A similar mechanism with a very different set of partners has been identified in *S. flexneri*. In this pathogen, the translocon proteins IpaB and IpaC are bound by the chaperone IpgC. The activator MxiE is inactivated through interaction with the anti-activator OspD1 and Spa15 protein, which acts as OspD1 chaperone under non-secretion conditions. IpaB, IpaC and OspD1 translocation into host cell allows released MxiE and IpgC to interact and activate T3SS gene expression (Parsot et al. 2005).

4.2 Modulation or Modification of T3SS Regulatory Components

Besides interaction with other regulatory proteins, activity of T3SS master regulators can be manipulated either by small compounds or through modification, such as phosphorylation. An exciting recent work demonstrated that long-chain fatty acids (e.g., oleate) prevent expression of the T3SS-1 genes in *S. enterica*. This inhibition is independent of the long-chain fatty acid degradation pathway and occurs solely through direct binding and inhibition of the DNA-binding activity of the T3SS activator HilD (Golubeva et al. 2016). Long-chain fatty acids are present in the intestinal tract and inhibit HilD-mediated T3SS gene expression until the bacteria reach the distal ileum of the small intestine. There, the metabolites are absorbed and the concentration falls below a critical threshold, which relieves the blockage in order to activate T3SS gene expression.

Another mechanism used to change the activity of T3SS regulators is through protein modification, i.e., tyrosine phosphorylation. However, only a few examples are known, and how the modification influences the activity of the modified proteins is often unclear. One example is *Shigella*, in which the second master regulator VirB (besides MxiE) is unable to stimulate T3SS gene expression when phosphorylated at residue 247 (Standish et al. 2016).

5 Conclusions and Future Perspectives

The long list of transcriptional, post-transcriptional and post-translational control mechanisms implicated in T3SS control in the different pathogens illustrates that T3SS regulation is highly complex. A large intervening network has been evolved that integrates various environmental signals, nutrient/ion availability and physiological conditions to control regulators, structural components and secreted effectors of the T3SS in a fine-tuned and concerted manner. This occurs through a plethora of sensory and regulatory factors, which can be RNA or proteins that modulate specifically or globally, and the players in the network can be modulated on different regulatory levels.

It is evident, that common players, e.g., conserved AraC/XylS regulators and transcriptional modulators, sensory and regulatory RNA elements, RNases, and common two-component systems are part of the control circuit. However, their interactions, targets and arrangements within the regulatory cascades and feedback cycles can vary significantly between the different pathogens. Moreover, even very small changes in the genetic information were found to provoke fundamental changes of T3SS control. This set of possible variations defines the different T3SS control variants, which adapt and optimize the expression of the secretion system to the distinct needs and lifestyles of the different pathogens.

Ongoing discovery of new regulatory factors and different T3SS control variants further illustrates that our understanding of T3SS regulation is still far from being complete. There is a dire need to explore the molecular mechanisms and the role of regulatory RNAs, controlled RNA degradation and translation changes under non-inducing and secretion conditions during the infection. However, this is challenging due to the complexity of the controlling signals and components. One major difficulty in the analysis of T3SS expression is the identification of appropriate in vitro growth conditions mimicking the different steps of T3SS production under infection conditions.

Another unsolved question is, how the detected molecular mechanisms of one representative strain extend to other strains or members of the family or species. In particular, post-transcriptional control elements such as sensory and regulatory RNAs appear more malleable to intrinsic changes and regulatory rewiring than transcriptional regulators. This promotes individual, strain-specific variations—a trait that is particularly advantageous for the bacteria to adapt to frequently changing host niches. Thus, it would be ill-advised to automatically extrapolate their role to others. For instance, it is known that GlmY/GlmZ promoted control of *espF* expression, which is essential for EHEC to form pedestals, is not encoded in the genome of closely related EPEC, and the DsrA RNA that activates transcription of *ler* in EHEC does not affect LEE in EPEC (Bhatt et al. 2016). Moreover, Hfq has a very strong influence on the Ysc/Yop T3SS in *Yersinia pestis*, but not in *Y. enterocolitica*, suggesting that T3SS regulation relies on different post-transcriptional mechanisms (Kakoschke et al. 2014, 2016; Schiano et al. 2010). Consequently, the function of these control elements must be experimentally investigated in additional members of the species and family.

An additional future challenge is to unravel the dynamics of the multicomponent and multilayered network of T3SS regulation during different phases of the process. This includes the transition from the repressed stage (e.g., 25 °C, outside host), to the preparing phase (host entry, 37 °C), and the different secretion phases upon host cell contact that allows the bacteria to carefully balance nutrient and energy use to maintain their biological fitness and competiveness.

References

Ahuja U, Shokeen B, Cheng N, Cho Y, Blum C, Coppola G et al (2016) Differential regulation of type III secretion and virulence genes in *Bordetella pertussis* and B*ordetella bronchiseptica* by a secreted anti-sigma factor. Proc Natl Acad Sci USA 113(9):2341–2348. https://doi.org/10.1073/pnas.1600320113. PubMed PMID: 26884180; PubMed Central PMCID: PMCPMC4780644

Altier C, Suyemoto M, Lawhon SD (2000) Regulation of *Salmonella enterica* serovar typhimurium invasion genes by *csrA*. Infect Immun 68(12):6790–6797. PubMed PMID: 11083797

Anderson DM, Schneewind O (1999) *Yersinia enterocolitica* type III secretion: an mRNA signal that couples translation and secretion of YopQ. Mol Microbiol 31(4):1139–1148. PubMed PMID: 10096081

Anderson DM, Ramamurthi KS, Tam C, Schneewind O (2002) YopD and LcrH regulate expression of *Yersinia enterocolitica* YopQ by a posttranscriptional mechanism and bind to *yopQ* RNA. J Bacteriol 184(5):1287–1295. PubMed PMID: 11844757; PubMed Central PMCID: PMC134855

Bent ZW, Poorey K, Brazel DM, LaBauve AE, Sinha A, Curtis DJ et al (2015) Transcriptomic analysis of *Yersinia enterocolitica* Biovar 1B infecting murine macrophages reveals new mechanisms of extracellular and intracellular survival. Infect Immun 83(7):2672–2685. https://doi.org/10.1128/iai.02922-14. PubMed PMID: 25895974; PubMed Central PMCID: PMCPMC4468540

Bhatt S, Edwards AN, Nguyen HT, Merlin D, Romeo T, Kalman D (2009) The RNA binding protein CsrA is a pleiotropic regulator of the locus of enterocyte effacement pathogenicity island of enteropathogenic *Escherichia coli*. Infect Immun 77(9):3552–3568. https://doi.org/10.1128/iai.00418-09. PubMed PMID: 19581394; PubMed Central PMCID: PMCPMC2737987

Bhatt S, Egan M, Jenkins V, Muche S, El-Fenej J (2016) The Tip of the iceberg: on the roles of regulatory small RNAs in the virulence of enterohemorrhagic and enteropathogenic *Escherichia coli*. Front Cell Infect Microbiol 6:105. https://doi.org/10.3389/fcimb.2016.00105. PubMed PMID: 27709103; PubMed Central PMCID: PMCPMC5030294

Böhme K, Steinmann R, Kortmann J, Seekircher S, Heroven AK, Berger E et al (2012) Concerted actions of a thermo-labile regulator and a unique intergenic RNA thermosensor control *Yersinia* virulence. PLoS Pathog 8(2):e1002518. Epub 2012/02/24. https://doi.org/10.1371/journal.ppat.1002518. PubMed PMID: 22359501; PubMed Central PMCID: PMC3280987

Brutinel ED, Yahr TL (2008) Control of gene expression by type III secretory activity. Curr Opin Microbiol 11(2):128–133. https://doi.org/10.1016/j.mib.2008.02.010. Epub 2008/04/09. doi:S1369-5274(08)00021-0 [pii]. PubMed PMID: 18396449; PubMed Central PMCID: PMC2387186

Brutinel ED, Vakulskas CA, Yahr TL (2010) ExsD inhibits expression of the *Pseudomonas aeruginosa* type III secretion system by disrupting ExsA self-association and DNA binding activity. J Bacteriol 192(6):1479–1486. https://doi.org/10.1128/jb.01457-09. PubMed PMID: 20008065; PubMed Central PMCID: PMC2832532

Bustos SA, Schleif RF (1993) Functional domains of the AraC protein. Proc Natl Acad Sci USA 90(12):5638–5642. PubMed PMID: 8516313; PubMed Central PMCID: PMCPMC46776

Büttner D (2012) Protein export according to schedule: architecture, assembly, and regulation of type III secretion systems from plant- and animal-pathogenic bacteria. Microbiol Mol Biol Rev 76(2):262–310. https://doi.org/10.1128/mmbr.05017-11. PubMed PMID: 22688814; PubMed Central PMCID: PMCPMC3372255

Cambronne ED, Schneewind O (2002) *Yersinia enterocolitica* type III secretion: *yscM1* and *yscM2* regulate *yop* gene expression by a posttranscriptional mechanism that targets the 5′ untranslated region of *yop* mRNA. J Bacteriol 184(21):5880–5893. PubMed PMID: 12374821; PubMed Central PMCID: PMC135404

Chakravarty S, Melton CN, Bailin A, Yahr TL, Anderson GG (2017) *Pseudomonas aeruginosa* magnesium transporter MgtE inhibits type III secretion system gene expression by stimulating *rsmYZ* transcription. J Bacteriol 199(23). https://doi.org/10.1128/jb.00268-17. PubMed PMID: 28847924; PubMed Central PMCID: PMCPMC5686585

Chao Y, Papenfort K, Reinhardt R, Sharma CM, Vogel J (2012) An atlas of Hfq-bound transcripts reveals 3' UTRs as a genomic reservoir of regulatory small RNAs. EMBO J 31 (20):4005–4019. https://doi.org/10.1038/emboj.2012.229. PubMed PMID: 22922465; PubMed Central PMCID: PMCPMC3474919

Chen Y, Anderson DM (2011) Expression hierarchy in the *Yersinia* type III secretion system established through YopD recognition of RNA. Mol Microbiol 80(4):966–980. https://doi.org/10.1111/j.1365-2958.2011.07623.x. Epub 2011/04/13. PubMed PMID: 21481017

Chevance FF, Hughes KT (2017) Coupling of flagellar gene expression with assembly in *Salmonella enterica*. Methods Mol Biol 1593:47–71. https://doi.org/10.1007/978-1-4939-6927-2_4. PubMed PMID: 28389944

Choi SM, Jeong JH, Choy HE, Shin M (2016)Amino acid residues in the Ler protein critical for derepression of the LEE5 promoter in enteropathogenic *E. coli*. J Microbiol 54(8):559–564. https://doi.org/10.1007/s12275-016-6027-6. PubMed PMID: 27480636

Dasgupta N, Lykken GL, Wolfgang MC, Yahr TL (2004) A novel anti-anti-activator mechanism regulates expression of the *Pseudomonas aeruginosa* type III secretion system. Mol Microbiol 53(1):297–308. https://doi.org/10.1111/j.1365-2958.2004.04128.x. PubMed PMID: 15225323

Dasgupta N, Ashare A, Hunninghake GW, Yahr TL (2006) Transcriptional induction of the *Pseudomonas aeruginosa* type III secretion system by low Ca^{2+} and host cell contact proceeds through two distinct signaling pathways. Infect Immun 74(6):3334–3341. https://doi.org/10.1128/iai.00090-06. PubMed PMID: 16714561; PubMed Central PMCID: PMCPMC1479281

Deng W, Marshall NC, Rowland JL, McCoy JM, Worrall LJ, Santos AS et al (2017) Assembly, structure, function and regulation of type III secretion systems. Nat Rev Microbiol 15(6):323–337. https://doi.org/10.1038/nrmicro.2017.20. PubMed PMID: 28392566

Diepold A, Wagner S (2014) Assembly of the bacterial type III secretion machinery. FEMS Microbiol Rev 38(4):802–822. https://doi.org/10.1111/1574-6976.12061. PubMed PMID: 24484471

Dorman CJ (2007) H-NS, the genome sentinel. Nat Rev Microbiol 5(2):157–161. https://doi.org/10.1038/nrmicro1598. Epub 2006/12/28. doi:nrmicro1598 [pii]. PubMed PMID: 17191074

Dubey AK, Baker CS, Romeo T, Babitzke P (2005) RNA sequence and secondary structure participate in high-affinity CsrA-RNA interaction. RNA 11(10):1579–1587. https://doi.org/10.1261/rna.2990205. PubMed PMID: 16131593; PubMed Central PMCID: PMCPMC1370842

Ellermeier JR, Slauch JM (2007) Adaptation to the host environment: regulation of the SPI1 type III secretion system in *Salmonella enterica* serovar Typhimurium. Curr Opin Microbiol 10(1):24–29. https://doi.org/10.1016/j.mib.2006.12.002. PubMed PMID: 17208038

Ellermeier JR, Slauch JM (2008) Fur regulates expression of the *Salmonella* pathogenicity island 1 type III secretion system through HilD. J Bacteriol 190(2):476–486. https://doi.org/10.1128/jb.00926-07. PubMed PMID: 17993530; PubMed Central PMCID: PMC2223717

Ellermeier CD, Ellermeier JR, Slauch JM (2005) HilD, HilC and RtsA constitute a feed forward loop that controls expression of the SPI1 type three secretion system regulator *hilA* in *Salmonella enterica* serovar Typhimurium. Mol Microbiol 57(3):691–705. https://doi.org/10.1111/j.1365-2958.2005.04737.x. PubMed PMID: 16045614

Erhardt M, Dersch P (2015) Regulatory principles governing *Salmonella* and *Yersinia* virulence. Front Microbiol 6:949. https://doi.org/10.3389/fmicb.2015.00949. PubMed PMID: 26441883; PubMed Central PMCID: PMCPMC4563271

Francis MS, Wolf-Watz H, Forsberg A (2002) Regulation of type III secretion systems. Curr Opin Microbiol 5(2):166–172. PubMed PMID: 11934613

Galan JE, Lara-Tejero M, Marlovits TC, Wagner S (2014) Bacterial type III secretion systems: specialized nanomachines for protein delivery into target cells. Annu Rev Microbiol 68:415–438. https://doi.org/10.1146/annurev-micro-092412-155725. PubMed PMID: 25002086; PubMed Central PMCID: PMCPMC4388319

Golubeva YA, Ellermeier JR, Cott Chubiz JE, Slauch JM (2016). Intestinal long-chain fatty acids act as a direct signal to modulate expression of the *Salmonella* Pathogenicity Island 1 type III secretion system. MBio 7(1):e02170–e02175. https://doi.org/10.1128/mbio.02170-15. PubMed PMID: 26884427; PubMed Central PMCID: PMCPMC4752608

Görke B, Stülke J (2008) Carbon catabolite repression in bacteria: many ways to make the most out of nutrients. Nat Rev Microbiol 6(8):613–624. https://doi.org/10.1038/nrmicro1932. PubMed PMID: 18628769

Gruber CC, Sperandio V (2014) Posttranscriptional control of microbe-induced rearrangement of host cell actin. MBio 5(1):e01025-13. https://doi.org/10.1128/mbio.01025-13. PubMed PMID: 24425733; PubMed Central PMCID: PMCPMC3903284

Gruber CC, Sperandio V (2015) Global analysis of posttranscriptional regulation by GlmY and GlmZ in enterohemorrhagic *Escherichia coli* O157:H7. Infect Immun 83(4):1286–1295. https://doi.org/10.1128/iai.02918-14. PubMed PMID: 25605763; PubMed Central PMCID: PMCPMC4363437

Heroven A, Bohme K, Rohde M, Dersch P (2008) A Csr-type regulatory system, including small non-coding RNAs, regulates the global virulence regulator RovA of *Yersinia pseudotuberculosis* through RovM. Mol Microbiol 68(5):1179–1195. https://doi.org/10.1111/j.1365-2958.2008.06218.x. Epub 2008/04/24. doi:MMI6218 [pii] . PubMed PMID: 18430141

Hoe NP, Goguen JD (1993) Temperature sensing in *Yersinia pestis*: translation of the LcrF activator protein is thermally regulated. J Bacteriol 175(24):7901–7909. PubMed PMID: 7504666

Hueck CJ (1998) Type III protein secretion systems in bacterial pathogens of animals and plants. Microbiol Mol Biol Rev 62(2):379–433

Intile PJ, Diaz MR, Urbanowski ML, Wolfgang MC, Yahr TL (2014) The AlgZR two-component system recalibrates the RsmAYZ posttranscriptional regulatory system to inhibit expression of the *Pseudomonas aeruginosa* type III secretion system. J Bacteriol 196(2):357–366. https://doi.org/10.1128/jb.01199-13. PubMed PMID: 24187093; PubMed Central PMCID: PMCPMC3911257

Intile PJ, Balzer GJ, Wolfgang MC, Yahr TL (2015) The RNA Helicase DeaD stimulates ExsA translation to promote expression of the *Pseudomonas aeruginosa* type III secretion system. J Bacteriol 197(16):2664–2674. https://doi.org/10.1128/jb.00231-15. PubMed PMID: 26055113; PubMed Central PMCID: PMCPMC4507347

Jackson M, Silva-Herzog E, Plano GV (2004) The ATP-dependent ClpXP and Lon proteases regulate expression of the *Yersinia pestis* type III secretion system via regulated proteolysis of YmoA, a small histone-like protein. Mol Microbiol 54(5):1364–1378. https://doi.org/10.1111/j.1365-2958.2004.04353.x. Epub 2004/11/24. doi:MMI4353 [pii]. PubMed PMID: 15554975

Kakoschke T, Kakoschke S, Magistro G, Schubert S, Borath M, Heesemann J et al (2014) The RNA chaperone Hfq impacts growth, metabolism and production of virulence factors in *Yersinia enterocolitica*. PLoS One 9(1):e86113. https://doi.org/10.1371/journal.pone.0086113. PubMed PMID: 24454955; PubMed Central PMCID: PMCPMC3893282

Kakoschke TK, Kakoschke SC, Zeuzem C, Bouabe H, Adler K, Heesemann J et al (2016) The RNA chaperone Hfq is essential for virulence and modulates the expression of four adhesins in *Yersinia enterocolitica*. Sci Rep 6:29275. https://doi.org/10.1038/srep29275. PubMed PMID: 27387855; PubMed Central PMCID: PMCPMC4937351

Katsowich N, Elbaz N, Pal RR, Mills E, Kobi S, Kahan T et al (2017) Host cell attachment elicits posttranscriptional regulation in infecting enteropathogenic bacteria. Science 355(6326):735–739. https://doi.org/10.1126/science.aah4886. PubMed PMID: 28209897

King JM, Schesser Bartra S, Plano G, Yahr TL (2013) ExsA and LcrF recognize similar consensus binding sites, but differences in their oligomeric state influence interactions with promoter DNA. J Bacteriol 195(24):5639–5650. https://doi.org/10.1128/jb.00990-13. PubMed PMID: 24142246; PubMed Central PMCID: PMC3889609

Kopaskie KS, Ligtenberg KG, Schneewind O (2013) Translational regulation of *Yersinia enterocolitica* mRNA encoding a type III secretion substrate. J Biol Chem 288(49):35478–35488. https://doi.org/10.1074/jbc.m113.504811. PubMed PMID: 24158443; PubMed Central PMCID: PMC3853294

Kusmierek M, Dersch P (2017) Regulation of host-pathogen interactions via the post-transcriptional Csr/Rsm system. Curr Opin Microbiol 41:58–67. https://doi.org/10.1016/j.mib.2017.11.022. PubMed PMID: 29207313

Laaberki MH, Janabi N, Oswald E, Repoila F (2006) Concert of regulators to switch on LEE expression in enterohemorrhagic *Escherichia coli* O157:H7: interplay between Ler, GrlA, HNS and RpoS. Int J Med Microbiol 296(4–5):197–210. https://doi.org/10.1016/j.ijmm.2006.02.017. PubMed PMID: 16618552

Lapouge K, Perozzo R, Iwaszkiewicz J, Bertelli C, Zoete V, Michielin O et al (2013) RNA pentaloop structures as effective targets of regulators belonging to the RsmA/CsrA protein family. RNA Biol 10(6):1031–1041. https://doi.org/10.4161/rna.24771. PubMed PMID: 23635605; PubMed Central PMCID: PMCPMC4111731

Leskinen K, Varjosalo M, Skurnik M (2015) Absence of YbeY RNase compromises the growth and enhances the virulence plasmid gene expression of *Yersinia enterocolitica* O:3. Microbiology 161(Pt 2):285–299. https://doi.org/10.1099/mic.0.083097-0. PubMed PMID: 25416689

Li L, Yan H, Feng L, Li Y, Lu P, Hu Y et al (2014) LcrQ blocks the role of LcrF in regulating the Ysc-Yop type III secretion genes in *Yersinia pseudotuberculosis*. PLoS One 9(3):e92243. https://doi.org/10.1371/journal.pone.0092243. PubMed PMID: 24658611; PubMed Central PMCID: PMCPMC3962397

Lodato PB, Hsieh PK, Belasco JG, Kaper JB (2012) The ribosome binding site of a mini-ORF protects a T3SS mRNA from degradation by RNase E. Mol Microbiol 86(5):1167–1182. https://doi.org/10.1111/mmi.12050. PubMed PMID: 23043360; PubMed Central PMCID: PMCPMC3537511

Lodato PB, Thuraisamy T, Richards J, Belasco JG (2017) Effect of RNase E deficiency on translocon protein synthesis in an RNase E-inducible strain of enterohemorrhagic *Escherichia coli* O157:H7. FEMS Microbiol Lett 364(13). https://doi.org/10.1093/femsle/fnx131. PubMed PMID: 28854682

Lopez-Garrido J, Puerta-Fernandez E, Casadesus J (2014) A eukaryotic-like 3' untranslated region in *Salmonella enterica hilD* mRNA. Nucleic Acids Res 42(9):5894–5906. https://doi.org/10.1093/nar/gku222. PubMed PMID: 24682814; PubMed Central PMCID: PMC4027200

Madrid C, Nieto JM, Juarez A (2002) Role of the Hha/YmoA family of proteins in the thermoregulation of the expression of virulence factors. Int J Med Microbiol 291(6–7):425–432. PubMed PMID: 11890540

Madrid C, Balsalobre C, Garcia J, Juarez A (2007) The novel Hha/YmoA family of nucleoid-associated proteins: use of structural mimicry to modulate the activity of the H-NS family of proteins. Mol Microbiol 63(1):7–14. PubMed PMID: 17116239

Martinez LC, Yakhnin H, Camacho MI, Georgellis D, Babitzke P, Puente JL et al (2011) Integration of a complex regulatory cascade involving the SirA/BarA and Csr global regulatory systems that controls expression of the *Salmonella* SPI-1 and SPI-2 virulence regulons through HilD. Mol Microbiol 80(6):1637–1656. https://doi.org/10.1111/j.1365-2958.2011.07674.x. PubMed PMID: 21518393; PubMed Central PMCID: PMC3116662

McCaw ML, Lykken GL, Singh PK, Yahr TL (2002) ExsD is a negative regulator of the *Pseudomonas aeruginosa* type III secretion regulon. Mol Microbiol 46(4):1123–1133. PubMed PMID: 12421316

Mercante J, Edwards AN, Dubey AK, Babitzke P, Romeo T (2009) Molecular geometry of CsrA (RsmA) binding to RNA and its implications for regulated expression. J Mol Biol 392(2):511–528. https://doi.org/10.1016/j.jmb.2009.07.034. PubMed PMID: 19619561; PubMed Central PMCID: PMC2735826

Miller HK, Kwuan L, Schwiesow L, Bernick DL, Mettert E, Ramirez HA et al (2014) IscR is essential for *Yersinia pseudotuberculosis* type III secretion and virulence. PLoS Pathog 10(6): e1004194. https://doi.org/10.1371/journal.ppat.1004194. PubMed PMID: 24945271; PubMed Central PMCID: PMC4055776

Mohanty BK, Kushner SR (2016)Regulation of mRNA decay in bacteria. Annu Rev Microbiol 70:25–44. https://doi.org/10.1146/annurev-micro-091014-104515. PubMed PMID: 27297126

Navarre WW, Porwollik S, Wang Y, McClelland M, Rosen H, Libby SJ et al (2006) Selective silencing of foreign DNA with low GC content by the H-NS protein in *Salmonella*. Science 313(5784):236–238. PubMed PMID: 16763111

Nuss AM, Beckstette M, Pimenova M, Schmühl C, Opitz W, Pisano F et al (2017) Tissue dual RNA-seq: a fast discovery path for infection-specific functions and riboregulators shaping host-pathogen transcriptomes. Proc Natl Acad Sci USA 114(5):E791–E800

Okan NA, Bliska JB, Karzai AW (2006) A Role for the SmpB-SsrA system in *Yersinia pseudotuberculosis* pathogenesis. PLoS Pathog 2(1):e6. https://doi.org/10.1371/journal.ppat.0020006. Epub 2006/02/02. PubMed PMID: 16450010; PubMed Central PMCID: PMC1358943

Olekhnovich IN, Kadner RJ (2007) Role of nucleoid-associated proteins Hha and H-NS in expression of *Salmonella enterica* activators HilD, HilC, and RtsA required for cell invasion. J Bacteriol 89(19):6882–6890. https://doi.org/10.1128/jb.00905-07. PubMed PMID: 17675384; PubMed Central PMCID: PMC2045230

Ozturk G, LeGrand K, Zheng Y, Young GM (2017) *Yersinia enterocolitica* CsrA regulates expression of the Ysa and Ysc type 3 secretion system in unique ways. FEMS Microbiol Lett 364(20). https://doi.org/10.1093/femsle/fnx204. PubMed PMID: 29044402

Parsot C, Ageron E, Penno C, Mavris M, Jamoussi K, d'Hauteville H et al (2005) A secreted anti-activator, OspD1, and its chaperone, Spa15, are involved in the control of transcription by the type III secretion apparatus activity in *Shigella flexneri*. Mol Microbiol 56(6):1627–1635. https://doi.org/10.1111/j.1365-2958.2005.04645.x. PubMed PMID: 15916611

Pettersson J, Nordfelth R, Dubinina E, Bergman T, Gustafsson M, Magnusson KE et al (1996) Modulation of virulence factor expression by pathogen target cell contact. Science 273 (5279):1231–1233

Poncet S, Milohanic E, Maze A, Abdallah JN, Ake FML et al (2009) Correlations between carbon metabolism and virulence in bacteria. Contrib Microbiol 16:88–102

Portaliou AG, Tsolis KC, Loos MS, Zorzini V, Economou A (2016) Type III secretion: building and operating a remarkable nanomachine. Trends Biochem Sci 41(2):175–189. https://doi.org/10.1016/j.tibs.2015.09.005. PubMed PMID: 26520801

Ren B, Shen H, Lu ZJ, Liu H, Xu Y (2014) The *phzA2-G2* transcript exhibits direct RsmA-mediated activation in *Pseudomonas aeruginosa* M18. PLoS One 9(2):e89653. https://doi.org/10.1371/journal.pone.0089653. PubMed PMID: 24586939; PubMed Central PMCID: PMCPMC3933668

Rhigetti F, Nuss AM, Twittenhoff C, Beele S, Urban K, Will S et al (2016) Temperature-responsive *in vitro* RNA structurome of *Yersinia pseudotuberculosis*. Proc Natl Acad Sci 113(26):7237–7242

Rietsch A, Vallet-Gely I, Dove SL, Mekalanos JJ (2005) ExsE, a secreted regulator of type III secretion genes in *Pseudomonas aeruginosa*. Proc Natl Acad Sci USA 102(22):8006–8011. https://doi.org/10.1073/pnas.0503005102. PubMed PMID: 15911752; PubMed Central PMCID: PMCPMC1142391

Rosenzweig JA, Chopra AK (2013) The exoribonuclease polynucleotide phosphorylase influences the virulence and stress responses of *yersiniae* and many other pathogens. Front Cell Infect Microbiol 3:81. https://doi.org/10.3389/fcimb.2013.00081. PubMed PMID: 24312901; PubMed Central PMCID: PMCPMC3832800

Rosenzweig JA, Weltman G, Plano GV, Schesser K (2005) Modulation of *Yersinia* type three secretion system by the S1 domain of polynucleotide phosphorylase. J Biol Chem 280(1):156–163. https://doi.org/10.1074/jbc.m405662200. PubMed PMID: 15509583

Rosenzweig JA, Chromy B, Echeverry A, Yang J, Adkins B, Plano GV et al (2007) Polynucleotide phosphorylase independently controls virulence factor expression levels and export in *Yersinia* spp. FEMS Microbiol Lett 270(2):255–264. https://doi.org/10.1111/j.1574-6968.2007.00689.x. PubMed PMID: 17391372

Schiano CA, Bellows LE, Lathem WW (2010) The small RNA chaperone Hfq is required for the virulence of *Yersinia pseudotuberculosis*. Infect Immun 78(5):2034–2044. https://doi.org/10.1128/iai.01046-09. PubMed PMID: 20231416; PubMed Central PMCID: PMC2863511

Schiano CA, Koo JT, Schipma MJ, Caulfield AJ, Jafari N, Lathem WW (2014) Genome-wide analysis of small RNAs expressed by *Yersinia pestis* identifies a regulator of the Yop-Ysc type III secretion system. J Bacteriol 196(9):1659–1670. Epub 2014/02/18. https://doi.org/10.1128/jb.01456-13. PubMed PMID: 24532772; PubMed Central PMCID: PMC3993326

Schleif R (2010) AraC protein, regulation of the l-arabinose operon in *Escherichia coli*, and the light switch mechanism of AraC action. FEMS Microbiol Rev 34(5):779–796. https://doi.org/10.1111/j.1574-6976.2010.00226.x. PubMed PMID: 20491933

Schulmeyer KH, Yahr TL (2017) Post-transcriptional regulation of type III secretion in plant and animal pathogens. Curr Opin Microbiol 36:30–36. https://doi.org/10.1016/j.mib.2017.01.009. PubMed PMID: 28189908; PubMed Central PMCID: PMCPMC5534366

Schwiesow L, Lam H, Dersch P, Auerbuch V (2015) *Yersinia* type III secretion system master regulator LcrF. J Bacteriol 198(4):604–614. https://doi.org/10.1128/jb.00686-15. PubMed PMID: 26644429; PubMed Central PMCID: PMCPMC4751813

Shimizu T, Ichimura K, Noda M (2015) The surface sensor NlpE of enterohemorrhagic *Escherichia coli* contributes to regulation of the type III secretion system and flagella by the Cpx response to adhesion. Infect Immun 84(2):537–549. https://doi.org/10.1128/iai.00881-15. PubMed PMID: 26644384; PubMed Central PMCID: PMCPMC4730559

Sittka A, Pfeiffer V, Tedin K, Vogel J (2007) The RNA chaperone Hfq is essential for the virulence of *Salmonella typhimurium*. Mol Microbiol 63(1):193–217. https://doi.org/10.1111/j.1365-2958.2006.05489.x. PubMed PMID: 17163975; PubMed Central PMCID: PMCPMC1810395

Standish AJ, Teh MY, Tran ENH, Doyle MT, Baker PJ, Morona R (2016) Unprecedented abundance of protein tyrosine phosphorylation modulates *Shigella flexneri* virulence. J Mol Biol 428(20):4197–4208. https://doi.org/10.1016/j.jmb.2016.06.016. PubMed PMID: 27380737

Tanabe T, Miyamoto K, Tsujibo H, Yamamoto S, Funahashi T (2015) The small RNA Spot 42 regulates the expression of the type III secretion system 1 (T3SS1) chaperone protein VP1682 in *Vibrio parahaemolyticus*. FEMS Microbiol Lett 362(21). https://doi.org/10.1093/femsle/fnv173. PubMed PMID: 26394644

Teixido L, Carrasco B, Alonso JC, Barbe J, Campoy S (2011) Fur activates the expression of *Salmonella enterica* pathogenicity island 1 by directly interacting with the *hilD* operator *in vivo* and *in vitro*. PLoS One 6(5):e19711. https://doi.org/10.1371/journal.pone.0019711. PubMed PMID: 21573071; PubMed Central PMCID: PMC3089636

Tobe T, Yen H, Takahashi H, Kagayama Y, Ogasawara N, Oshima T (2014)Antisense transcription regulates the expression of the enterohemorrhagic *Escherichia coli* virulence regulatory gene ler in response to the intracellular iron concentration. PLoS One 9(7):e101582.

https://doi.org/10.1371/journal.pone.0101582. PubMed PMID: 25006810; PubMed Central PMCID: PMCPMC4090186

Urbanowski ML, Lykken GL, Yahr TL (2005) A secreted regulatory protein couples transcription to the secretory activity of the *Pseudomonas aeruginosa* type III secretion system. Proc Natl Acad Sci USA 102(28):9930–9935. https://doi.org/10.1073/pnas.0504405102. PubMed PMID: 15985546; PubMed Central PMCID: PMC1175016

Vakulskas CA, Brady KM, Yahr TL (2009) Mechanism of transcriptional activation by *Pseudomonas aeruginosa* ExsA. J Bacteriol 191(21):6654–6664. https://doi.org/10.1128/jb.00902-09. PubMed PMID: 19717612; PubMed Central PMCID: PMC2795306

Vakulskas CA, Potts AH, Babitzke P, Ahmer BM, Romeo T (2015) Regulation of bacterial virulence by Csr (Rsm) systems. Microbiol Mol Biol Rev 79(2):193–224. https://doi.org/10.1128/mmbr.00052-14. PubMed PMID: 25833324

Vergnes A, Viala JP, Ouadah-Tsabet R, Pocachard B, Loiseau L, Meresse S et al (2017) The iron-sulfur cluster sensor IscR is a negative regulator of Spi1 type III secretion system in *Salmonella enterica*. Cell Microbiol 19(4). https://doi.org/10.1111/cmi.12680. PubMed PMID: 27704705

Wang H, Avican K, Fahlgren A, Erttmann SF, Nuss AM, Dersch P et al (2016) Increased plasmid copy number is essential for *Yersinia* T3SS function and virulence. Science 353(6298):492–495. https://doi.org/10.1126/science.aaf7501. PubMed PMID: 27365311

Yakhnin AV, Baker CS, Vakulskas CA, Yakhnin H, Berezin I, Romeo T et al (2013) CsrA activates *flhDC* expression by protecting *flhDC* mRNA from RNase E-mediated cleavage. Mol Microbiol 87(4):851–866. https://doi.org/10.1111/mmi.12136. PubMed PMID: 23305111; PubMed Central PMCID: PMCPMC3567230

Yang J, Jain C, Schesser K (2008) RNase E regulates the *Yersinia* type 3 secretion system. J Bacteriol 190(10):3774–3778. https://doi.org/10.1128/jb.00147-08. PubMed PMID: 18359811; PubMed Central PMCID: PMCPMC2395017

Zheng Z, Chen G, Joshi S, Brutinel ED, Yahr TL, Chen L (2007) Biochemical characterization of a regulatory cascade controlling transcription of the *Pseudomonas aeruginosa* type III secretion system. J Biol Chem 282(9):6136–6142. https://doi.org/10.1074/jbc.m611664200. PubMed PMID: 17197437

Zhou X, Konkel ME, Call DR (2010) Regulation of type III secretion system 1 gene expression in *Vibrio parahaemolyticus* is dependent on interactions between ExsA, ExsC, and ExsD. Virulence 1(4):260–272. https://doi.org/10.4161/viru.1.4.12318. PubMed PMID: 21178451; PubMed Central PMCID: PMCPMC3073295

Zierler MK, Galan JE (1995) Contact with cultured epithelial cells stimulates secretion of *Salmonella typhimurium* invasion protein InvJ. Infect Immun 63(10):4024–4028

Assembly and Post-assembly Turnover and Dynamics in the Type III Secretion System

Andreas Diepold

Contents

Abstract The type III secretion system (T3SS) is one of the largest transmembrane complexes in bacteria, comprising several intricately linked and embedded substructures. The assembly of this nanomachine is a hierarchical process which is regulated and controlled by internal and external cues at several critical points. Recently, it has become obvious that the assembly of the T3SS is not a unidirectional and deterministic process, but that parts of the T3SS constantly exchange or rearrange. This article aims to give an overview on the assembly and post-assembly dynamics of the T3SS, with a focus on emerging general concepts and adaptations of the general assembly pathway.

A. Diepold (✉)
Department of Ecophysiology, Max Planck Institute for Terrestrial Microbiology, Karl-von-Frisch-Straße 10, 35043 Marburg, Germany
e-mail: andreas.diepold@mpi-marburg.mpg.de

Current Topics in Microbiology and Immunology (2020) 427: 35–66
https://doi.org/10.1007/82_2019_164
© Springer Nature Switzerland AG 2019
Published Online: 20 June 2019

1 Introduction

Bacteria have evolved an astounding variety of ways to manipulate the behavior of eukaryotic host cells to their benefit. One of the most complex and intriguing examples is the type III secretion system (T3SS), a molecular machine that injects effector proteins into host cells (Cornelis 2006; Galán and Wolf-Watz 2006; Büttner 2012; Burkinshaw and Strynadka 2014; Deng et al. 2017; Wagner et al. 2018). The effect of these translocated proteins on the host can range from mutually beneficial in symbiosis to deleterious in pathogenesis (Galán 2009)—*see review by David Guttman in this edition for more details*. In fact, many important Gram-negative bacterial plant and animal pathogens such as *Salmonella*, *Shigella*, *Yersinia*, *Pseudomonas*, *Xanthomonas*, and *pathogenic Escherichia coli* employ a T3SS for infection, which is often essential for pathogenicity. While the translocated effector proteins greatly vary between different bacteria and reflect their different infection strategies, the secretion machinery, also called injectisome, is strongly conserved. A better understanding of the assembly, structure, and function of the system is therefore an important step to find new ways to interfere with this conserved virulence factor.

The T3SS injectisome consists of over 300 single copies of more than 15 different proteins that assemble to form a multi-membrane-spanning complex of >7 MDa (Table 1). The injectisome can be separated into distinct substructures: (i) the extracellular needle capped by a tip complex on its distal side, (ii) the membrane-spanning rings traversing both bacterial membranes and the periplasm, (iii) the export apparatus in and above the inner membrane (IM), and (iv) a cytosolic complex on the proximal interface of the injectisome (Fig. 1).

The injectisome is evolutionarily related to the bacterial flagellum (Abby and Rocha 2012; Pallen and Gophna 2007), which harbors an integral T3SS for the export of the subunits of its extracellular filament. Homologous parts of the machineries include the IM ring, export apparatus, and the cytosolic complex (Erhardt et al. 2010; Diepold and Armitage 2015).

In this review, I will summarize the general assembly pathway of the T3SS (see also Diepold and Wagner 2014; Deng et al. 2017; Wagner et al. 2018) and focus on recent studies that have shed light on open questions in the assembly and post-assembly dynamics of the T3SS: is there a universal assembly pathway, or do certain bacterial species deviate from the norm—and if yes, how and why? Is the resulting structure stable or adaptive? How can it respond to external signals and changes in the environment? And how are assembly and dynamics linked to the function of the T3SS? While this review focuses on the well-studied T3SS of animal pathogens, most of the described principles will apply to plant pathogens or symbionts (reviewed in Büttner and He (2009), Tampakaki (2014)) as well.

Table 1 The building blocks and substructures of the T3SS

	General Sct protein name (Hueck 1998)	Functional description	Stoichiometry	MW single protein* (kDa)	MW total* (kDa)	MW sub-structure* (MDa)	References for stoichiometry
Membrane rings	SctC	Secretin	12–15**	67.2	807	**2.58**	Kowal et al. (2013)
	SctD	Outer MS ring protein	24	46.8	1123		Schraidt et al. (2010)
	SctJ	Inner MS ring protein	24	27.1	649		Yip et al. (2005), Schraidt et al. (2010)
Export apparatus	SctR	Minor export apparatus protein	5	24.4	122	**0.93**	Zilkenat et al. (2016)
	SctS	Minor export apparatus protein	4	9.6	38		Zilkenat et al. (2016), Kuhlen et al. (2018)
	SctT	Minor export apparatus protein	1	28.6	29		Zilkenat et al. (2016)
	SctU	Export apparatus switch protein	1	40.3	40		Zilkenat et al. (2016)
	SctV	Major export apparatus protein	9	77.9	701		Abrusci et al. (2014)
Cytosolic complex	SctK	Accessory cytosolic protein	6	24.0	144	**2.05/2.34**	Diepold et al. (2017), Zhang et al. (2017)
	SctQ	C-ring protein	24	34.4	826		Diepold et al. (2017), Zhang et al. (2017)
	$SctQ_C$		48***	10.0	480		

(continued)

Table 1 (continued)

	General Sct protein name (Hueck 1998)	Functional description	Stoichiometry	MW single protein* (kDa)	MW total* (kDa)	MW sub-structure* (MDa)	References for stoichiometry
		Additional C-term. SctQ fragment					Bzymek et al. (2012), McDowell et al. (2016), Diepold et al. (2017), Zhang et al. (2017)
	SctL	Stator	12	25.0	300		Minamino and Macnab (2000b), Diepold et al. (2017)
	SctN	ATPase	6/12	47.8	287/ 573		Pozidis et al. (2003), Müller et al. (2006), Akeda and Galán (2005), Diepold et al. (2017)
	SctO	Stalk	1	18.9	19		
Needle, rod, tip	SctF	Needle filament protein	~140****	9.4	1323	**1.59**	Broz et al. (2007)
	SctI	Inner rod protein	6	12.7	76		Zilkenat et al. (2016), Dietsche et al. (2016)
	SctA*****	hydrophilic translocator, tip protein	5	37.3	186		Mueller et al. (2005)
Total*			328/334			**7.15/7.44**	

Only conserved proteins present in the steady-state structure (prior to host cell contact) are included. For other components, such as the hydrophobic translocators or species-specific T3SS proteins, see Table S1

*Molecular weight and total stoichiometry are listed for *Yersinia enterocolitica* pYVe227, Genbank accession number AF102990

**Stoichiometry of 15 for *Salmonella* (Schraidt and Marlovits 2011; Worrall et al. 2016)

***Conflicting evidence on presence in injectisome (Diepold et al. 2015; Lara-Tejero et al. 2019)

****Needle length and stoichiometry significantly differ between species and subspecies (Journet et al. 2003; Wee and Hughes 2015)

*****Sct name not assigned by Hueck (1998), "SctA" assigned by Portaliou et al. (2016), *Yersinia*/*Salmonella*/*Shigella*-specific names are LcrV/SipD/IpaD

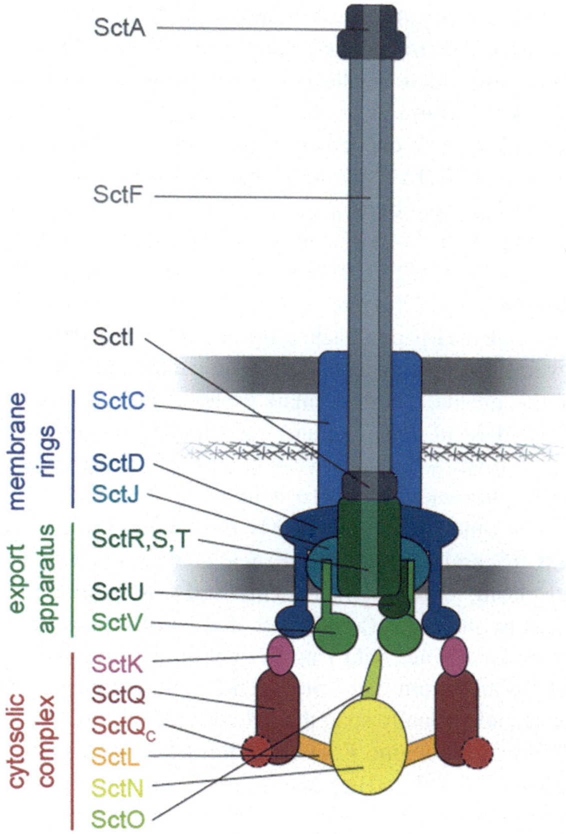

Fig. 1 Overall structure of the T3SS injectisome. Schematic structural overview of the T3SS injectisome, based on the *Salmonella* SPI-1 injectisome in situ structure (Hu et al. 2017) and the cryo-EM structure of the export apparatus (Kuhlen et al. 2018). Dashed lines indicate conflicting results about the presence of SctQ$_C$ at the injectisome

2 Structure and Function of the Injectisome

In several organisms, large parts of the intact injectisome can be purified (Kubori et al. 1998; Tamano et al. 2000; Daniell et al. 2001; Blocker et al. 2001; Sekiya et al. 2001). These so-called needle complexes comprising the needle, membrane rings and, at least in some cases, the export apparatus, but not the cytosolic components, have been structurally characterized to a molecular level (Schraidt and Marlovits 2011; Bergeron et al. 2013; Worrall et al. 2016; Hu et al. 2018). The recent developments in cryo-electron tomography have allowed to also visualize the injectisome in situ (Kudryashev et al. 2013; Kawamoto et al. 2013; Nans et al. 2015; Hu et al. 2015; Makino et al. 2016; Hu et al. 2017; Park et al. 2018), revealing for the first time the structure and arrangement at the cytosolic interface of the injectisome (Hu et al. 2015; Nans et al. 2015; Makino et al. 2016; Hu et al. 2017; Park et al. 2018) (Fig. 1). *See review by Thomas Marlovits in this edition for more details on the structure of the T3SS.*

The injectisome is anchored in the peptidoglycan (PG) layer and spans both bacterial membranes, extending into both the extracellular and the cytosolic space. The extracellular needle is formed through helical polymerization of a small protein, **SctF**[1] (Fujii et al. 2012)—*see review by Ariel Blocker in this edition*. At its distal side, it is capped by a pentameric tip structure formed by the hydrophilic translocator **SctA** (Mueller et al. 2005), which in some cases includes a copy of one of the hydrophobic translocators (Johnson et al. 2006; Veenendaal et al. 2007; Blocker et al. 2008; Cheung et al. 2014). In enteropathogenic and enterohemorrhagic *E. coli* (EPEC/EHEC), the tip protein extends into a long flexible filament, thought to span the intestinal mucus layer, whereas the needles of plant pathogens, also called Hrp pili, lack a tip structure and instead are greatly elongated (up to several μm, Weber et al. 2005) to penetrate the plant cell wall. At the proximal end of the needle, another small protein, **SctI**, forms an inner rod (Marlovits et al. 2004). Recent structural and cross-linking results indicate that the rod structure is a rather small adapter between the needle and the export apparatus (Zilkenat et al. 2016; Dietsche et al. 2016; Kuhlen et al. 2018).

The outer membrane (OM) is spanned by a 12–15-mer of a secretin protein, **SctC** (Schraidt et al. 2010; Kowal et al. 2013), which also penetrates the PG. In the periplasm, SctC interacts with **SctD**, a bitopic protein forming a 24-mer ring structure in the IM (Schraidt et al. 2010; Bergeron et al. 2013), which interacts with a smaller 24-mer **SctJ** ring (Yip et al. 2005; Schraidt et al. 2010), anchored in the IM by an *N*-terminal lipid anchor (Allaoui et al. 1992). While most SctJ homologues additionally span the IM with a C-terminal TMH, this region is absent in the T3SS of pathogenic *E. coli* and has been shown not to be required for secretion in *Salmonella* SPI-1 (Bergeron et al. 2015).

Within the ring formed by SctDJ, five membrane proteins, **SctR**, **S**, **T**, **U**, **and V**, form the so-called export apparatus (Wagner et al. 2010). The composition and assembly of the export apparatus has only been uncovered in recent years. Structural and mass spectrometry-based studies defined the stoichiometry of the participating proteins as $SctR_5S_4T_1U_1V_9$ (Abrusci et al. 2013; Zilkenat et al. 2016; Kuhlen et al. 2018) with the SctR pentamer forming an IM pore with a diameter of 1.5 nm (Dietsche et al. 2016). The core export apparatus proteins, $SctR_5S_4T$, form a unique helical structure that is located above the IM plane in the final structure, and define the helical structure of the adjacent rod and needle (Kuhlen et al. 2018). Two of the export apparatus proteins have large cytosolic domains, the "substrate switch" protein SctU, which undergoes autocleavage at a conserved NPTH motif (Minamino and Macnab 2000a; Lavander et al. 2002; Sorg et al. 2007), and SctV,

[1]Throughout this review, the general Sct nomenclature for T3SS components (Hueck 1998; Portaliou et al. 2016) is used. Where no Sct name has been assigned, the *Yersinia/Salmonella/Shigella*-specific names are listed. See Table S1 for species-specific names and flagellar homologues.

whose C-terminal domains form a suspended toroid structure at the cytosolic side of the export apparatus (Abrusci et al. 2013).[2] *See review by Tohru Minamino in this edition for more details.*

On the cytosolic side, five soluble proteins are required for T3SS export: **SctN** is an oligomerization-activated ATPase that was shown to detach chaperones and unfold effectors prior to export through the T3SS (Akeda and Galán 2005). In addition to homohexamerization, SctN can form 1:2 complexes with **SctL**, which negatively regulates ATPase activity (Pallen et al. 2006; Blaylock et al. 2006). **SctQ** is a homologue of the flagellar C-ring proteins FliM and FliN. SctQ has an internal translation initiation site (Yu et al. 2011; Bzymek et al. 2012) leading to the additional expression of its C-terminal fragment, **SctQ$_C$**, which is highly homologous to FliN. Additional SctQ$_C$ is required for function of most, but not all, tested T3SS (Bzymek et al. 2012; Diepold et al. 2015; Notti et al. 2015; McDowell et al. 2016; Song et al. 2017; Lara-Tejero et al. 2019). Full-length SctQ (SctQ$_{FL}$) and SctQ$_C$ form a 1:2 complex (Bzymek et al. 2012), which structurally resembles the flagellar FliM:FliN complex (McDowell et al. 2016). While the flagellar C-ring is part of the switch complex used to reverse or stop the rotation of the flagellum in chemotaxis, the exact role of SctQ in the injectisome is unknown. **SctK** is an injectisome-specific protein associated with the IM (Soto et al. 2016). The four proteins interact in a linear SctK-Q-L-N chain (Jackson and Plano 2000). In *Salmonella*, a complex of SctK, SctQ, and SctL was shown to bind to cytosolic chaperone-effector complexes in a sequential manner and termed "sorting platform" (Lara-Tejero et al. 2011)—*see review by Maria Lara-Tejero in this edition*. In *Shigella* and *Yersinia*, similar cytosolic complexes include (and depend on) the ATPase SctN (Johnson and Blocker 2008; Diepold et al. 2017). A last cytosolic protein, **SctO**, binds to the ATPase complex as a monomer (Majewski et al. 2019). SctO or its flagellar homologue have been shown to promote ATPase oligomerization (Ibuki et al. 2011), bind chaperones of effector proteins (Evans and Hughes 2009), and modulate PMF conversion together with a cytosolic regulator (PcrG in *Pseudomonas aeruginosa*, Lee et al. 2014). If and how these functions are coordinated remains unclear at this point. Notably, SctN, O, L, and parts of the large export apparatus component SctV display homology to the α/β, the γ, the β/δ, and the E subunits of F/V-type ATPases, respectively (Ibuki et al. 2011; Pallen et al. 2006; Imada et al. 2007; Worrall et al. 2010). This similarity has been largely confirmed by recent structural data (Hu et al. 2015, 2017; Majewski et al. 2019). While the cytosolic components display clear homology to flagellar counterparts, the resulting structure is strikingly different: instead of the continuous ring structure of the flagellar switch complex, cryo-ET revealed a sixfold symmetry with six large "pod" structures (Hu et al. 2015; Makino et al. 2016; Hu et al. 2017). Each pod is estimated to contain one SctK (connecting the pod to four SctD proteins each at the

[2]In *Yersinia* and *Pseudomonas*, additional species-specific essential T3SS components YscX/Y and Pcr3/4 (Iriarte and Cornelis 1999; Day and Plano 2000; Bröms et al. 2005; Yang et al. 2007) have been shown to tightly bind to SctV (Diepold et al. 2011). Whether their—currently unknown—function is taken over by different proteins in other T3SS is unknown at the moment.

IM), four SctQ and possibly eight additional $SctQ_C$, and two SctL, which in turn connect to the multimeric ATPase below the export apparatus (Hu et al. 2017; Diepold et al. 2017; Zhang et al. 2017; McDowell et al. 2016).

Two hydrophobic translocators, **SctB** and **SctE**, which interact with the hydrophilic translocator at the needle tip, form pores in the host cell membrane. Recent in situ structures of injectisomes in contact to the host cell (Park et al. 2018; Nauth et al. 2018) are strongly suggestive of a direct injection of effector proteins from the bacterial cytosol to the host cytosol in a one-step mechanism. Indeed, the structures of needle complexes with trapped translocator or effector substrates (Radics et al. 2014; Dohlich et al. 2014) highlight that substrates do travel through the needle. However, there is evidence that at least under certain circumstances, a binary AB-toxin-like two-step translocation mechanism can also lead to translocation. This pathway relies on a functional T3SS and the presence of pores in the host cell membrane, but features an intermediate extracellular presence of the effectors, which are then internalized through interactions with the hydrophobic translocators (Akopyan et al. 2011; Edgren et al. 2012; Tejeda-Dominguez et al. 2017). Whether this pathway significantly contributes to T3SS protein translocation in vivo remains to be determined.

3 What Drives Protein Assemblies?

Pairwise protein interactions are governed by binding affinities and the local concentrations of the respective binding partners. However, in the formation of protein complexes, additional factors come into place: Binding of one protein might change the confirmation of its partner, thus altering its affinity for a third binding partner, or directly sterically influence binding of additional proteins. Additionally, two or more (identical or different) proteins might offer a joint binding surface. To prevent non-functional or dead-end interactions, chaperones may be required (Daley 2008). Membrane protein complexes have yet different binding constraints: Integration into the membrane restricts movement and orientation and increases the chance of rendering an interacting protein inaccessible for additional binding partners. In Gram-negative bacteria, the OM and IM are largely separated by the PG layer, and in many cases, specific or housekeeping PG-modifying enzymes must be utilized to connect proteins on either side.

In recent years, it has become obvious that protein complexes, including some membrane-bound complexes, can form co-translationally (Duncan and Mata 2011; Wells et al. 2015; Natan et al. 2017; Shiber et al. 2018). The prokaryotic organization of genes in operons can conceivably strongly enhance the efficiency of such co-translational complex formation. Indeed, complex assembly was found to be co-translational for the *E. coli* luciferase complex (Shieh et al. 2015), and structure-based assembly predictions for most protein complexes matched the gene order within operons (Wells et al. 2016).

Bacterial secretion systems and related systems display a remarkable variety of assembly processes (see Box 1). Focusing on the T3SS, one of the most complex protein assemblies in prokaryotes, steric limitations of interactions are obvious. Also, the integration of some parts, especially the export apparatus, into closed membrane structures suggests an ordered rather than random assembly. Interestingly, the gene order of the members of the export apparatus is among the most strongly conserved among T3SS (Abby and Rocha 2012). Along those lines, complementation of T3SS components *in trans* often requires a higher expression level of the respective protein for full functionality, or even leads to toxic effects (Wagner et al. (2010) and own unpublished results). A recent study by Song et al. showed that a recoded and reordered set of T3SS genes (that were still arranged on two operons) can lead to the formation of functional T3SS, albeit at a lower efficiency (Song et al. 2017). This indicates that while the above-mentioned mechanisms are not completely essential in vitro, they lead to a more efficient, and possibly faster assembly of the T3SS during infection.

4 Assembly of the IM Export Apparatus

A quick look at the schematic structure of the membrane-spanning part of the T3SS (Fig. 1) reveals a clear candidate for a nucleation point of T3SS assembly: the export apparatus, embedded in the SctJ ring in the IM, which itself is associated with the SctD ring. At least in a static structure, it is hard to envision how this arrangement of multi-transmembrane helix proteins could be integrated into the membrane rings at a later time point. And indeed, it was shown that the export apparatus can form independently of the remaining T3SS (Wagner et al. 2010; Diepold et al. 2011).

Recent advances in determining the structure and composition of the export apparatus have led to a better understanding of the assembly of this early substructure. Based on the stability of different subcomplexes and the cryo-electron microscopy structure of the core *Salmonella* SPI-1 export apparatus, the start of the assembly of the export apparatus is the formation of a SctR pentamer, which is stabilized by the addition of one SctT molecule (see Fig. 2 for a schematic representation of the assembly process). After this, four SctS molecules and subsequently one SctU bind to this complex with weaker affinity (Wagner et al. 2010; Dietsche et al. 2016; Kuhlen et al. 2018). The formation of this complex allows nine SctV molecules (Abrusci et al. 2013) to associate with the other export apparatus proteins (Wagner et al. 2010; Dietsche et al. 2016; Kuhlen et al. 2018). The structural arrangement of these proteins already implies an assembly order, as the $SctR_5S_4T$ subcomplex, formed by unambiguously predicted transmembrane proteins, forms a pseudohexameric helical arrangement that is ultimately displaced from the IM toward the periplasm, probably at the addition of the SctV nonamer (Kuhlen et al. 2018). Notably, while SctV was not required for the formation of the SctRSTU complex in *Salmonella* (Wagner et al. 2010; Dietsche et al. 2016), SctU

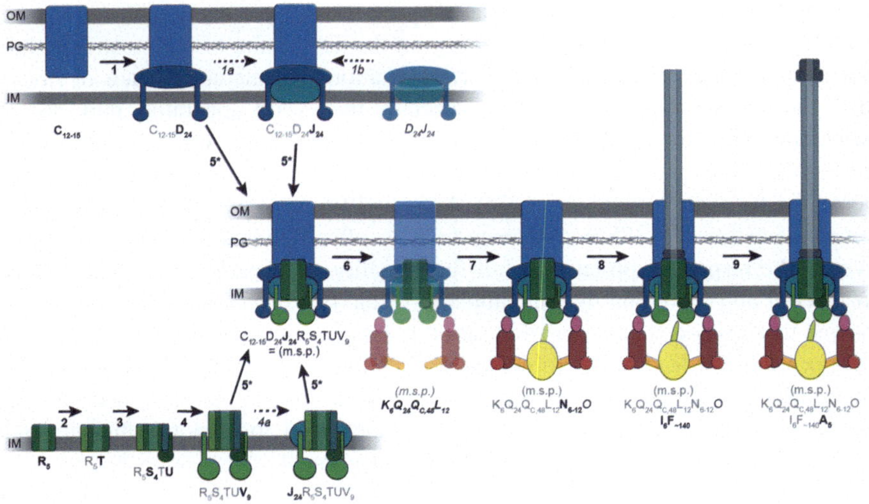

Fig. 2 Schematic overview of the assembly of the T3SS injectisome. The composition of the complexes is indicated below the depicted assembly step (letter according to Sct nomenclature, m.s.p. = membrane-spanning part $SctC_{12\text{-}15}D_{24}J_{24}R_5S_4TUV_9$). Newly added proteins are highlighted in bold font. Assembly steps that were shown to occur, but are not thought to be part of the canonical assembly process, are indicated by dashed arrows. Complexes that are instable or absent from some organisms are indicated by dashed lines, opaque fills, and italic letters. Assembly can start independently in the OM and the IM. In the OM (top branch), the secretin SctC forms a stable ring structure with the sole help of a pilotin protein in many organisms. SctD can add on to this structure (1). In the IM (bottom branch, slightly enlarged), a $SctR_5$ complex is stabilized by SctT (2). Next, $SctS_4$ and SctU are added (3), and the nonameric SctV ring assembles around this patch of membrane proteins (4). At this point, the membrane ring structures merge, possibly through the integration and multimerization of SctJ (central branch, 5). Asterisks denote the various possible combinations of subcomplexes at this step. In *Salmonella*, addition of SctK, SctQ, $SctQ_C$, and SctL (6) leads to a stable structure, whereas in *Shigella* and *Yersinia*, SctN also needs to be present for the next step (7). The time point of SctO integration is unclear; however, it is required for the formation of the inner rod and needle (8). Once the correct length of the needle is sensed, which probably involves the ruler protein SctP, the hydrophilic translocator protein SctA forms the tip of the needle (9). The SctCD substructure allows the integration of SctJ (1a) and, to a certain degree, the cytosolic complex (not shown), although this might represent a dead-end assembly which is of minor importance in vivo. SctD and SctJ can form a complex in some organisms (1b). Like the SctCD complex, the export apparatus can bind to SctJ (4a), but it is unclear if this is part of the assembly process in vivo

(but not SctRST) was dispensable for SctV oligomerization and its interaction with the IM ring in *Yersinia* (Diepold et al. 2011), possibly indicating different protein affinities which might be reflected in the respective assembly orders.

In the absence of any of the three membrane ring components SctC, D, J, the export apparatus moves within the IM (Diepold et al. 2011), suggesting that the export apparatus needs to be incorporated into (or synthesized within) the IM ring, and that its final location is determined by the distribution of the secretin SctC in the PG and OM.

5 Assembly of the Membrane Rings

Given the independent assembly of the export apparatus, the most straightforward assembly path would be a linear inside-out extension through subsequent addition of SctJ and SctD in the IM, and finally SctC in the OM. However, a variety of studies has shown that in different organisms, the complete SctCDJ complex can form in the absence of the export apparatus and other T3SS components (Kubori et al. 2000; Kimbrough and Miller 2000; Sukhan et al. 2001; Ogino et al. 2006). Where assessed, assembly kinetics and efficiency were reduced in these studies. This indicates while it is possible that these SctCDJ complexes are intermediates in parallel assembly pathways, they might also represent dead ends of assembly, possibly favored by the used expression or purification methods (see below).

The assembly order of the membrane rings themselves might differ between species. The OM ring formed by the secretin SctC is stable and assembles independently with the sole help of a pilotin lipoprotein, YscW/InvH/MxiM (Koster et al. 1997; Crago and Koronakis 1998; Daefler and Russel 1998; Burghout et al. 2004; Okon et al. 2008; Perdu et al. 2015). Interestingly, in EPEC, where no pilotin is known (Gaytán et al. 2016), a recent study showed the IM ring component SctD was required for oligomerization of SctC in the OM, suggesting that SctD can partially take over pilotin function in this case (Tseytin et al. 2017). In contrast, in *Yersinia*, the secretin SctC is required for assembly of SctD and its subsequent interaction with SctJ (Diepold et al. 2010). This outside-in assembly may ensure penetration of the PG at the site of the injectisome assembly, which is especially important for the *Yersinia* T3SS, as it has no dedicated PG hydrolase associated with the T3SS (see Box 2). The assembly order of the membrane rings in other species is unknown. SctD and SctJ can independently form a ring structure in some species (Kimbrough and Miller 2000; Schraidt et al. 2010), which is stabilized in presence of SctC, indicating an inside-out assembly of the membrane rings in these cases.

Given the possible formation of SctDJ and SctCDJ membrane rings in the absence of export apparatus components, it is particularly interesting how the formation of "empty" ring structures (which, if stable, most likely represent dead ends) is prevented. A structural study by Burkinshaw et al. (2015) showed a surprisingly small interaction interface between SctD and SctJ and indicated the presence of two SctJ confirmations, a monomeric, assembly-incompetent confirmation and an assembly-competent form, which is stabilized by contact to SctD. It is tempting to speculate that despite the closer interaction between SctD and SctJ shown by Worrall et al. (2016) this mechanism might also ensure integration of the export apparatus prior to SctJ oligomerization; however, this concept currently lacks experimental validation.

6 Assembly of the Cytosolic Complex

At the cytosolic interface of the injectisome, five proteins are required for the function of the T3SS: SctK, Q, L, N, O (see Fig. 1). Little is known about the role in T3SS formation of the "stalk" protein SctO, which is not required for the assembly of the other cytosolic components (Diepold et al. 2010, 2012; Zhang et al. 2017; Hu et al. 2017). In contrast, the interactions and assembly of the other four cytosolic T3SS components have been studied in detail. Live fluorescence microscopy experiments showed that the assembly of each of the four proteins SctK, Q, L, N, including the additionally expressed $SctQ_C$, requires the presence all four proteins, as well as the SctCDJ membrane rings (Diepold et al. 2010; Zhang et al. 2017).[3] The presence of the export apparatus at least stabilizes the assembly of the cytosolic proteins (Diepold et al. 2010; Zhang et al. 2017). These results were corroborated by cryo-ET studies and biochemical analyses for *Salmonella* and *Shigella* (Hu et al. 2015, 2017; Lara-Tejero et al. 2019), with the notable exception that in *Salmonella* SPI-1, assembly of the remaining cytosolic complex still occurred to some extent in strains lacking SctN or $SctQ_C$ (Hu et al. 2017; Lara-Tejero et al. 2019). Absence of single cytosolic proteins can even decrease the stability of the other interacting proteins (Makino et al. 2016; Lara-Tejero et al. 2019). Together, these results suggest the formation of one large interdependent cytosolic complex at the proximal interface of the injectisome. How exactly this complex (which matches the membrane rings in size, see Table 1) attaches to the cytosolic interface is unknown. Possibilities range from the addition (and mutual stabilization) of single proteins to addition of single pods or attachment of the whole cytosolic complex. The finding that the binding of SctK to the cytosolic domain of four SctD molecules (Hu et al. 2017) does not occur in the absence of SctQ or SctN (Diepold et al. 2017) supports the idea of a pod-wise assembly or strongly cooperative binding of the cytosolic components at the injectisome.

7 Assembly of the Needle

After assembly of the cytosolic components to the export apparatus/membrane rings complex, the T3SS is ready for secretion of the early substrates, including the needle subunit SctF, the ruler protein SctP, and the inner rod subunit SctI. Interestingly, these three proteins themselves are required for different steps in export: While SctI was already essential for export of an early substrate across the IM, SctF was only required for complete secretion across both membranes and SctP was dispensable for secretion of early export substrates (Diepold et al. 2012).

[3]Notably, large, partially membrane-associated complexes of the cytosolic components can still form in the absence of the membrane rings (Johnson and Blocker 2008; Lara-Tejero et al. 2011; Diepold et al. 2017)—see "Stability and dynamics of the injectisome" for more details.

Formation of the needle opens the previously closed periplasmic gate of the SctC secretin ring, which leads to significant structural rearrangements in the membrane ring structures (Marlovits et al. 2004; Hu et al. 2018).

The location within the injectisome of the inner rod formed by SctI suggests that is formed prior to the needle. However, export of SctI occurs at the same time as needle formation. Mutants in the ruler protein, SctP, lead to parallel changes in the export levels of both of these proteins (Wood et al. 2008) and the expression ratio of SctF and SctI influenced needle length in *Salmonella* (Marlovits et al. 2006), although the exact mechanism remains unclear, especially in the light of the low stoichiometry of SctI proposed in recent studies (Zilkenat et al. 2016; Kuhlen et al. 2018).

In at least some T3SS, SctF secretion requires the presence of cytosolic needle chaperones (Quinaud et al. 2005; Sun et al. 2008; Sal-Man et al. 2013) thought to prevent premature oligomerization, which occurs spontaneously in vitro (Poyraz et al. 2010). Needle elongation occurs by addition of single secreted SctF proteins at the distal end of the needle (Poyraz et al. 2010), similar to the elongation of the flagellum. Attachment of the needle subunits has long been thought to occur without the assistance of cap proteins required for elongation of the flagellum (Blocker et al. 2008). However, it was recently shown that in *Salmonella* SPI-1, a protein previously characterized as effector assists needle elongation in a very similar way (Kato et al. 2018).

8 The First Substrate Switch and Assembly of the Needle Tip

Needles have a defined length that varies between different bacterial species and subspecies. Length control probably ensures that the distance between the bacterium and the host cell membrane is bridged, as has been elegantly shown by manipulating the length of the adhesin, YadA, which determines this distance in *Yersinia enterocolitica* (Mota et al. 2005). Three proteins have been proposed to be involved in regulating needle length and the subsequent switch of substrate specificity to stop elongation of the needle and allow formation of the needle tip: the largely unstructured "ruler" protein SctP, the substrate switch protein SctU in the export apparatus, and the inner rod protein, SctI.

Different models for how these proteins measure needle length and switch the substrate specificity of the T3SS have been proposed. Based on the lack of an inner rod structure in ruler mutants in *Salmonella* SPI-1 and effects of the expression ratio of needle and inner rod subunit on needle length, an inner rod needle length control model was proposed, according to which complete assembly of the inner rod terminates needle growth and switches substrate specificity (Marlovits et al. 2006; Lefebre and Galán 2013). Another model, the ruler model, is based on the observation that the length of the ruler protein is directly correlated to needle length in *Yersinia* (Journet et al. 2003; Mota et al. 2005; Wagner et al. 2009), as was shown for the flagellar hook (Shibata et al. 2007; Erhardt et al. 2011) and recently for SPI-1 (Wee and Hughes 2015). In line with this hypothesis, the recent structural characterization of the *P. aeruginosa* ruler defined the interaction between the C-terminus of SctP (Bergeron et al. 2016) and SctU. How the mechanism(s) that sense completion of the needle lead to switching of the substrate specificity is still under debate. Release of the C-terminal domain of the switch protein might lead to a functional rearrangement of the export apparatus or linked components (Frost et al. 2012), allowing the export of the hydrophilic translocator, whose assembly at the needle tip prevents further elongation of the needle (Poyraz et al. 2010). A point mutant in *Y. enterocolitica* SctU which is unable to undergo autocleavage specifically prevents the export of the translocators (Sorg et al. 2007; Zarivach et al. 2008). This phenotype is linked to the specific recognition of export signals, as export can be restored by adding an effector export signal to LcrV (Sorg et al. 2007). Notably, needle length control in this strain is unaffected if the export of the ruler protein is ensured (Sorg et al. 2007; Monjarás Feria et al. 2015), suggesting that length measurement and switching can be uncoupled. *See review by Shin-Ichi Aizawa in this edition for more details.*

9 The Second Substrate Switch and Assembly of the Translocon

Once the formation of the needle has been finished by the assembly of the tip, the T3SS enters a steady state in which secretion of the late substrates (hydrophobic translocators and virulence effectors) can be induced by external signals, most prominently by contact to a host cell. The two hydrophobic translocators SctB and SctE can destabilize or disrupt the host cell membrane and form the host membrane-spanning translocon, thought to ensure a direct conduit for effector translocation. The hydrophobic translocators form both homo- and hetero-multimeric structures in varying sizes in membranes (Ide et al. 2001; Schoehn et al. 2003; Faudry et al. 2006; Montagner et al. 2011; Myeni et al. 2013), which makes them difficult study objects. Romano et al. showed the formation of a defined heterohexadecameric $SctB_8$-$SctE_8$ pore that depended on the interaction of both hydrophobic translocators during translocon assembly (Romano et al. 2016). Recent in situ visualizations of translocons at the interface between needle and host membrane showed that translocon formation was strongly induced by host cell cues (Nauth et al. 2018) and was accompanied by significant distortions in the host membrane, which may contribute to efficient insertion of the pore or delivery of effector proteins (Park et al. 2018).

But how is the second substrate switch initiated? Host cell contact leads to conformational changes in the translocon proteins (Veenendaal et al. 2007; Roehrich et al. 2013; Armentrout and Rietsch 2016), which are transmitted to the base by conformational rearrangements of the needle (Kenjale et al. 2005; Davis and Mecsas 2006; Torruellas et al. 2005) and lead to release of the cytosolic gatekeeper protein SctW (Martinez-Argudo and Blocker 2010). The exact mechanism of gatekeeper release is currently unclear; interactions with the hydrophilic translocator (Roehrich et al. 2013), as well as a calcium-dependent interaction with the ruler (Shaulov et al. 2017), have been implied to participate. A key player in the second substrate switch is the large export apparatus protein SctV. Its cytosolic domain, $SctV_C$, binds gatekeeper proteins (Portaliou et al. 2017), as well as substrate-chaperone complexes, and appears to regulate the switch from translocator-chaperone to effector-chaperone complexes in response to external signals activating effector secretion (Büttner et al. 2006; Bange et al. 2010; Hartmann and Büttner 2013; Gaytán et al. 2018; Portaliou et al. 2017; Xing et al. 2018). *See review by William Picking in this edition.* Chemical activation of the T3SS by chelation of Ca^{2+}, which has been proposed to mimic the intracellular environment within host cells (Fowler and Brubaker 1994), may activate effector secretion differently (Gaytán et al. 2018), and has been shown to alter protein interactions among the cytosolic T3SS components even in the absence of needles (Diepold et al. 2017). If and when these two pathways converge is unclear at the moment. Notably, co-IP experiments with the export apparatus switch protein SctU showed a different subset of interacting cytosolic components between the wild-type and non-cleavable versions of SctU (Riordan and Schneewind 2008; Botteaux et al. 2010), suggesting a link between the substrate switch and the arrangement of the export apparatus and the cytosolic complex.

Despite—or maybe because of—the dazzling number of reported interactions between substrate-chaperone complexes, different proteins of the cytosolic sorting platform, the export apparatus, the plug complex and the ruler protein, the path and regulation of protein export, and the establishment of a secretion hierarchy by the T3SS remains one of the most important open research questions in the field.

10 Stability and Dynamics of the Injectisome

The needle complex (NC) of the *Salmonella* SPI-1 T3SS was first isolated and visualized by Kubori et al. (1998) (Kubori et al. 1998), and since then NCs from various bacteria have been isolated and described. Purification steps frequently include basic pH, addition of detergents, and ultracentrifugation, indicating a high post-assembly stability of the NCs. On the other hand, *Yersinia* needles can be easily detached by mechanical shearing (Hoiczyk and Blobel 2001; Mueller et al. 2005), possibly a reason for the lack of purified NC structures from this organism. In contrast to their flagellar equivalents, hook basal bodies, purified NCs do not contain cytosolic components and only recently, these became accessible to structural analysis by cryo-ET (Hu et al. 2015; Nans et al. 2015; Hu et al. 2017; Makino et al. 2016), showing surprising differences in the arrangement of the homologous proteins between flagella and injectisomes. This may indicate that the cytosolic components are less tightly bound to the NC in the injectisome, and indeed, it has been shown that the main cytosolic T3SS components SctK, SctL,

SctN, and SctQ exist both in a membrane-bound and a cytosolic state (Johnson and Blocker 2008; Lara-Tejero et al. 2011; Diepold et al. 2017; Zhang et al. 2017). A combination of superresolution and fluorescence recovery microscopy revealed that at least for SctQ, the injectisome-bound and the cytosolic pool exchange subunits between each other (Diepold et al. 2015), similar to what was shown for the flagellar homologue (Leake et al. 2006). Notably, the exchange rate was reduced under non-secreting conditions, suggesting a link between protein dynamics and effector export (Diepold et al. 2015). In line with these data, recent cryo-ET studies showed changes in the electron density of the cytosolic complex between secreting and non-secreting conditions (Nans et al. 2015), and in certain tomogram subclasses (Makino et al. 2016), although no effect of host cell binding was observed for *Salmonella* SPI-1 (Park et al. 2018), and at any given time, most *Yersinia* basal bodies harbor cytosolic complexes including all four components, SctK, L, N, and Q (Diepold et al. 2017). Taken together, these studies strongly suggest that the T3SS displays subunit exchange and rearrangements of the cytosolic complex in response to external cues, a phenomenon that may be more common than thought in biological assemblies (Tusk et al. 2018).

The interactions between the cytosolic components also play an important role in the formation and behavior of soluble complexes. Johnson and Blocker (2008) and Lara-Tejero et al. (2011) demonstrated the presence of large complexes in the cytosol, containing SctK, Q, L, and—in *Shigella*, but not in *Salmonella* SPI-1—SctN. Measurements in live *Y. enterocolitica* also confirmed the presence of soluble SctKQLN complexes and showed a wide range of diffusion coefficients for these complexes, indicating rearrangement or turnover of the participating proteins. Both protein interaction and diffusion of complexes change upon chemical induction of T3SS secretion, again suggesting that these cytosolic complexes are directly involved in (or react to) effector secretion (Diepold et al. 2017). Similar concepts have been brought up for the T3SS ATPase in the flagellar motor (Bai et al. 2014), and the injectisome (Case and Dickenson 2018), where oligomerization, ATPase activity and potentially binding to the injectisome of SctN were influenced by its regulator SctL. A recent study indicated that in *Salmonella* SPI-1, the $SctQ_{FL}$ binding sites for $SctQ_C$ and SctL overlap (Lara-Tejero et al. 2019), and the authors proposed that $SctQ_C$ may stabilize cytosolic sorting platform components or subcomplexes, rather than take a structural role in the injectisome. As the localization of $SctQ_C$ in injectisomes, shown for *Yersinia* (Diepold et al. 2015), was not tested in this study, it is difficult to estimate whether this "handover" model may represent a species-specific adaptation.

What could be the reason for this remarkable conformational and binding flexibility of the cytosolic components, especially SctQ? Recently, it was found that the presence of two SPOA[4] domains in SctQ—one on each side of the translation initiation site leading to expression of additional $SctQ_C$—promotes both

[4]The surface presentation of antigens (SPOA) domain (Pfam 01052) (Suvarnapunya et al. 2003) has been mainly found in the cytosolic components of T3SS including flagella; the archetype is based on the flagellar C-ring component FliN.

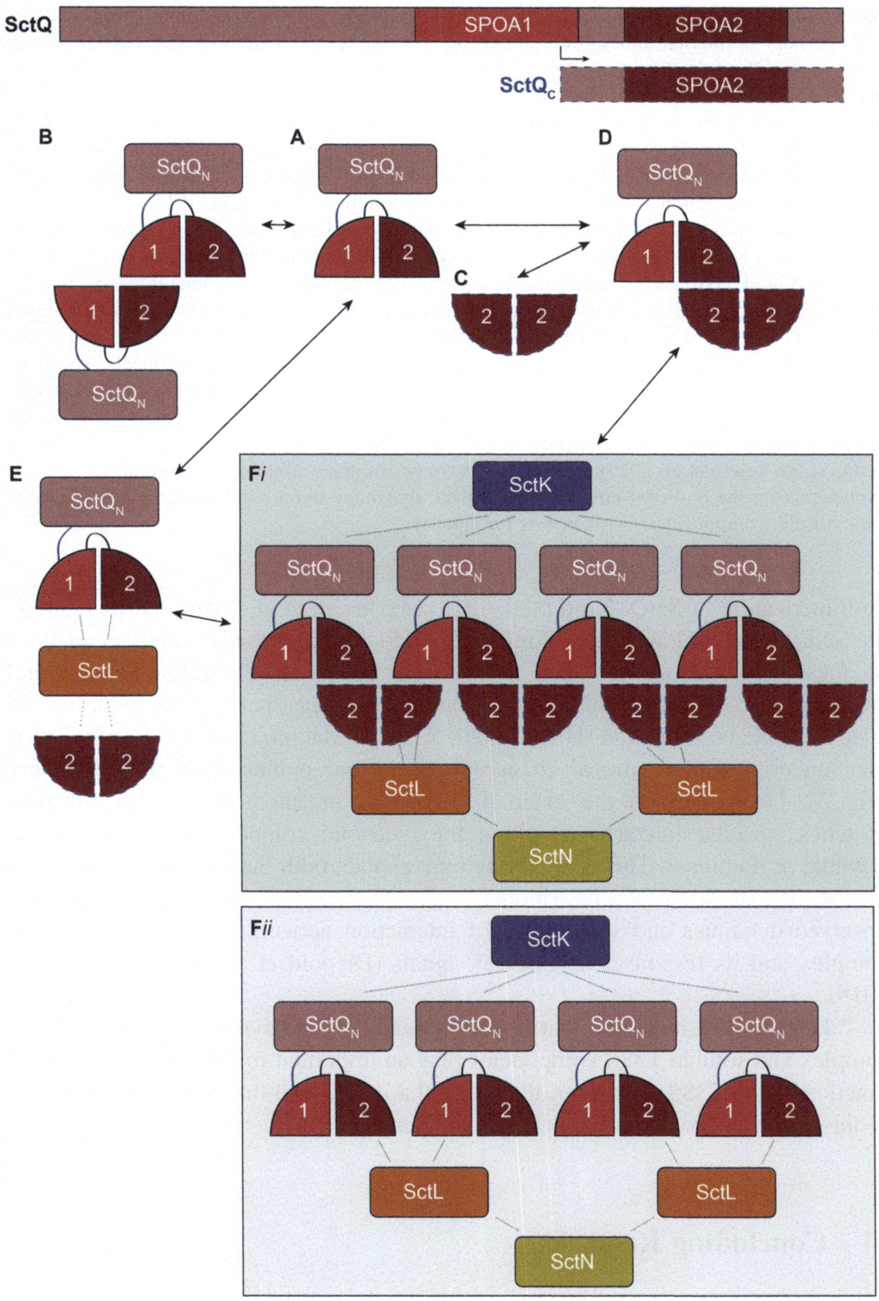
SctQ
SPOA1
SPOA2
SctQ$_C$
SPOA2
B
A
D
C
E
SctQ$_N$
SctL
SctK
SctN
F*i*
F*ii*

◂**Fig. 3 Possible pathways for the assembly and adaptation of the C-terminal complex**. *Top*: SctQ harbors an internal translation initiation site, leading to the additional expression of $SctQ_C$ (blue, dashed borders) (Yu et al. 2011; Bzymek et al. 2012). SctQ contains two SPOA domains, the SPOA1 domain (light red, only present in the full-length protein $SctQ_{FL}$), and the SPOA2 domain (dark red, present in both $SctQ_{FL}$ and $SctQ_C$). *Below*: Several possible interactions of SctQ and its SPOA domains have been shown: *A*, the two SPOA domains interact intramolecularly (Notti et al. 2015; McDowell et al. 2016). *B*, two $SctQ_{FL}$ proteins can interact intermolecularly; however, this interaction is not required for protein secretion (Notti et al. 2015). *C*, two $SctQ_C$ proteins can form a stable dimer (Bzymek et al. 2012; McDowell et al. 2016). *D*, this dimer can interact with a $SctQ_{FL}$ protein, forming the $SctQ{:}SctQ_{C,2}$ "building block" (Bzymek et al. 2012; McDowell et al. 2016). This interaction is predominant over the $SctQ_{FL}$ homodimer in *Shigella* (McDowell et al. 2016). *E*, SctL can bind both $SctQ_{FL}$ and, with lower affinity, $SctQ_C$ dimers (Notti et al. 2015). *F*, the $SctQ{:}SctQ_{C,2}$ "building block" can oligomerize into elongated arrays (McDowell et al. 2016). Putative interactions with the other cytosolic components (based on Jackson and Plano 2000; Notti et al. 2015; Hu et al. 2017) are indicated by dotted lines (left side, SctL interaction with SPOA1 + 2, right side, interaction with SPOA2 dimer). *i*, possible assembly of single pod including $SctQ_C$ (Diepold et al. 2015); *ii*, possible assembly of single pod excluding $SctQ_C$ (Lara-Tejero et al. 2019). Notably, the variety of different interactions provides a possible explanation for the observed protein exchange and dynamics and possible structural flexibility of the cytosolic components (see main text for details)

multimerization of SctQ (Notti et al. 2015; McDowell et al. 2016) and its binding, via SctL, to the ATPase SctN (Notti et al. 2015). The proposed flexible interactions of these SPOA domains have a profound effect on the possible structural arrangements of the cytosolic complex: although at least parts of the complex can adapt a stable confirmation (Hu et al. 2015, 2017; Makino et al. 2016), other parts may undergo conformational changes or bind and unbind from the machinery (Fig. 3). Depending on the external conditions or interactions upon substrate switches, specific interactions within the cytosolic components might become possible or dominant. This adaptability may explain both the multitude of detected interactions of SctQ (Morita-ishihara et al. 2005; Spaeth et al. 2009), and the observed dynamics and wide range of interaction networks within the cytosolic complex and its responses to external signals (Diepold et al. 2017; Rocha et al. 2018).

Taken together, it appears that structural flexibility and dynamics of the cytosolic complex and soluble T3SS components play an important role in the assembly and function of the T3SS. However, the molecular details and the exact functional role of this feature remain unknown at this moment.

11 Concluding Remarks

Our knowledge on the assembly of the T3SS has considerably advanced over the last years and is a showcase for how structural, biochemical, and microscopy-based studies can lead to a holistic picture of such a complex process. We now can both describe the common assembly pathway and identify species-specific deviations, often linked to unique properties of the respective T3SS. An interesting question for future studies is whether and how the assembly pathways—both the canonical and the specific ones—are linked to the function of the machinery.

It now seems clear that beyond the assembly, ongoing dynamics and adaptation of the T3SS, especially its cytosolic parts, influence the function of the T3SS in vivo. While this is an exciting field of research, the link of the phenomenological observations to the functional role of these findings still remains to be defined in many cases. Together, these studies will provide a much clearer picture of not only the assembly, but also the function of the T3SS during the infection process.

An important and active field of research is the search for inhibitors of the T3SS. Therapeutics that inhibit the assembly or dynamics of the T3SS, a crucial virulence factor for many important pathogens, have the potential to prevent or attenuate a huge number of cases of illnesses, and various reports highlight the potency of such assembly inhibitors (Veenendaal et al. 2009; Duncan et al. 2012; Marshall and Finlay 2014; Morgan et al. 2017). The identification of common assembly steps and bottlenecks during the assembly process will undoubtedly yield valuable targets for structure-based drug design.

Box 1: There's more than one way to skin a cat—communalities and differences in the assembly of secretion systems and the flagellum

Protein complexes spanning both bacterial membranes in Gram-negative bacteria face similar conceptual challenges during the assembly process: They need to penetrate the peptidoglycan layer, ensure proper association of the outer and inner membrane-spanning rings, prevent leakage, and regulate the formation and function of the system in response to—often external—signals. Is the assembly of bacterial secretion systems thus governed by certain common pathways or concepts?

Various bacterial secretion systems (discussed in more detail by Costa et al. (2015), Green and Mecsas (2016), Galán and Waksman (2018)) and related machines such as the flagellum span both bacterial membranes. The type 1 secretion system (T1SS), which secretes diverse, often nutrient-scavenging proteins into the extracellular space; the T2SS, which exports folded proteins from the periplasm, and the related type IV pili; the T3SS injectisome and the related flagellum; the T4SS, which can translocate DNA and proteins into a host cell in a one-step mechanism; the T6SS, which shares evolutionary and structural similarities with an inverted contractile phage tail and can translocate proteins into eukaryotic or prokaryotic hosts, and the T9SS, another protein secretion pathway additionally involved in motility, which is restricted to the *Bacteroidetes* phylum.

The **T1SS** consists of three main parts, an ABC transporter in the IM, a periplasmic membrane fusion protein anchored in the IM, and an OM porin (Masi and Wandersman 2010; Thomas et al. 2014). The two IM components can independently assemble stable complexes. However, only upon engagement of the substrate, a transient bridge formed by the membrane fusion protein links to the porin present in the OM (Létoffé et al. 1996; Thanabalu et al. 1998).

In the more complex **T2SS** and related **type IV pili** (T4P), distinct networks of interaction have been observed for the IM components, the pseudopilins, and the

OM secretin and pilotin (Howard 2014). However, the localization of IM proteins depends on the presence of the secretin in the OM (Lybarger et al. 2009). Starting from the secretin, an outside-in assembly is initiated via a tripartite IM complex, which subsequently binds to a cytoplasmic component (Friedrich et al. 2014). The symmetry mismatch between the OM ring and the IM platform may lead to formation of a metastable complex that specifically allows entry of *bona fide* T2SS substrates into the export channel (Hay et al. 2017). The cytosolic ATPase complexes responsible for extension and retraction of the T4P are recruited—mutually exclusively—on demand (Chang et al. 2016). In several of the assembly steps, disordered regions of proteins and their structural transitions have been shown to play crucial roles (Gu et al. 2017).

Despite a long research history, the assembly of the **T3SS** part of the **flagellum** is not fully elucidated yet. Based on stable assembly intermediates, a largely inside-out assembly starting at the MS ring in the IM, with subsequent additions of the C-ring, ATPase and export apparatus, and the periplasmic and OM rings was established (Kubori et al. 1992, 1997; Lux et al. 2000; Macnab 2003). Likewise, the periplasmic rods assemble in an inside-out order (Burrage et al. 2018), with the length of the outermost distal rod determined by the distance to the OM (Cohen et al. 2017). There is conflicting evidence on whether the export apparatus can serve as a nucleation point for assembly of the flagellar motor (Li and Sourjik 2011; Morimoto et al. 2014).

Based on structure and interaction networks, Fronzes et al. (2009a) concluded that the first step in **T4SS** assembly is the formation of a stable core of three proteins (VirB7, 9, 10) already connecting the OM and the IM (Fronzes et al. 2009b; Chandran et al. 2009). In the next steps, IM components including the ATPase (Yuan et al. 2005; Walldén et al. 2012) and subsequently the periplasmic proteins forming the pilus can attach (Yuan et al. 2005; Fronzes et al. 2009a; Low et al. 2014). Interestingly, in *Legionella pneumophila*, the polar localization of the T4SS is conferred by two T6SS baseplate protein homologs, followed by stable insertion of a lipoprotein in the OM, and further additions in a largely outside-in direction, with the cytosolic ATPase DotB only recruited to the system at a late stage (Jeong et al. 2018; Chetrit et al. 2018; Ghosal et al. 2018).

The **T6SS** is a dynamic assembly, controlled by different signals in various organisms (Zoued et al. 2014; Basler 2015). During its multi-step formation, complexes of two to three IM components assemble at the IM, which form potential assembly sites for the baseplate and the sheath (Durand et al. 2015; Gerc et al. 2015; Brunet et al. 2015; Cherrak et al. 2018). The sheath then assembles at its distal end in the cytosol until contact to the opposing membrane has been established, and a regulatory protein stops and maintains the assembly (Vettiger et al. 2017; Santin et al. 2018). The T6SS uses a "domesticated" lytic transglycosylase whose activity is stimulated by an IM T6SS component to pass the PG layer (Santin and Cascales 2017).

Major aspects of the assembly of the **T9SS**, the latest addition to the secretion pathway collection, are still unknown. One of the earliest steps is most likely the formation of a core complex spanning both membranes (Vincent et al. 2017), which might be initiated at the IM, based on a comparison of pairwise protein affinities (Leone et al. 2018). Interestingly, the T9SS structure seems to be dynamic or at least can switch between two states, as its large OM translocon was found in two mutually exclusive conformations with a lateral or periplasmic opening, respectively (Lauber et al. 2018).

Looking at the impressive variety of methods used by these secretion systems to span both membranes and secrete proteins, it is obvious that assembly of multi-membrane-spanning complexes can occur in different fashions, both with respect to the assembly order and the dynamics of the secretion system. It will be interesting to see which tricks newly discovered secretion systems (such as the T9SS, and the T7SS in Gram-positive bacteria) have in store.

Secretion system	Assembly of membrane-spanning rings	Stability of core system
T1SS	Inside-out	On-demand assembly
T2SS	Outside-in	Stable
T3SS (injectisome)	Species-specific?	Stable (parts dynamic)
T3SS (flagellum)	Inside-out	Stable (parts dynamic)
T4SS	Simultaneous/outside in	Stable (parts and localization dynamic)
T6SS	Inside-out?	On-demand/dynamic assembly
T9SS	Unknown	Dynamic/various states

Box 2: Traversing the peptidoglycan and distribution of the T3SS

Any multi-membrane-spanning protein complex needs to penetrate the peptidoglycan (PG) layer, a mesh with an average pore size of only around 2 nm (Demchick and Koch 1996; Vollmer et al. 2008), much smaller than the diameter of the T3SS at this place (the minimal diameter of the membrane-spanning rings is at least 12 nm (Schraidt and Marlovits 2011; Bergeron et al. 2013; Hu et al. 2015, 2017)).

To this aim, the PG-spanning part of the T3SS, most likely the secretin ring (see Fig. 1), has to be inserted into the PG either by exploiting the action of a cellular PG-modifying enzyme (e.g. during cell growth or division), or by applying a dedicated PG-modifying enzyme. Such enzymes associated with the T3SS are widespread (Miras et al. 1995; Koraimann 2003), but are missing in some organisms, like *Yersinia*. In these cases, colocalization of the secretin and/or its pilotin protein with PG-modifying enzymes (as has been observed for the *E. coli* T6SS (Santin and Cascales 2017)) is likely to determine the localization of the final T3SS complex, which might explain the strict outside-in assembly observed for the *Yersinia* membrane rings (Diepold et al. 2010). In contrast, the availability of dedicated T3SS PG-modifying enzymes could facilitate the integration of the secretin in the OM at the locus of the IM rings after their completion. Indeed, Burkinshaw et al. (2015) showed the interaction and activation of a PG lytic enzyme, EtgA, by the T3SS inner rod component SctI in EPEC. Recently, Hausner et al. observed a similar association in *Xanthomonas campestris* (Hausner et al. 2017). If SctI is already recruited to the growing injectisome in the absence of a secretin ring, such a mechanism could account for penetration of the PG layer and connection of the IM ring (which at this point would already need to harbor a functional secretion machinery) and the secretin, putting the secretin at a much later point of the EPEC T3SS assembly. Further experimental validation of this interesting hypothesis in various bacteria could greatly increase our knowledge about the integration, and overall assembly, of the T3SS.

Notably, despite the possibly dissimilar ways to position their T3SS, the distribution of at least the C-ring component SctQ in *Yersinia* (where injectisomes have been shown to form clusters in the membrane (Kudryashev et al. 2015)) and *Salmonella* SPI-1 is strikingly similar (Diepold et al. 2010; Notti et al. 2015; Zhang et al. 2017). Whether the integration of systems such as the *Salmonella* SPI-2 system which only requires few injectisomes to establish contact with the vacuolar membrane (Chakravortty et al. 2005) or the *Shigella* injectisome present in 50–100 copies per cell (Blocker et al. 1999) is reflected in different PG integration and positioning pathways remains to be tested.

References

Abby SS, Rocha EPC (2012) The non-flagellar type III secretion system evolved from the bacterial flagellum and diversified into host-cell adapted systems. PLoS Genet 8:e1002983

Abrusci P, Vergara-Irigaray M, Johnson S, Beeby MD, Hendrixson DR, Roversi P, Friede ME, Deane JE, Jensen GJ, Tang CM, Lea SM (2013) Architecture of the major component of the type III secretion system export apparatus. Nat Struct Mol Biol 20:99–104

Abrusci P, McDowell MA, Lea SM, Johnson S (2014) Building a secreting nanomachine: a structural overview of the T3SS. Curr Opin Struct Biol 25C:111–117

Akeda Y, Galán JE (2005) Chaperone release and unfolding of substrates in type III secretion. Nature 437:911–915

Akopyan K, Edgren T, Wang-Edgren H, Rosqvist R, Fahlgren A, Wolf-Watz H, Fällman M (2011) Translocation of surface-localized effectors in type III secretion. Proc Natl Acad Sci U S A 108:1639–1644

Allaoui A, Sansonetti PJ, Parsot C (1992) MxiJ, a lipoprotein involved in secretion of *Shigella* Ipa invasins, is homologous to YscJ, a secretion factor of the *Yersinia* Yop proteins. J Bacteriol 174:7661–7669

Armentrout EI, Rietsch A (2016) The type III secretion translocation pore senses host cell contact. PLoS Pathog 12:e1005530

Bai F, Morimoto YV, Yoshimura SDJ, Hara N, Kami-Ike N, Namba K, Minamino T (2014) Assembly dynamics and the roles of FliI ATPase of the bacterial flagellar export apparatus. Sci. Rep 4:6528

Bange G, Kümmerer N, Engel C, Bozkurt G, Wild K, Sinning I (2010) FlhA provides the adaptor for coordinated delivery of late flagella building blocks to the type III secretion system. Proc Natl Acad Sci U S A 107:11295–11300

Basler M (2015) Type VI secretion system: secretion by a contractile nanomachine. Philos Trans R Soc Lond B Biol, Sci, p 370

Bergeron JRC, Worrall LJ, Sgourakis NG, DiMaio F, Pfuetzner RA, Felise HB, Vuckovic M, Yu AC, Miller SI, Baker D, Strynadka NCJ (2013) A refined model of the prototypical *Salmonella* SPI-1 T3SS basal body reveals the molecular basis for its assembly. PLoS Pathog 9:e1003307

Bergeron JRC, Worrall LJ, De S, Sgourakis NG, Cheung AH, Lameignère E, Okon M, Wasney GA, Baker D, McIntosh LP, Strynadka NCJ, Lameignere E, Okon M, Wasney GA, Baker D, McIntosh LP, Strynadka NCJ (2015) The modular structure of the inner-membrane ring component PrgK facilitates assembly of the type III secretion system basal body. Structure 23:161–172

Bergeron JRC, Fernández L, Wasney GA, Vuckovic M, Reffuveille F, Hancock REW, Strynadka NCJ (2016) The structure of a type 3 secretion system (T3SS) ruler protein suggests a molecular mechanism for needle length sensing. J Biol Chem 291:1676–1691

Blaylock B, Riordan KE, Missiakas D, Schneewind O (2006) Characterization of the *Yersinia enterocolitica* type III secretion ATPase YscN and its regulator, YscL. J Bacteriol 188:3525–3534

Blocker AJ, Gounon P, Larquet E, Niebuhr K, Cabiaux V, Parsot C, Sansonetti P (1999) The tripartite type III secreton of *Shigella flexneri* inserts IpaB and IpaC into host membranes. J Cell Biol 147:683–693

Blocker AJ, Jouihri NN, Larquet E, Gounon P, Ebel F, Parsot C, Sansonetti P, Allaoui A (2001) Structure and composition of the *Shigella flexneri* 'needle complex', a part of its type III secreton. Mol Microbiol 39:652–663

Blocker AJ, Deane JE, Veenendaal AKJ, Roversi P, Hodgkinson J, Johnson S, Lea SM (2008) What's the point of the type III secretion system needle? Proc Natl Acad Sci U S A 105:6507–6513

Botteaux A, Kayath CA, Page A-L, Jouihri NN, Sani M, Boekema EJ, Biskri L, Parsot C, Allaoui A (2010) The 33 carboxyl terminal residues of Spa40 orchestrate the multi-step assembly process of the type III secretion needle complex in *Shigella flexneri*. Microbiology 156:2807–2817

Bröms JE, Edqvist PJ, Carlsson KE, Forsberg Å, Francis MS, Bro JE (2005) Mapping of a YscY binding domain within the LcrH chaperone that is required for regulation of *Yersinia* type III secretion. J Bacteriol 187:7738–7752

Broz P, Mueller CA, Müller SA, Philippsen A, Sorg I, Engel A, Cornelis GR (2007) Function and molecular architecture of the *Yersinia* injectisome tip complex. Mol Microbiol 65:1311–1320

Brunet YR, Zoued A, Boyer F, Douzi B, Cascales E (2015) The type VI secretion TssEFGK-VgrG phage-like baseplate is recruited to the TssJLM membrane complex via multiple contacts and serves as assembly platform for tail tube/sheath polymerization. PLoS Genet 11:e1005545

Burghout P, Beckers F, de Wit E, van Boxtel R, Cornelis GR, Tommassen J, Koster M (2004) Role of the pilot protein YscW in the biogenesis of the YscC secretin in *Yersinia enterocolitica*. J Bacteriol 186:5366–5375

Burkinshaw BJ, Strynadka NCJ (2014) Assembly and structure of the T3SS. Biochim Biophys Acta Mol Cell Res 1843:1648–1693

Burkinshaw BJ, Deng W, Lameignère E, Wasney GA, Zhu H, Worrall LJ, Finlay BB, Strynadka NCJ (2015) Structural analysis of a specialized type III secretion system peptidoglycan-cleaving enzyme. J Biol Chem 290:10406–10417

Burrage AM, Vanderpool E, Kearns DB (2018) The assembly order of flagellar rod subunits in Bacillus subtilis. J Bacteriol JB.00425-18

Büttner D (2012) Protein export according to schedule: architecture, assembly, and regulation of type III secretion systems from plant- and animal-pathogenic bacteria. Microbiol Mol Biol Rev 76:262–310

Büttner D, He SY (2009) Type III protein secretion in plant pathogenic bacteria. Plant Physiol 150:1656–1664

Büttner D, Lorenz C, Weber E, Bonas U (2006) Targeting of two effector protein classes to the type III secretion system by a HpaC- and HpaB-dependent protein complex from *Xanthomonas campestris* pv. *vesicatoria*. Mol Microbiol 59:513–527

Bzymek KP, Hamaoka BY, Ghosh P (2012) Two translation products of *Yersinia yscQ* assemble to form a complex essential to type III secretion. Biochemistry 51:1669–1677

Case HB, Dickenson NE (2018) MxiN differentially regulates monomeric and oligomeric species of the *Shigella* type three secretion system ATPase Spa47. Biochemistry. acs.biochem.8b00070

Chakravortty D, Rohde M, Jäger L, Deiwick J, Hensel M (2005) Formation of a novel surface structure encoded by *Salmonella* Pathogenicity Island 2. EMBO J 24:2043–2052

Chandran V, Fronzes R, Duquerroy S, Cronin N, Navaza J, Waksman G (2009) Structure of the outer membrane complex of a type IV secretion system. Nature 462:1011–1015

Chang Y-W, Rettberg LA, Treuner-Lange A, Iwasa J, Sogaard-Andersen L, Jensen GJ (2016) Architecture of the type IVa pilus machine. Science (80) 351:aad2001-aad2001

Cherrak Y, Rapisarda C, Pellarin R, Bouvier G, Bardiaux B, Allain F, Malosse C, Rey M, Chamot-Rooke J, Cascales E, Fronzes R, Durand E (2018) Biogenesis and structure of a type VI secretion baseplate. Nat Microbiol 3:1404–1416

Chetrit D, Hu B, Christie PJ, Roy CR, Liu J (2018) A unique cytoplasmic ATPase complex defines the *Legionella pneumophila* type IV secretion channel. Nat Microbiol 3:678–686

Cheung M, Shen D-K, Makino F, Kato T, Roehrich AD, Martinez-Argudo I, Walker ML, Murillo I, Liu X, Pain M, Brown J, Frazer G, Mantell J, Mina P, Todd T, Sessions RB, Namba K, Blocker AJ (2014) Three-dimensional electron microscopy reconstruction and cysteine-mediated crosslinking provide a model of the type III secretion system needle tip complex. Mol Microbiol 95:31–50

Cohen EJ, Ferreira JL, Ladinsky MS, Beeby M, Hughes KT (2017) Nanoscale-length control of the flagellar driveshaft requires hitting the tethered outer membrane. Science 356:197–200

Cornelis GR (2006) The type III secretion injectisome. Nat Rev Microbiol 4:811–825

Costa TRD, Felisberto-Rodrigues C, Meir A, Prevost MS, Redzej A, Trokter M, Waksman G (2015) Secretion systems in Gram-negative bacteria: structural and mechanistic insights. Nat Rev Microbiol 13:343–359

Crago A, Koronakis V (1998) *Salmonella* InvG forms a ring-like multimer that requires the InvH lipoprotein for outer membrane localization. Mol Microbiol 30:47–56

Daefler S, Russel M (1998) The *Salmonella typhimurium* InvH protein is an outer membrane lipoprotein required for the proper localization of InvG. Mol Microbiol 28:1367–1380

Daley D (2008) The assembly of membrane proteins into complexes. Curr Opin Struct Biol 18:420–424

Daniell S, Takahashi N, Wilson R, Friedberg D, Rosenshine I, Booy F, Shaw R, Knutton S, Frankel G, Aizawa S-I (2001) The filamentous type III secretion translocon of enteropathogenic *Escherichia coli*. Cell Microbiol 3:865–871

Davis AJ, Mecsas J (2006) Mutations in the *Yersinia pseudotuberculosis* type III secretion system needle protein, YscF, that specifically abrogate effector translocation into host cells. J Bacteriol 189:83–97

Day JB, Plano GV (2000) The *Yersinia pestis* YscY protein directly binds YscX, a secreted component of the type III secretion machinery. J Bacteriol 182:1834–1843

Demchick P, Koch AL (1996) The permeability of the wall fabric of *Escherichia coli* and *Bacillus subtilis*. J Bacteriol 178:768–773

Deng W, Marshall NC, Rowland JL, McCoy JM, Worrall LJ, Santos AS, Strynadka NCJ, Finlay BB (2017) Assembly, structure, function and regulation of type III secretion systems. Nat. Rev, Microbiol

Diepold A, Armitage JP (2015) Type III secretion systems: the bacterial flagellum and the injectisome. Philos Trans R Soc B Biol Sci 370:20150020

Diepold A, Wagner S (2014) Assembly of the bacterial type III secretion machinery. FEMS Microbiol Rev 38:802–822

Diepold A, Amstutz M, Abel S, Sorg I, Jenal U, Cornelis GR (2010) Deciphering the assembly of the *Yersinia* type III secretion injectisome. EMBO J 29:1928–1940

Diepold A, Wiesand U, Cornelis GR (2011) The assembly of the export apparatus (YscR, S, T, U, V) of the *Yersinia* type III secretion apparatus occurs independently of other structural components and involves the formation of an YscV oligomer. Mol Microbiol 82:502–514

Diepold A, Wiesand U, Amstutz M, Cornelis GR (2012) Assembly of the *Yersinia* injectisome: the missing pieces. Mol Microbiol 85:878–892
Diepold A, Kudryashev M, Delalez NJ, Berry RM, Armitage JP (2015) Composition, formation, and regulation of the cytosolic C-ring, a dynamic component of the type III secretion injectisome. PLoS Biol 13:e1002039
Diepold A, Sezgin E, Huseyin M, Mortimer T, Eggeling C, Armitage JP (2017) A dynamic and adaptive network of cytosolic interactions governs protein export by the T3SS injectisome. Nat Commun 8:15940
Dietsche T, Tesfazgi Mebrhatu M, Brunner MJ, Abrusci P, Yan J, Franz-Wachtel M, Schärfe C, Zilkenat S, Grin I, Galán JE, Kohlbacher O, Lea SM, Macek B, Marlovits TC, Robinson CV, Wagner S (2016) Structural and functional characterization of the bacterial type III secretion export apparatus. PLoS Pathog 12:e1006071
Dohlich K, Zumsteg AB, Goosmann C, Kolbe M (2014) A substrate-fusion protein is trapped inside the type III secretion system channel in *Shigella flexneri*. PLoS Pathog 10:e1003881
Duncan CDS, Mata J (2011) Widespread cotranslational formation of protein complexes. PLoS Genet 7:e1002398
Duncan MC, Linington RG, Auerbuch V (2012) Chemical inhibitors of the type three secretion system: disarming bacterial pathogens. Antimicrob Agents Chemother 56:5433–5441
Durand E, Nguyen VS, Zoued A, Logger L, Péhau-Arnaudet G, Aschtgen M-S, Spinelli S, Desmyter A, Bardiaux B, Dujeancourt A, Roussel A, Cambillau C, Cascales E, Fronzes R (2015) Biogenesis and structure of a type VI secretion membrane core complex. Nature
Edgren T, Forsberg Å, Rosqvist R, Wolf-Watz H (2012) Type III secretion in *Yersinia*: injectisome or not? PLoS Pathog 8:e1002669
Erhardt M, Namba K, Hughes KT (2010) Bacterial nanomachines: the flagellum and type III injectisome. Cold Spring Harb Perspect Biol 2:a000299
Erhardt M, Singer HM, Wee DH, Keener JP, Hughes KT (2011) An infrequent molecular ruler controls flagellar hook length in *Salmonella enterica*. EMBO J 30:2948–2961
Evans LDB, Hughes C (2009) Selective binding of virulence type III export chaperones by FliJ escort orthologues InvI and YscO. FEMS Microbiol Lett 293:292–297
Faudry E, Vernier G, Neumann E, Forge V, Attree I (2006) Synergistic pore formation by type III toxin translocators of *Pseudomonas aeruginosa*. Biochemistry 45:8117–8123
Fowler JM, Brubaker RR (1994) Physiological basis of the low calcium response in *Yersinia pestis*. Am Soc Microbiol J
Friedrich C, Bulyha I, Sogaard-Andersen L (2014) Outside-in assembly pathway of the type IV pilus system in *Myxococcus xanthus*. J Bacteriol 196:378–390
Fronzes R, Christie PJ, Waksman G (2009a) The structural biology of type IV secretion systems. Nat Rev Microbiol 7:703–714
Fronzes R, Schäfer E, Wang L, Saibil HR, Orlova E, Waksman G (2009b) Structure of a type IV secretion system core complex. Science (80) 323:266–268
Frost S, Ho O, Login FH, Weise CF, Wolf-Watz H, Wolf-Watz M (2012) Autoproteolysis and intramolecular dissociation of *Yersinia* YscU precedes secretion of its C-terminal polypeptide YscU(CC). PLoS ONE 7:e49349
Fujii T, Cheung M, Blanco A, Kato T, Blocker AJ, Namba K (2012) Structure of a type III secretion needle at 7-Å resolution provides insights into its assembly and signaling mechanisms. Proc Natl Acad Sci U S A 109:4461–4466
Galán JE (2009) Common themes in the design and function of bacterial effectors. Cell Host Microbe 5:571–579
Galán JE, Waksman G (2018) Protein-injection machines in bacteria. Cell 172:1306–1318
Galán JE, Wolf-Watz H (2006) Protein delivery into eukaryotic cells by type III secretion machines. Nature 444:567–573
Gaytán MO, Martínez-Santos VI, Soto E, González-Pedrajo B (2016) Type three secretion system in attaching and effacing pathogens. Front Cell Infect Microbiol 6:129

Gaytán MO, Monjarás Feria J, Soto E, Espinosa N, Benítez JM, Georgellis D, González-Pedrajo B (2018) Novel insights into the mechanism of SepL-mediated control of effector secretion in enteropathogenic *Escherichia coli*. Microbiologyopen 7:e00571

Gerc AJAJ, Diepold A, Trunk K, Porter M, Rickman C, Armitage JP, Stanley-Wall NRNR, Coulthurst SJ (2015) Visualization of the Serratia type VI secretion system reveals unprovoked attacks and dynamic assembly. Cell Rep 12:2131–2142

Ghosal D, Chang Y-W, Jeong KC, Vogel JP, Jensen G (2018) Molecular architecture of the *Legionella* Dot/Icm type IV secretion system. bioRxiv 312009

Green ER, Mecsas J (2016) Bacterial secretion systems: an overview. Microbiol Spectr 4

Gu S, Shevchik VE, Shaw R, Pickersgill RW, Garnett JA (2017) The role of intrinsic disorder and dynamics in the assembly and function of the type II secretion system. Biochim Biophys Acta Proteins Proteomics 1865:1255–1266

Hartmann N, Büttner D (2013) The inner membrane protein HrcV from *Xanthomonas* spp. Is involved in substrate docking during type III secretion. Mol Plant-Microbe Interact 26:1176–1189

Hausner J, Hartmann N, Jordan M, Büttner D (2017) The predicted lytic transglycosylase HpaH from *Xanthomonas campestris* pv. *vesicatoria* associates with the type III secretion system and promotes effector protein translocation. Infect Immun 85

Hay ID, Belousoff MJ, Lithgow T (2017) Structural basis of type 2 secretion system engagement between the inner and outer bacterial membranes. MBio 8

Hoiczyk E, Blobel G (2001) Polymerization of a single protein of the pathogen *Yersinia enterocolitica* into needles punctures eukaryotic cells. Proc Natl Acad Sci U S A 98:4669–4674

Hu B, Morado DR, Margolin W, Rohde JR, Arizmendi O, Picking WL, Picking WD, Liu J (2015) Visualization of the type III secretion sorting platform of *Shigella flexneri*. Proc Natl Acad Sci 112:1047–1052

Hu B, Lara-Tejero M, Kong Q, Galán JE, Liu J (2017) In situ molecular architecture of the *Salmonella* type III secretion machine. Cell 168:1065–1074.e10

Hu J, Worrall LJ, Hong C, Vuckovic M, Atkinson CE, Caveney N, Yu Z, Strynadka NCJ (2018) Cryo-EM analysis of the T3S injectisome reveals the structure of the needle and open secretin. Nat Commun 9:3840

Hueck CJ (1998) Type III protein secretion systems in bacterial pathogens of animals and plants. Microbiol Mol Biol Rev 62:379–433

Ibuki T, Imada K, Minamino T, Kato T, Miyata T, Namba K (2011) Common architecture of the flagellar type III protein export apparatus and F- and V-type ATPases. Nat Struct Mol Biol 18:277–282

Ide T, Laarmann S, Greune L, Schillers H, Oberleithner H, Schmidt MA (2001) Characterization of translocation pores inserted into plasma membranes by type III-secreted Esp proteins of enteropathogenic *Escherichia coli*. Cell Microbiol 3:669–679

Imada K, Minamino T, Tahara A, Namba K (2007) Structural similarity between the flagellar type III ATPase FliI and F1-ATPase subunits. Proc Natl Acad Sci U S A 104:485–490

Iriarte M, Cornelis GR (1999) Identification of SycN, YscX, and YscY, three new elements of the *Yersinia* yop virulon. J Bacteriol 181:675–680

Jackson MW, Plano GV (2000) Interactions between type III secretion apparatus components from *Yersinia pestis* detected using the yeast two-hybrid system. FEMS Microbiol Lett 186:85–90

Jeong KC, Gyore J, Teng L, Ghosal D, Jensen GJ, Vogel JP (2018) Polar targeting and assembly of the *Legionella* Dot/Icm type IV secretion system (T4SS) by T6SS-related components. bioRxiv 315721

Johnson S, Blocker AJ (2008) Characterization of soluble complexes of the *Shigella flexneri* type III secretion system ATPase. FEMS Microbiol Lett 286:274–278

Johnson S, Roversi P, Espina M, Olive A, Deane JE, Birket S, Field T, Picking WL, Blocker AJ, Galyov EE, Lea SM (2006) Self-chaperoning of the type III secretion system needle tip proteins IpaD and BipD. J Biol Chem 282:4035–4044

Journet L, Agrain C, Broz P, Cornelis GR (2003) The needle length of bacterial injectisomes is determined by a molecular ruler. Science (80) 302:1757–1760

Kato J, Dey S, Soto JE, Butan C, Wilkinson MC, De Guzman RN, Galán JE (2018) A protein secreted by the *Salmonella* type III secretion system controls needle filament assembly. Elife 7

Kawamoto A, Morimoto YV, Miyata T, Minamino T, Hughes KT, Kato T, Namba K (2013) Common and distinct structural features of *Salmonella* injectisome and flagellar basal body. Sci Rep 3:3369

Kenjale R, Wilson J, Zenk S, Saurya S, Picking WL, Blocker AJ (2005) The needle component of the type III secreton of *Shigella* regulates the activity of the secretion apparatus. J Biol Chem 280:42929–42937

Kimbrough TG, Miller SI (2000) Contribution of *Salmonella* typhimurium type III secretion components to needle complex formation. Proc Natl Acad Sci U S A 97:11008–11013

Koraimann G (2003) Lytic transglycosylases in macromolecular transport systems of Gram-negative bacteria. Cell Mol Life Sci 60:2371–2388

Koster M, Bitter W, de Cock H, Allaoui A, Cornelis GR, Tommassen J (1997) The outer membrane component, YscC, of the Yop secretion machinery of *Yersinia enterocolitica* forms a ring-shaped multimeric complex. Mol Microbiol 26:789–797

Kowal J, Chami M, Ringler P, Müller SA, Kudryashev M, Castaño-Díez D, Amstutz M, Cornelis GR, Stahlberg H, Engel A (2013) Structure of the dodecameric *Yersinia enterocolitica* secretin YscC and Its trypsin-resistant core. Structure 21:2152–2161

Kubori T, Shimamoto N, Yamaguchi S, Namba K, Aizawa S-I (1992) Morphological pathway of flagellar assembly in *Salmonella* typhimurium. J Mol Biol 226:433–446

Kubori T, Yamaguchi S, Aizawa S-I (1997) Assembly of the switch complex onto the MS ring complex of *Salmonella* typhimurium does not require any other flagellar proteins. J Bacteriol 179:813–817

Kubori T, Matsushima Y, Nakamura D, Uralil J, Lara-Tejero M, Sukhan A, Galán JE, Aizawa S-I (1998) Supramolecular structure of the *Salmonella* typhimurium type III protein secretion system. Science (80) 280:602–605

Kubori T, Sukhan A, Aizawa S-I, Galán JE (2000) Molecular characterization and assembly of the needle complex of the *Salmonella* typhimurium type III protein secretion system. Proc Natl Acad Sci U S A 97:10225–10230

Kudryashev M, Stenta M, Schmelz S, Amstutz M, Wiesand U, Castaño-Díez D, Degiacomi MTM, Münnich S, Bleck CKC, Kowal J, Diepold A, Heinz DWD, Dal Peraro M, Cornelis GR, Stahlberg H (2013) In situ structural analysis of the *Yersinia enterocolitica* injectisome. Elife 2: e00792

Kudryashev M, Diepold A, Amstutz M, Armitage JP, Stahlberg H, Cornelis GR (2015) *Yersinia enterocolitica* type III secretion injectisomes form regularly spaced clusters, which incorporate new machines upon activation. Mol Microbiol 95:875–884

Kuhlen L, Abrusci P, Johnson S, Gault J, Deme J, Caesar J, Dietsche T, Mebrhatu MT, Ganief T, Macek B, Wagner S, Robinson CV, Lea SM (2018) Structure of the core of the type III secretion system export apparatus. Nat Struct Mol Biol 25:583–590

Lara-Tejero M, Kato J, Wagner S, Liu X, Galán JE (2011) A sorting platform determines the order of protein secretion in bacterial type III systems. Science (80) 331:1188–1191

Lara-Tejero M, Qin Z, Hu B, Butan C, Liu J, Galán JE (2019) Role of SpaO in the assembly of the sorting platform of a *Salmonella* type III secretion system. PLoS Pathog 15:e1007565

Lauber F, Deme JC, Lea SM, Berks BC (2018) Type 9 secretion system structures reveal a new protein transport mechanism. Nature 564:77–82

Lavander M, Sundberg L, Edqvist PJ, Lloyd SA, Wolf-Watz H, Forsberg Å (2002) Proteolytic cleavage of the FlhB homologue YscU of *Yersinia pseudotuberculosis* is essential for bacterial survival but not for type III secretion. J Bacteriol 184:4500–4509

Leake MC, Chandler J, Wadhams GH, Bai F, Berry RM, Armitage JP (2006) Stoichiometry and turnover in single, functioning membrane protein complexes. Nature 443:355–358

Lee P-C, Zmina SE, Stopford CM, Toska J, Rietsch A (2014) Control of type III secretion activity and substrate specificity by the cytoplasmic regulator PcrG. Proc Natl Acad Sci U S A 111: E2027-36

Lefebre MD, Galán JE (2013) The inner rod protein controls substrate switching and needle length in a *Salmonella* type III secretion system. Proc Natl Acad Sci U S A 2013:1–6

Leone P, Roche J, Vincent MS, Tran QH, Desmyter A, Cascales E, Kellenberger C, Cambillau C, Roussel A (2018) Type IX secretion system PorM and gliding machinery GldM form arches spanning the periplasmic space. Nat Commun 9:429

Létoffé S, Delepelaire P, Wandersman C (1996) Protein secretion in gram-negative bacteria: assembly of the three components of ABC protein-mediated exporters is ordered and promoted by substrate binding. EMBO J 15:5804–5811

Li H, Sourjik V (2011) Assembly and stability of flagellar motor in *Escherichia coli*. Mol Microbiol 80:886–899

Low HH, Gubellini F, Rivera-Calzada A, Braun N, Connery S, Dujeancourt A, Lu F, Redzej A, Fronzes R, Orlova EV, Waksman G (2014) Structure of a type IV secretion system. Nature 508:550–553

Lux R, Kar N, Khan S (2000) Overproduced *Salmonella* typhimurium flagellar motor switch complexes. J Mol Biol 298:577–583

Lybarger S, Johnson TL, Gray M, Sikora A, Sandkvist M (2009) Docking and assembly of the type II secretion complex of *Vibrio cholerae*. J Bacteriol 191:3149–3161

Macnab RM (2003) How bacteria assemble flagella. Annu Rev Microbiol 57:77–100

Majewski DDD, Worrall LJ, Hong C, Atkinson CEE, Vuckovic M, Watanabe N, Yu Z, Strynadka NCJ (2019) Cryo-EM structure of the homohexameric T3SS ATPase-central stalk complex reveals rotary ATPase-like asymmetry. Nat Commun 10:626

Makino F, Shen D, Kajimura N, Kawamoto A, Pissaridou P, Oswin H, Pain M, Murillo I, Namba K, Blocker AJ (2016) The architecture of the cytoplasmic region of type III secretion systems. Sci Rep 6:33341

Marlovits TC, Kubori T, Sukhan A, Thomas DR, Galán JE, Unger VM (2004) Structural insights into the assembly of the type III secretion needle complex. Science (80) 306:1040–1042

Marlovits TC, Kubori T, Lara-Tejero M, Thomas D, Unger VM, Galán JE (2006) Assembly of the inner rod determines needle length in the type III secretion injectisome. Nature 441:637–640

Marshall NC, Finlay BB (2014) Targeting the type III secretion system to treat bacterial infections. Expert Opin Ther Targets 18:137–152

Martinez-Argudo I, Blocker AJ (2010) The *Shigella* T3SS needle transmits a signal for MxiC release, which controls secretion of effectors. Mol Microbiol 78:1365–1378

Masi M, Wandersman C (2010) Multiple signals direct the assembly and function of a type 1 secretion system. J Bacteriol 192:3861–3869

McDowell MA, Marcoux J, McVicker G, Johnson S, Fong YH, Stevens R, Bowman LAH, Degiacomi MT, Yan J, Wise A, Friede ME, Benesch JLP, Deane JE, Tang CM, Robinson CV, Lea SM (2016) Characterisation of *Shigella* Spa33 and *Thermotoga* FliM/N reveals a new model for C-ring assembly in T3SS. Mol Microbiol 99:749–766

Minamino T, Macnab RM (2000a) Domain structure of *Salmonella* FlhB, a flagellar export component responsible for substrate specificity switching. J Bacteriol 182:4906–4914

Minamino T, Macnab RM (2000b) FliH, a soluble component of the type III flagellar export apparatus of *Salmonella*, forms a complex with FliI and inhibits its ATPase activity. Mol Microbiol 37:1494–1503

Miras I, Hermant D, Arricau N, Popoff MY (1995) Nucleotide sequence of iagA and iagB genes involved in invasion of HeLa cells by *Salmonella enterica* subsp. enterica ser. Typhi. Res Microbiol 146:17–20

Monjarás Feria JV, Lefebre MD, Stierhof YD, Galán JE, Wagner S (2015) Role of autocleavage in the function of a type iii secretion specificity switch protein in *Salmonella enterica* serovar typhimurium. MBio 6

Montagner C, Arquint C, Cornelis GR (2011) Translocators YopB and YopD from *Yersinia enterocolitica* form a multimeric integral membrane complex in eukaryotic cell membranes. J Bacteriol 193:6923–6928

Morgan JM, Duncan MC, Johnson KS, Diepold A, Lam H, Dupzyk AJ, Martin LR, Wong WR, Armitage JP, Linington RG, Auerbuch V (2017) Piericidin A1 Blocks *Yersinia* Ysc type III secretion system needle assembly. mSphere 2:e00030-17

Morimoto YV, Ito M, Hiraoka KD, Che YS, Bai F, Kami-ike N, Namba K, Minamino T (2014) Assembly and stoichiometry of FliF and FlhA in *Salmonella* flagellar basal body. Mol Microbiol 91:1214–1226

Morita-ishihara T, Ogawa M, Sagara H, Yoshida M, Katayama E, Sasakawa C (2005) *Shigella* Spa33 is an essential C-ring component of type III secretion machinery. J Biol Chem 281:599–607

Mota LJ, Journet L, Sorg I, Agrain C, Cornelis GR (2005) Bacterial injectisomes: needle length does matter. Science (80) 307:2005

Mueller CA, Broz P, Müller SA, Ringler P, Erne-Brand F, Sorg I, Kuhn M, Engel A, Cornelis GR (2005) The V-antigen of *Yersinia* forms a distinct structure at the tip of injectisome needles. Science (80) 310:674–676

Müller SA, Pozidis C, Stone R, Meesters C, Chami M, Engel A, Economou A, Stahlberg H (2006) Double hexameric ring assembly of the type III protein translocase ATPase HrcN. Mol Microbiol 61:119–125

Myeni SK, Wang L, Zhou D (2013) SipB-SipC complex is essential for translocon formation. PLoS ONE 8:e60499

Nans A, Kudryashev M, Saibil HR, Hayward RD (2015) Structure of a bacterial type III secretion system in contact with a host membrane in situ. Nat Commun 6:10114

Natan E, Wells JN, Teichmann SA, Marsh JA (2017) Regulation, evolution and consequences of cotranslational protein complex assembly. Curr Opin Struct Biol 42:90–97

Nauth T, Huschka F, Schweizer M, Bosse JB, Diepold A, Failla AV, Steffen A, Stradal TEB, Wolters M, Aepfelbacher M (2018) Visualization of translocons in *Yersinia* type III protein secretion machines during host cell infection. PLoS Pathog 14:e1007527

Notti RQ, Bhattacharya S, Lilic M, Stebbins CE (2015) A common assembly module in injectisome and flagellar type III secretion sorting platforms. Nat Commun 6:7125

Ogino T, Ohno R, Sekiya K, Kuwae A, Matsuzawa T, Nonaka T, Fukuda H, Imajoh-Ohmi S, Abe A (2006) Assembly of the type III secretion apparatus of enteropathogenic *Escherichia coli*. J Bacteriol 188:2801–2811

Okon M, Moraes T, Lario P, Creagh A, Haynes C, Strynadka NCJ, McIntosh LP (2008) Structural characterization of the type-III pilot-secretin complex from *Shigella flexneri*. Structure 16:1544–1554

Pallen MJ, Gophna U (2007) Bacterial flagella and type III secretion: case studies in the evolution of complexity. Genome Dyn 3:30–47

Pallen MJ, Bailey CM, Beatson SA (2006) Evolutionary links between FliH/YscL-like proteins from bacterial type III secretion systems and second-stalk components of the FoF1 and vacuolar ATPases. Protein Sci 15:935–941

Park D, Lara-Tejero M, Waxham MN, Li W, Hu B, Galán JE, Liu J (2018) Visualization of the type III secretion mediated *Salmonella*-host cell interface using cryo-electron tomography. Elife 7

Perdu C, Huber P, Bouillot S, Blocker AJ, Elsen S, Attree I, Faudry E (2015) ExsB is required for correct assembly of *Pseudomonas aeruginosa* type III secretion apparatus in the bacterial membrane and full virulence in vivo. Infect Immun IAI.00048-15

Portaliou AG, Tsolis KC, Loos MS, Zorzini V, Economou A (2016) Type III secretion: building and operating a remarkable nanomachine. Trends Biochem Sci 41:175–189

Portaliou AG, Tsolis KC, Loos MS, Balabanidou V, Rayo J, Tsirigotaki A, Crepin VF, Frankel G, Kalodimos CG, Karamanou S, Economou A (2017) Hierarchical protein targeting and secretion is controlled by an affinity switch in the type III secretion system of enteropathogenic *Escherichia coli*. EMBO J 36:3517–3531

Poyraz O, Schmidt H, Seidel K, Delissen F, Ader C, Tenenboim H, Goosmann C, Laube B, Thünemann A, Zychlinsky A, Baldus M, Lange A, Griesinger C, Kolbe M (2010) Protein refolding is required for assembly of the type three secretion needle. Nat Struct Mol Biol 17:788–792

Pozidis C, Chalkiadaki A, Gomez-Serrano A, Stahlberg H, Brown I, Tampakaki AP, Lustig A, Sianidis G, Politou AS, Engel A, Panopoulos NJ, Mansfield J, Pugsley AP, Karamanou S, Economou A (2003) Type III protein translocase: HrcN is a peripheral ATPase that is activated by oligomerization. J Biol Chem 278:25816–25824

Quinaud M, Chabert J, Faudry E, Neumann E, Lemaire D, Pastor A, Elsen S, Dessen A, Attree I (2005) The PscE-PscF-PscG complex controls type III secretion needle biogenesis in *Pseudomonas aeruginosa*. J Biol Chem 280:36293–36300

Radics J, Königsmaier L, Marlovits TC (2014) Structure of a pathogenic type 3 secretion system in action. Nat Struct Mol Biol 21:82–87

Riordan KE, Schneewind O (2008) YscU cleavage and the assembly of *Yersinia* type III secretion machine complexes. Mol Microbiol 68:1485–1501

Rocha JM, Richardson CJ, Zhang M, Darch CM, Cai E, Diepold A, Gahlmann A (2018) Single-molecule tracking in live *Yersinia enterocolitica* reveals distinct cytosolic complexes of injectisome subunits. Integr Biol 10:502–515

Roehrich AD, Guillossou E, Blocker AJ, Martinez-Argudo I (2013) *Shigella* IpaD has a dual role: signal transduction from the type III secretion system needle tip and intracellular secretion regulation. Mol Microbiol 87:690–706

Romano FB, Tang Y, Rossi KC, Monopoli KR, Ross JL, Heuck AP (2016) Type 3 Secretion translocators spontaneously assemble a hexadecameric transmembrane complex. J Biol Chem

Sal-Man N, Setiaputra D, Scholz R, Deng W, Yu ACY, Strynadka NCJ, Finlay BB (2013) EscE and EscG are cochaperones for the type III needle protein EscF of enteropathogenic *Escherichia coli*. J Bacteriol 195:2481–2489

Santin YG, Cascales E (2017) Domestication of a housekeeping transglycosylase for assembly of a Type VI secretion system. EMBO Rep 18:138–149

Santin YG, Doan T, Lebrun R, Espinosa L, Journet L, Cascales E (2018) In vivo TssA proximity labelling during type VI secretion biogenesis reveals TagA as a protein that stops and holds the sheath. Nat Microbiol 3:1304–1313

Schoehn G, Di Guilmi AM, Lemaire D, Attree I, Weissenhorn W, Dessen A (2003) Oligomerization of type III secretion proteins PopB and PopD precedes pore formation in *Pseudomonas*. EMBO J 22:4957–4967

Schraidt O, Marlovits TC (2011) Three-dimensional model of *Salmonella's* needle complex at subnanometer resolution. Science (80) 331:1192–1195

Schraidt O, Lefebre MD, Brunner MJ, Schmied WH, Schmidt A, Radics J, Mechtler K, Galán JE, Marlovits TC (2010) Topology and organization of the *Salmonella* typhimurium type III secretion needle complex components. PLoS Pathog 6:e1000824

Sekiya K, Ohishi M, Ogino T, Tamano K, Sasakawa C, Abe A (2001) Supermolecular structure of the enteropathogenic *Escherichia coli* type III secretion system and its direct interaction with the EspA-sheath-like structure. Proc Natl Acad Sci U S A 98:11638–11643

Shaulov L, Gershberg J, Deng W, Finlay BB, Sal-Man N (2017) The ruler protein EscP of the enteropathogenic *Escherichia coli* type III secretion system is involved in calcium sensing and secretion hierarchy regulation by interacting with the gatekeeper protein SepL. MBio 8: e01733-16

Shibata S, Takahashi N, Chevance FFV, Karlinsey JE, Hughes KT, Aizawa S-I (2007) FliK regulates flagellar hook length as an internal ruler. Mol Microbiol 64:1404–1415

Shiber A, Döring K, Friedrich U, Klann K, Merker D, Zedan M, Tippmann F, Kramer G, Bukau B (2018) Cotranslational assembly of protein complexes in eukaryotes revealed by ribosome profiling. Nature 561:268–272

Shieh Y-W, Minguez P, Bork P, Attree I, Guilbride DL, Kramer G, Bukau B (2015) Operon structure and cotranslational subunit association direct protein assembly in bacteria. Science (80) 350:678–680

Song M, Sukovich DJ, Ciccarelli L, Mayr J, Fernandez-Rodriguez J, Mirsky EA, Tucker AC, Gordon DB, Marlovits TC, Voigt CA (2017) Control of type III protein secretion using a minimal genetic system. Nat Commun 8:14737

Sorg I, Wagner S, Amstutz M, Müller SA, Broz P, Lussi Y, Engel A, Cornelis GR (2007) YscU recognizes translocators as export substrates of the *Yersinia* injectisome. EMBO J 26:3015–3024

Soto E, Espinosa N, Díaz-Guerrero M, Gaytán MO, Puente JL, González-Pedrajo B (2016) Functional characterization of EscK (Orf4), a sorting platform component of the enteropathogenic *Escherichia coli* injectisome. J Bacteriol JB.00538-16

Spaeth K, Chen Y-S, Valdivia R (2009) The *Chlamydia* type III secretion system C-ring engages a chaperone-effector protein complex. PLoS Pathog 5:e1000579

Sukhan A, Kubori T, Wilson J, Galán JE (2001) Genetic analysis of assembly of the *Salmonella enterica* serovar Typhimurium type III secretion-associated needle complex. J Bacteriol 183:1159–1167

Sun P, Tropea JE, Austin BP, Cherry S, Waugh DS (2008) Structural characterization of the *Yersinia pestis* type III secretion system needle protein YscF in complex with its heterodimeric chaperone YscE/YscG. 377

Suvarnapunya AE, Zurawski DV, Guy RL, Stein MA (2003) Molecular characterization of the prototrophic *Salmonella* mutants defective for intraepithelial replication. Infect Immun 71:2247–2252

Tamano K, Aizawa S-I, Katayama E, Nonaka T, Imajoh-Ohmi S, Kuwae A, Nagai S, Sasakawa C (2000) Supramolecular structure of the *Shigella* type III secretion machinery: the needle part is changeable in length and essential for delivery of effectors. EMBO J 19:3876–3887

Tampakaki AP (2014) Commonalities and differences of T3SSs in rhizobia and plant pathogenic bacteria. Front Plant Sci 5:114

Tejeda-Dominguez F, Huerta-Cantillo J, Chavez-Dueñas L, Navarro-Garcia F (2017) A novel mechanism for protein delivery by the type 3 secretion system for extracellularly secreted proteins. MBio 8:e00184-17

Thanabalu T, Koronakis E, Hughes C, Koronakis V (1998) Substrate-induced assembly of a contiguous channel for protein export from *E. coli*: reversible bridging of an inner-membrane translocase to an outer membrane exit pore. EMBO J 17:6487–6496

Thomas S, Holland IB, Schmitt L (2014) The type 1 secretion pathway—the hemolysin system and beyond. Biochim Biophys Acta Mol Cell Res 1843:1629–1641

Torruellas J, Jackson MW, Pennock JW, Plano GV (2005) The *Yersinia pestis* type III secretion needle plays a role in the regulation of Yop secretion. Mol Microbiol 57:1719–1733

Tseytin I, Dagan A, Oren S, Sal-Man N (2017) The role of EscD in supporting EscC polymerization in the type III secretion system of enteropathogenic *Escherichia coli*. Biochim Biophys Acta

Tusk SE, Delalez NJ, Berry RM (2018) Subunit exchange in protein complexes. J Mol Biol

Veenendaal AKJ, Hodgkinson J, Schwarzer L, Stabat D, Zenk S, Blocker AJ (2007) The type III secretion system needle tip complex mediates host cell sensing and translocon insertion. Mol Microbiol 63:1719–1730

Veenendaal AKJ, Sundin C, Blocker AJ (2009) Small-molecule type III secretion system inhibitors block assembly of the *Shigella* type III secreton. J Bacteriol 191:563–570

Vettiger A, Winter J, Lin L, Basler M (2017) The type VI secretion system sheath assembles at the end distal from the membrane anchor. Nat Commun 8:16088

Vincent MS, Canestrari MJ, Leone P, Stathopulos J, Ize B, Zoued A, Cambillau C, Kellenberger C, Roussel A, Cascales E (2017) Characterization of the *Porphyromonas gingivalis* type IX Secretion trans-envelope PorKLMNP core complex. J Biol Chem 292:3252–3261

Vollmer W, Blanot D, De Pedro MA (2008) Peptidoglycan structure and architecture. FEMS Microbiol Rev 32:149–167

Wagner S, Sorg I, Degiacomi MT, Journet L, Dal Peraro M, Cornelis GR (2009) The helical content of the YscP molecular ruler determines the length of the *Yersinia* injectisome. Mol Microbiol 71:692–701

Wagner S, Königsmaier L, Lara-Tejero M, Lefebre MD, Marlovits TC, Galán JE (2010) Organization and coordinated assembly of the type III secretion export apparatus. Proc Natl Acad Sci U S A 107:17745–17750

Wagner S, Grin I, Malmsheimer S, Singh N, Torres-Vargas CE, Westerhausen S (2018) Bacterial type III secretion systems: a complex device for delivery of bacterial effector proteins into eukaryotic host cells. FEMS Microbiol Lett

Walldén K, Williams R, Yan J, Lian PW, Wang L, Thalassinos K, Orlova EV, Waksman G (2012) Structure of the VirB4 ATPase, alone and bound to the core complex of a type IV secretion system. Proc Natl Acad Sci U S A 109:11348–11353

Weber E, Ojanen-Reuhs T, Huguet E, Hause G, Romantschuk M, Korhonen TK, Bonas U, Koebnik R (2005) The type III-dependent Hrp pilus is required for productive interaction of *Xanthomonas campestris* pv. *vesicatoria* with pepper host plants. J Bacteriol 187:2458–2468

Wee DH, Hughes KT (2015) Molecular ruler determines needle length for the *Salmonella* Spi-1 injectisome. Proc Natl Acad Sci 112:4098–4103

Wells JN, Bergendahl LT, Marsh JA (2015) Co-translational assembly of protein complexes. Biochem Soc Trans 43:1221–1226

Wells JN, Bergendahl LT, Marsh JA (2016) Operon gene order is optimized for ordered protein complex assembly. Cell Rep 14:679–685

Wood S, Jin J, Lloyd S (2008) YscP and YscU switch the substrate specificity of the *Yersinia* type III secretion system by regulating export of the inner rod protein YscI. J Bacteriol

Worrall LJ, Vuckovic M, Strynadka NCJ (2010) Crystal structure of the C-terminal domain of the *Salmonella* type III secretion system export apparatus protein InvA. Protein Sci 19:1091–1096

Worrall LJ, Hong C, Vuckovic M, Deng W, Bergeron JRC, Majewski DD, Huang RK, Spreter T, Finlay BB, Yu Z, Strynadka NCJ (2016) Near-atomic-resolution cryo-EM analysis of the *Salmonella* T3S injectisome basal body. Nature 540:597–601

Xing Q, Shi K, Portaliou AG, Rossi P, Economou A, Kalodimos CG (2018) Structures of chaperone-substrate complexes docked onto the export gate in a type III secretion system. Nat Commun 9:1773

Yang H, Shan Z, Kim J, Wu W, Lian W, Zeng L, Xing L, Jin S (2007) Regulatory role of PopN and its interacting partners in type III secretion of *Pseudomonas aeruginosa.* J Bacteriol 189:2599–2609

Yip CK, Kimbrough TG, Felise HB, Vuckovic M, Thomas N, Pfuetzner RA, Frey EA, Finlay BB, Miller SI, Strynadka NCJ (2005) Structural characterization of the molecular platform for type III secretion system assembly. Nature 435:702–707

Yu X-J, Liu M, Matthews S, Holden DW (2011) Tandem translation generates a chaperone for the *Salmonella* type III secretion system protein SsaQ. J Biol Chem 286:36098–36107

Yuan Q, Carle A, Gao C, Sivanesan D, Aly KA, Höppner C, Krall L, Domke N, Baron C (2005) Identification of the VirB4-VirB8-VirB5-VirB2 pilus assembly sequence of type IV secretion systems. J Biol Chem 280:26349–26359

Zarivach R, Deng W, Vuckovic M, Felise HB, Nguyen H, Miller SI, Finlay BB, Strynadka NCJ (2008) Structural analysis of the essential self-cleaving type III secretion proteins EscU and SpaS. Nature 453:124–127

Zhang Y, Lara-Tejero M, Bewersdorf J, Galán JE (2017) Visualization and characterization of individual type III protein secretion machines in live bacteria. Proc Natl Acad Sci U S A 114:6098–6103

Zilkenat S, Franz-Wachtel M, Stierhof Y-D, Galán JE, Macek B, Wagner S (2016) Determination of the stoichiometry of the complete bacterial type III secretion needle complex using a combined quantitative proteomic approach. Mol Cell Proteomics M115.056598

Zoued A, Brunet YR, Durand E, Aschtgen M-S, Logger L, Douzi B, Journet L, Cambillau C, Cascales E (2014) Architecture and assembly of the type VI secretion system. Biochim Biophys Acta Mol Cell Res 1843:1664–1673

The Structure of the Type III Secretion System Needle Complex

Sean Miletic, Nikolaus Goessweiner-Mohr and Thomas C. Marlovits

Contents

Abstract The type III secretion system (T3SS) is an essential virulence factor of many pathogenic bacterial species including *Salmonella, Yersinia, Shigella* and enteropathogenic *Escherichia coli* (EPEC). It is an intricate molecular machine that spans the bacterial membranes and injects effector proteins into target host cells, enabling bacterial infection. The T3SS needle complex comprises of proteinaceous rings supporting a needle filament which extends out into the extracellular environment. It serves as the central conduit for translocating effector proteins. Multiple laboratories have dedicated a remarkable effort to decipher the structure and function of the needle complex. A combination of structural biology techniques such as

S. Miletic · T. C. Marlovits (✉)
Center for Structural Systems Biology, Institute for Structural and Systems Biology, Universitätsklinikum Hamburg-Eppendorf, 85 Notkestraße, Hamburg 22607, Germany
e-mail: marlovits@marlovitslab.org

N. Goessweiner-Mohr
Institute of Biophysics, Johannes Kepler University Linz, Gruberstraße 40, 4020 Linz, Austria
e-mail: nikolaus.goessweiner-mohr@jku.at

Current Topics in Microbiology and Immunology (2020) 427: 67–90
https://doi.org/10.1007/82_2019_178
© Springer Nature Switzerland AG 2019
Published Online: 31 October 2019

cryo-electron microscopy (cryoEM), X-ray crystallography, nuclear magnetic resonance (NMR) spectroscopy and computer modelling have been utilized to study different structural components at progressively higher resolutions. This chapter will provide an overview of the structural details of the T3SS needle complex, shedding light on this essential component of this fascinating bacterial system.

1 Visualizing the Needle Complex

Research on the type III secretion system (T3SS) spans over two decades with increasingly complex experiments having been utilized to study the architecture of this molecular machine. In 1990, a novel secretion system in *Yersinia* was identified (Michiels et al. 1990) and was soon noted for its similarity to genes in the plant pathogen *Ralstonia solanacearum*, thus suggesting the discovery of a conserved secretion system utilized by different bacterial pathogens (Gough et al. 1992). Salmond and Reeves put these discoveries into context by coining the name, type III secretion system (1993). Since then, homologues of the T3SS have been identified in a variety of bacterial pathogens from *Yersinia* spp., *Shigella flexneri, Salmonella enterica,* enteropathogenic *Escherichia coli* (EPEC), *Pseudomonas aeruginosa, Chlamydia* spp. (Hueck 1998), *Burkholderia pseudomallei* (Vander Broek and Stevens 2017) and *Vibrio parahaemolyticus* (Makino et al. 2003), to the plant pathogens *Ralstonia solanacearum, Erwinia* spp., *Xanthomonas campestris, Pseudomonas syringae* and the plant symbionts from *Rhizobium* spp. (Hueck 1998).

Having been identified genetically, efforts by several laboratories focused on the protein structure of the T3SS. It was visualized in a partially purified state from *Salmonella* by negative-stain transmission electron microscopy (TEM; Kubori et al. 1998). It became apparent that the T3SS resembles the flagellar basal body—it is a large, supramolecular complex that spans both bacterial membranes, with two upper rings embedded in the outer membrane connected to two larger rings embedded in the inner membrane (Fig. 1a). From the smaller outer membrane rings extends a filament resembling a needle and hence the name, the needle complex (Kubori et al. 1998). Three-dimensional reconstructions from TEM negative-stain images of the needle complex in *Shigella flexneri* determined that the complex has a central channel (Blocker et al. 2001). The use of cryo-electron microscopy (cryoEM) led to a three-dimensional, 17 Å structure of the needle complex in *Salmonella* (Marlovits et al. 2004). Since then, persistent efforts aided by new technological developments have pushed the resolution of the needle complex structure further using cryoEM (Schraidt and Marlovits 2011) to recently, a near-atomic resolution of the needle complex (Worrall et al. 2016; Hu et al. 2018). The fundamental work accomplished using TEM, combined with an extraordinary effort using other structural techniques such as X-ray crystallography and nuclear magnetic resonance (NMR) spectroscopy, has provided an increasingly clear picture of this intricate molecular machine. As the complete atomic structure of the T3SS becomes closer to reality, new research

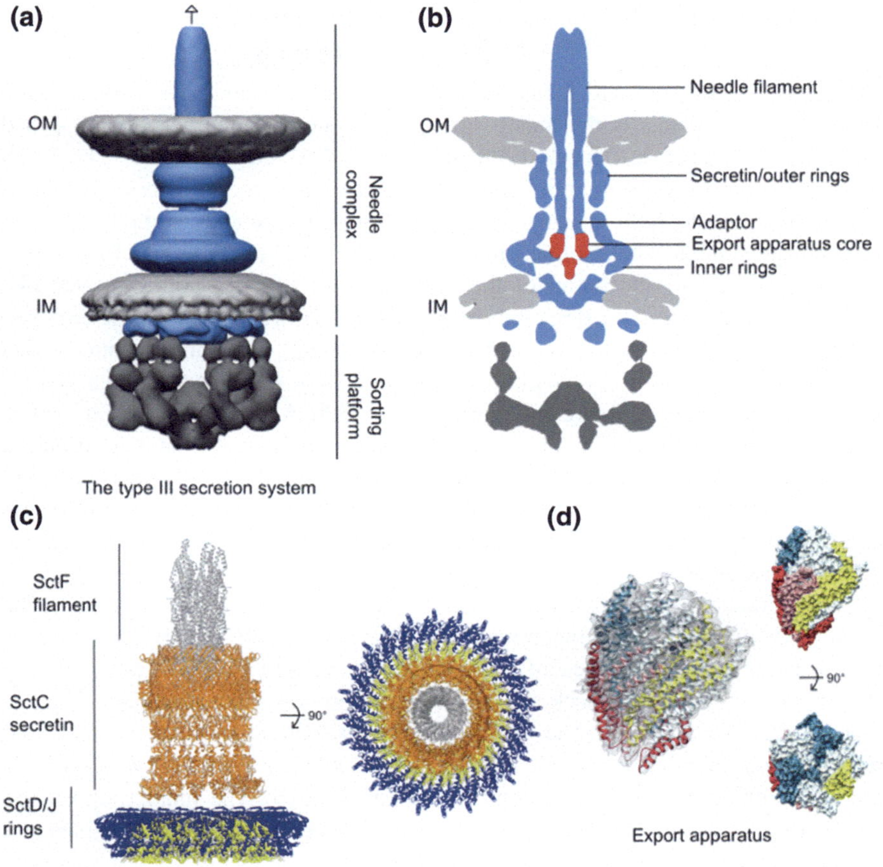

Fig. 1 **a** A cryoEM tomographic reconstruction of the *Salmonella* type III secretion system embedded in the inner (IM) and outer (OM) cell membranes. In blue is the needle complex and in dark grey is the cytoplasmic sorting platform complex (EMDB accession code EMD-8544; Hu et al. 2017). **b** A slice through the map shown in (**a**) showing the interior of the complex with labelled components. In red, is the approximate location of the export apparatus core complex. **c** Ribbon diagrams from high-resolution, single-particle cryoEM reconstructions of the inner membrane SctD (blue) and SctJ (yellow) rings, the outer membrane SctC (orange) secretin and the helical needle filament (light grey). The diagrams are shown in a 90° rotated view on the right (PDB, accession numbers 6DUZ, 6DV3 and 6DWB; Hu et al. 2018). **d** The export apparatus core complex. On the left, a ribbon diagram of the flagellar export apparatus is positioned in the corresponding cryoEM density map. On the top right is the corresponding space-filling representation which is also shown rotated 90° on the bottom. Depicted in light/dark blue is SctR, in yellow is SctT and in red in SctS flagellar homologues (PDB accession number 6F2D, EMDB accession code EMDB-4173; Kuhlen et al. 2018)

directions will undoubtedly arise, yielding a better understanding of bacterial pathogenesis.

2 A Structural Blueprint of the Needle Complex

This chapter will primarily focus on proteins from the *Salmonella* pathogenicity island-1 T3SS, as this is the most well-studied system to date and the proposed universal nomenclature (Hueck 1998) will be used to aid the reader. The T3SS needle complex can be broadly likened to a common medical syringe with a hypodermic needle, both in terms of its overall appearance and its ability to inject proteins into a variety of organisms. It comprises of a cylindrical basal body analogous to the plastic barrel of the syringe, from which a hollow needle-like filament extends into the extracellular environment. However, the needle complex is much more intricate both structurally and functionally, and not to mention much smaller, than a relatively simple plastic pump. In *Salmonella*, this molecular syringe is a supramolecular structure of approximately 300 Å (30 nm) tall and 240 Å (24 nm) wide (Marlovits et al. 2004). It resides in both of the two bacterial membranes, extending from the inner membrane, through the periplasmic space and peptidoglycan layer, to the outer membrane, and from there the helical needle extends out on average 800 Å (80 nm) into the extracellular environment (Fig. 1a; Kubori et al. 1998).

A multitude of proteins assemble intricately to form the needle complex. The basal body, which is the membrane-spanning base of the needle complex, is composed of multiple copies of three proteins arranged into two pairs of ring structures connected via a neck and an internal export apparatus complex (Fig. 1b; Kubori et al. 1998; Galán et al. 2014). The larger, bottom ring pair is integrated with the inner membrane and is comprised of multiple copies of two proteins SctD and SctJ with an inverted orientation (Schraidt et al. 2010). The smaller, upper ring pair is associated with the outer membrane and is comprised of multiple copies of a single protein, SctC (Kubori et al. 2000). This protein spans further down into the periplasmic space contacting the lower rings and forming a neck between the ring pairs (Fig. 1b–c). Associated in the centre of the inner membrane rings, are densities likely corresponding to the export apparatus complex comprised of five different proteins (Fig. 1d). They are SctV, SctR, SctS, SctT and SctU (Galán et al. 1992; Groisman and Ochman 1993; Fields et al. 1994; Allaoui et al. 1994; Wagner et al. 2010). Although their precise function remains to be elucidated, these five proteins are essential for needle filament assembly and host cell invasion (Sukhan et al. 2001).

Several additional proteins integrate with the basal body to form a functioning needle complex. SctI assembles inside (Marlovits et al. 2004), forming a structure referred to historically as the inner rod (Marlovits et al. 2006), or more recently the needle adaptor (Hu et al. 2018; Torres-Vargas et al. 2019) which anchors the needle filament. Extending from the adaptor out into the extracellular environment is the needle filament composed of a helical polymer of the protein SctF (Kubori et al. 2000). At the distal end of the filament is a tip complex of proteins which varies depending on the species (Mueller et al. 2005; Epler et al. 2012; Kaur et al. 2016). It likely controls secretion and serves as an environmental sensor (Espina et al. 2006; Blocker et al. 2008). There are additional proteins on the cytoplasmic side of the needle complex, SctQ, SctK, SctL and the ATPase SctN, which form the sorting platform complex (Lara-Tejero et al. 2011) that is homologous to the flagellar C ring complex (Francis et al. 1994). This is a separate complex of the T3SS and is believed to queue proteins for export (Lara-Tejero et al. 2011) and to energize transport (Eichelberg et al. 1994). From this overall picture of the needle complex, one can begin to appreciate the structural complexity of this system. The following sections will focus on the structural arrangement of individual proteins of the needle complex which, for the purposes of this chapter, is defined as the fully assembled basal body with filament and export apparatus proteins.

3 The Individual Building Blocks of the Needle Complex

3.1 The Inner Membrane Protein SctD

SctD ($\sim$45 kDa for animal pathogens) is one of the three main structural proteins of the needle complex basal body. Early on, it was determined that SctD forms part of the needle complex (Kubori et al. 1998) where it forms the majority of the inner membrane rings (Kimbrough and Miller 2000; Kubori et al. 2000; Blocker et al. 2001). SctD is 392 residues long in *Salmonella* and has four globular domains with a transmembrane domain connecting the first two (Fig. 2a). The crystal structures of fragments of SctD, residues 11–120, 170–362 and 170–392, have provided insight on the structure and multimeric assembly of this protein (Spreter et al. 2009; Bergeron et al. 2013). The first domain folds into a globular structure containing nine, mostly anti-parallel β sheets resembling a forkhead-associated domain (FHA; Bergeron et al. 2013). Domains 2–4 have a modular architecture, each containing two α helices and one 3-stranded β sheet and are connected by short linkers, giving an overall appearance similar to that of a boot (Spreter et al. 2009). This domain structure is seen in other ring-forming proteins from EPEC such as the secretin protein SctC and the inner ring protein SctJ. A wedge-shaped fold between two α helices and the β sheet in domains 2 and 3 is conserved between these proteins. This fold may be a ring-building motif (RBM) structurally conserved across different T3SS-containing bacterial species (Spreter et al. 2009). However, this fold is also observed in other proteins that do not assemble into ring-like structures (Mishima et al. 2005; Valverde et al. 2008; Korotkov et al. 2009).

SctD intricately assembles into two, stacked, 24-mer ring structures positioned in the inner bacterial membrane. It is oriented with the first domain in the cytoplasm, while domains 2–4 remain in the periplasm of the bacterial envelope (Fig. 2a; Schraidt et al. 2010). The periplasmic inner membrane ring (IR1) is about 250 Å in diameter at its widest and composed of SctD domains 2–4. Domain 1 forms the

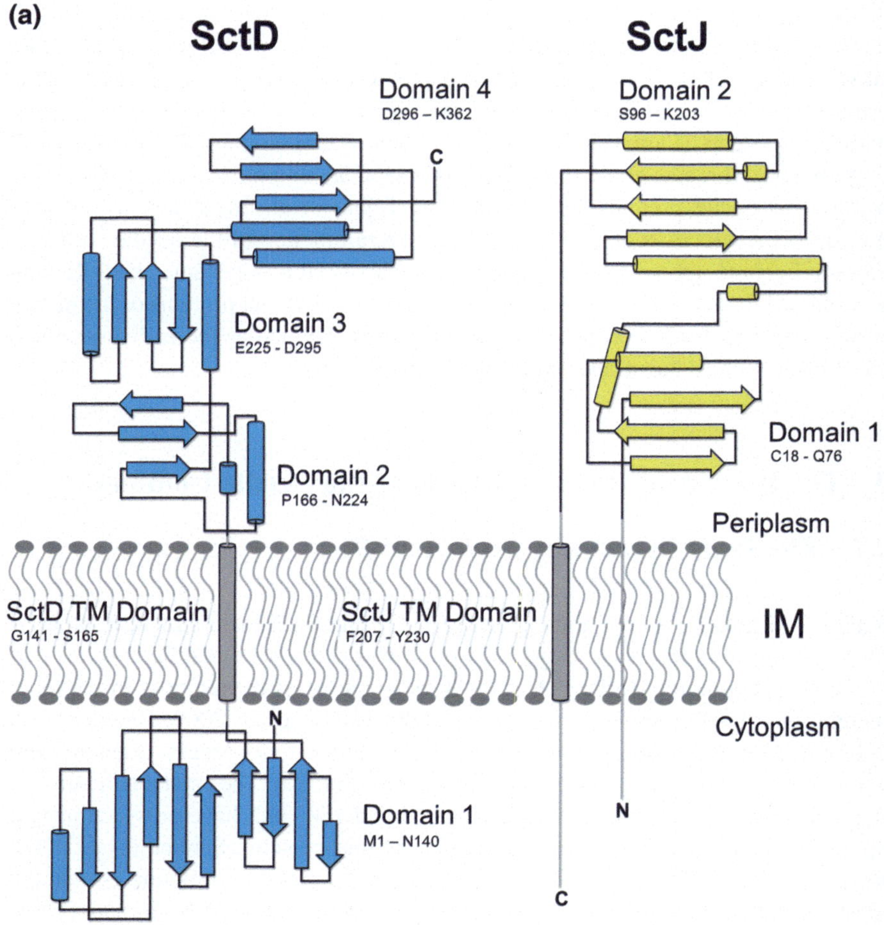

Fig. 2 **a** Topology diagrams of SctD and SctJ with labelled domains and positioned in the inner cell membrane (IM). TM = Transmembrane domain. **b** Molecular contacts in SctD and SctJ. Ribbon representations of SctD (blue) and SctJ (yellow) with shown surfaces and notable residues highlighted with red spheres. Close-ups of these residues are shown in the surrounding panels (PDB, accession numbers 6DUZ and 3J1W; Hu et al. 2018; Bergeron et al. 2013)

cytoplasmic inner membrane ring (IR2; Schraidt et al. 2010; Schraidt and Marlovits 2011). This domain is very flexible and undergoes a rearrangement from a continuous ring into hexamers of tetramers upon binding of the cytoplasmic sorting platform complex (Hu et al. 2017).

Intermolecular contacts between monomers are present in the SctD ring. A cryoEM structure of the inner ring predicted that a tyrosine at position 239 in one monomer of SctD closely associates with glutamic acid at position 252 from an adjacent monomer (Fig. 2b(I); Schraidt and Marlovits 2011). However, single mutations to either of these residues were insufficient to perturb T3SS function and

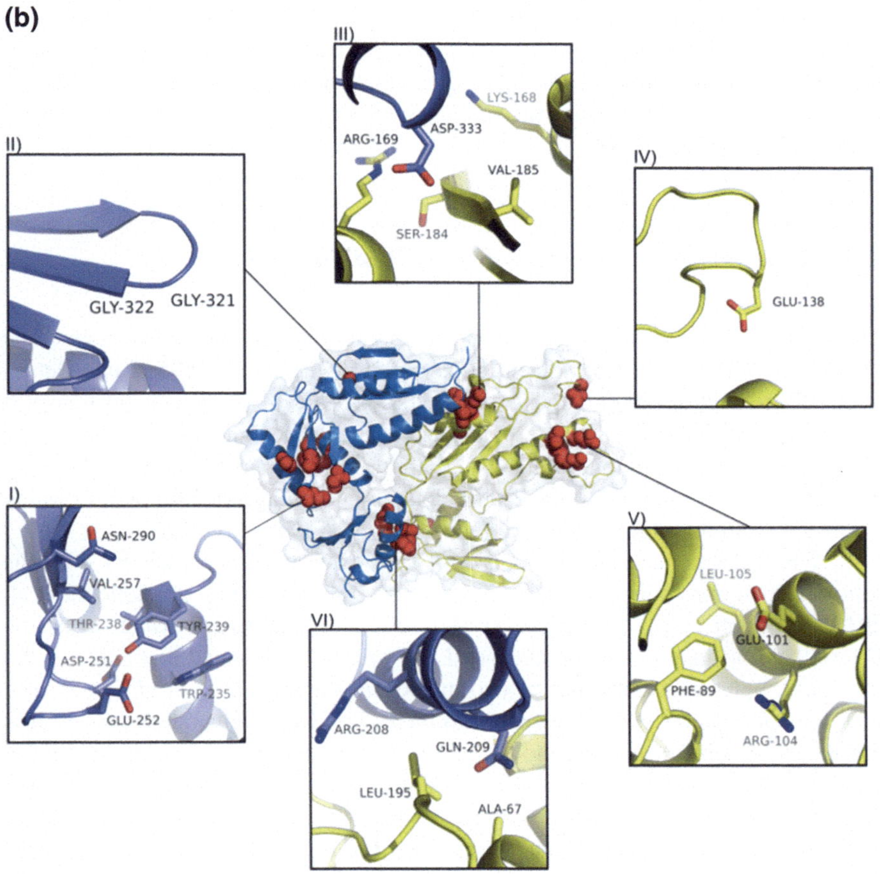

Fig. 2 (continued)

therefore needle complex assembly. Double mutants did, though, providing evidence that this is a contact point between monomers and it is at least partially responsible for SctD ring assembly (Schraidt and Marlovits 2011). In other work, secretion was abrogated and needle complex assembly was disrupted when a glycine at position 322 of a periplasmic loop was mutated (Fig. 2b(II); Bergeron et al. 2013), suggesting that there are likely multiple contact points between monomers of the SctD ring.

The SctD monomers likely contact the inner SctJ ring and the above SctC rings. Rosetta modelling of SctD and SctJ in the previous cryoEM map predicted that aspartic acid at position 333 in the second α helix of domain 4 may be an important residue of SctD for contacting SctJ (Bergeron et al. 2015). Its side chain is buried in a positively charged pocket formed by SctJ monomers (Fig. 2b(III)). This association was confirmed by mutating the residue to arginine which inhibited T3SS secretion (Bergeron et al. 2015). The C-terminus of SctD is in close contact with the secretin SctC positioned above in the periplasm. This occurs via a lysine at position 367, as demonstrated by chemical crosslinking (Schraidt et al. 2010). Deletion of the last four amino acids of the SctD C-terminus destabilized needle complexes, providing more evidence that this region is important for assembly of the basal body (Schraidt et al. 2010).

3.2 The Second Inner Membrane Protein SctJ

The second component of the inner membrane rings is the inner membrane lipoprotein SctJ ($\sim$27 kDa; Allaoui et al. 1992; Pegues et al. 1995; Kubori et al. 1998; Tamano et al. 2000; Blocker et al. 2001). It has a Sec-dependent transport signal sequence located at its N-terminus and is anchored to the inner membrane by a C-terminal transmembrane domain or a lipid anchor (Yip et al. 2005; Schraidt et al. 2010). SctJ is similar to the central domain of the flagellar FliF protein which forms a ring in the inner membrane (Yip et al. 2005; Bergeron 2016).

SctJ is 252 residues long in *Salmonella* and consists of two globular domains followed by a transmembrane domain and a C-terminal tail (Fig. 2a; Schraidt et al. 2010; Bergeron et al. 2015). The crystal structure from EPEC which lacks the C-terminal transmembrane domain found in *Salmonella*, revealed that SctJ is a flat triangular molecule comprising of two mixed α/β domains (Yip et al. 2005). The first domain contains two α helices and a three-stranded, anti-parallel β sheet. It is connected by a linker to a second domain of two α helices and a three-stranded β sheet. This linker has been postulated to fine-tune the angular orientation of the two domains with respect to each other (Yip et al. 2005). In *Salmonella*, the solution and crystal structures of the first domain and the crystal structure of the second domain were solved, revealing a structural similarity to EPEC (Fig. 2; Bergeron et al. 2015). Both of these domains contain the same structural RBM as seen in SctD, highlighting its importance throughout the assembly of the system.

SctJ forms a ring comprised of 24 monomers with a diameter of 180 Å and a height of 52 Å and has an inverted orientation compared to SctD as its N-terminus is located in the periplasm (Fig. 2a; Yip et al. 2005; Schraidt et al. 2010). Monomers may intrinsically oligomerize during ring formation. When crystallized, EPEC SctJ was found in a tetrameric conformation which packs into superhelical stacks of rings with each helix comprising of 24 monomers (Yip et al. 2005). Based on this data, the EPEC SctJ crystal structure was first modelled into a 24-ring structure (Yip et al. 2005), which was subsequently used in later studies to model *Salmonella* SctJ (Schraidt et al. 2010) until it was also shown to have 24-fold symmetry by cryoEM (Schraidt and Marlovits 2011).

Extensive interactions form between each monomer in the SctJ ring. In EPEC, monomers interact with both neighbouring molecules causing about a third of the total solvent-accessible surface to be buried. Each interface contains multiple hydrogen bonds between charged residues in both domains 1 and 2. As well, there are multiple van der Waals interactions and hydrophobic contacts dispersed across the intermolecular interface (Yip et al. 2005). Interestingly, it is hypothesized that the conformation of a linker between domains 1 and 2 influences oligomerization. When the linker is more strongly bound to the first domain, SctJ exists in an assembly incompetent state, whereas releasing the linker promotes oligomerization (Bergeron et al. 2015). When assembled in a ring, the linker region packs between neighbouring domain 2 helices via a phenylalanine residue at position 89, which packs between monomers like a zipper, an observation supported by mutational analysis (Fig. 2b(V); Bergeron et al. 2015; Worrall et al. 2016).

SctJ contacts the surrounding SctD ring and likely other protein components of the needle complex. Major interactions occur between the C-terminal region of SctD and the second domain of SctJ (Bergeron et al. 2015). As previously mentioned, the fourth α helix of two adjacent SctJ monomers forms a positively charged pocket which buries a conserved asparagine residue from SctD (Fig. 2b(III); Bergeron et al. 2015). The loop connecting the second domain with the transmembrane domain additionally forms extensive interactions with neighbouring SctJ monomers (Fig. 2b(VI); Worrall et al. 2016) and is supported by crosslinking data (Sanowar et al. 2010).

SctJ could make additional contacts in less resolved regions of the needle complex. A flexible loop between the fifth β strand and the fifth α helix of SctJ might be located in a cavity observed between the SctD and SctJ rings, providing an additional contact between the two ring structures (Bergeron et al. 2015). Although it cannot be resolved in cryoEM maps, the N-terminus of SctD also interacts with the C-terminus of SctJ as demonstrated by chemical crosslinking (Schraidt et al. 2010). Furthermore, the second domain contacts the export apparatus proteins, revealed by the crosslinking glutamic acid at position 138 (Fig. 2b(IV); Kuhlen et al. 2018).

3.3 The Outer Membrane Protein SctC

The next major ring-forming protein of the needle complex is SctC (~62 kDa) which forms both outer membrane rings. This protein assembles into a large pore or secretin in the outer bacterial membrane and is similar to other secretins found in the type II secretion system (T2SS), the type IV pili system (T4PS) and the PulD family of proteins (Kaniga et al. 1994; Collins et al. 2003; Korotkov et al. 2011). Early work reported that in *Salmonella,* SctC forms ring-shaped structures as visualized by negative-stain TEM (Crago and Koronakis 1998). The similarities seen between SctC and other secretins led to the early proposal that it forms the bacterial outer membrane rings. Due to its homology with PulD and pIV secretins, SctC was oriented with the N-terminus in the periplasm (Kubori et al. 2000). This was confirmed using cryoEM revealing that it extends from the outer membrane all the way to the inner ring complex (Schraidt et al. 2010).

SctC is 564 residues long in *Salmonella* and is much more structurally complex compared to the other ring-forming proteins discussed so far. It is comprised of three globular domains (N0, N1 and N3) situated underneath a double-barrelled secretin domain that has a periplasmic gate, a tilted lip on top, and an outer helix-turn-helix S domain (Fig. 3a; Worrall et al. 2016). The crystal structures of SctC in *Salmonella* (residues 22–178) and in EPEC (residues 21–174) were solved, revealing that the first two domains are tilted towards each other and form a modular fold (Spreter et al. 2009; Bergeron et al. 2013). The first domain comprises of two α helices sandwiched between two β sheets, and the second domain has two α helices alongside one β sheet. Interestingly, these domains have the same RBM found in SctD and SctJ (Spreter et al. 2009; Hu et al. 2018).

Recent, high-resolution cryoEM studies have provided a structural model of the entire SctC protein (Worrall et al. 2016; Hu et al. 2018). The third domain N3 contains an RBM, the two α helices and a β sheet, as well as another two α helices and a β hairpin (Worrall et al. 2016). Following this, is the secretin domain containing an outer β sheet of essentially four strands and an inner β sheet of also four strands forming a retractable periplasmic gate (Fig. 3a). Above, extending to the extracellular surface, is the secretin lip, formed by a double-stranded hairpin. On the C-terminal end is the S domain, comprising of two α helices connected to the rest of the protein by an extended loop which wraps around the surface of the outer β barrel, linking one SctC monomer to two successive protomers (Worrall et al. 2016).

This protein intricately assembles into a large ring structure of a different symmetry than the inner membrane rings. Earlier studies debated the symmetry of T3SS secretins, with reports claiming 12–14-fold symmetries depending on the species studied and techniques used (Bitter 2003; Burghout et al. 2004; Hodgkinson et al. 2009). Using cryoEM, it was determined that SctC has a 15-fold symmetry in *Salmonella*, giving the entire basal body, in respect to the rings, an overall C3 symmetry where 15 SctC subunits assemble with 24 SctJ and 24 SctD subunits (Schraidt and Marlovits 2011). To tolerate the symmetry mismatch between the different rings, presumably five monomers of SctC contact eight monomers of SctD or SctJ. It should, however, be noted that because of the presence of the helical extracellular needle filament in fully assembled needle complexes, and a pseudo-hexameric structure of the export apparatus (Kuhlen et al. 2018), the overall symmetry is asymmetric.

The 15-mer secretin domain is a double-barrelled pore formed by two β sheets sandwiching each other. The outer β sheet of SctC forms a 60-strand, anti-parallel β barrel positioned in the outer membrane. The second β-sheet forms an additional anti-parallel β barrel that slants towards the interior lumen of the complex forming the periplasmic gate (Worrall et al. 2016). Evidently, this gate is not complete as a central pore of 15 Å in diameter still exists which is large enough for small molecules to transverse (Fig. 3b). The gate is structurally supported underneath by a β hairpin in the third domain (Fig. 3b(I)). Mutants of the gate or the supporting β hairpin impede secretion, and it is proposed that the third domain stabilizes the gate and may also have a role in gating (Worrall et al. 2016). This periplasmic gate changes conformation to an open state which permits needle filament assembly (Marlovits et al. 2004; Hu et al. 2018), activating the system.

Contacts between the several domains of SctC help support its ring structure. The secretin domains in the barrel form an extensive hydrogen bond network (Worrall et al. 2016). The RBM in domain 3 forms interfaces with the β sandwich of the secretin domain which is supported by mutational analysis (Fig. 3b(II); Worrall et al. 2016). The loop region between the first and second domain makes contact between adjacent monomers, supported by mutational analysis (Fig. 3b(III); Bergeron et al. 2013). The outer S domain wraps around the outer surface of the secretin, spanning two protomers, and contacts the outer β sheet with a salt bridge forming between an asparagine residue at position 544 and lysine at 315 (Fig. 3b (IV) and is also supported by mutational data in this region (Worrall et al. 2016). This domain is believed to function as a molecular staple which stabilizes the β barrel structure.

The SctC rings make multiple contacts with other needle complex components. Using chemical crosslinking, the N-terminus was shown to make contact with the C-terminal domain of SctD (Schraidt et al. 2010). Indeed, modelling using available structures placed a lysine at residue 38 of the SctC N-terminus, closest to the inner ring proteins consistent with cryoEM maps (Fig. 3b(V); Schraidt and Marlovits 2011; Hu et al. 2018). As well, the lower SctC N0 ring has a basic surface and a positively charged collar in its interior, both of which may provide an initial assembly force for the ring structures (Bergeron et al. 2013). In the secretin, the outer S domain also interacts with another protein, the pilotin chaperone InvH in *Salmonella*, which is essential for T3SS assembly. This pilotin could recognize a common turn-helix motif present in the S domain (Worrall et al. 2016). The outer β-barrel lip is proposed to contact and span the outer bacterial membrane. It contains an amphipathic helix loop (AHL) facing the inner leaflet, known to associate with membranes (Fig. 3a). The lip is tilted in the purified structure, which matches well with in situ tomography data showing an inward bending of the outer membrane associated with the secretin (Hu et al. 2015, 2017).

Comparing the open and closed SctC cryoEM structures, structural differences are evident in the N3 domain, the inner β barrel and the upper lip of the secretin domain (Fig. 3b(VI–X); Worrall et al. 2016; Hu et al. 2018). The first β hairpin, inner gate is kinked at residues Asn386 and Gly407 in the closed structure; however in the open conformation, they are unkinked, moving the tip of the hairpin approximately 40 Å upwards (Fig. 3b(VI)). It is oriented in such a conformation to allow interactions to the outer β barrel via residues Lys392 and Ile394, which influence secretion when mutated (Worrall et al. 2016; Hu et al. 2018). The second β-hairpin gate is also unkinked at residues Gly430 and Gly451 (Fig. 3b(VII)), and the loop region is rotated 180° (Hu et al. 2018). As well, the secretin lip is pushed up and out increasing the pore diameter. The N3 domain is rotated slightly, and the β-hairpin strut is pulled away and rotated making contacts with Arg411 of the gate (Fig. 3VIII; Hu et al. 2018) supported by mutational analysis (Worrall et al. 2016). Previously unresolved residues 217–227 and 252–267 form additional interfaces which further stabilize the needle complex at the N1 and N3 domains (Hu et al. 2018).

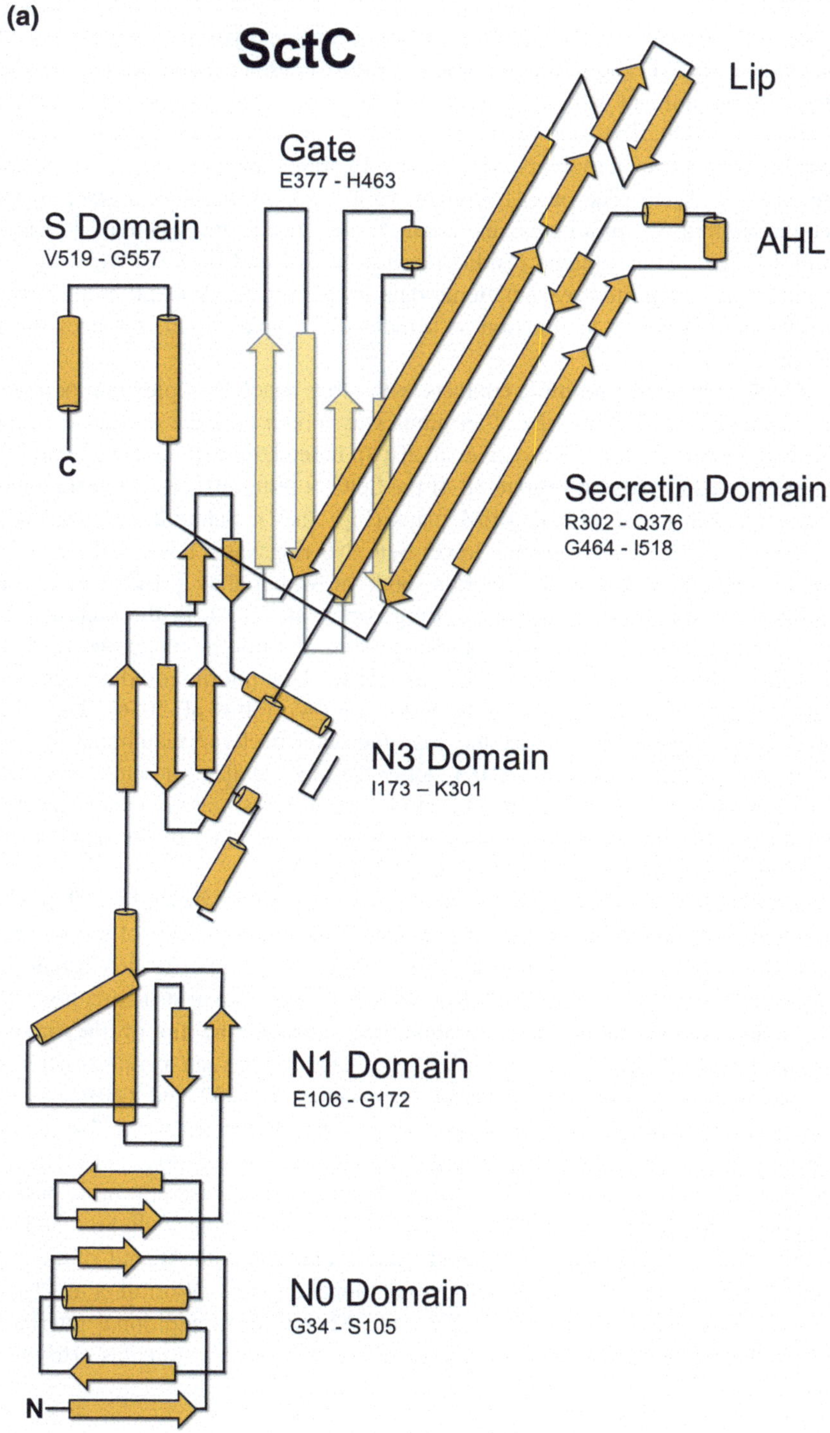
(a)
SctC
Lip
Gate
E377 - H463
S Domain
V519 - G557
AHL
C
Secretin Domain
R302 - Q376
G464 - I518
N3 Domain
I173 – K301
N1 Domain
E106 - G172
N0 Domain
G34 - S105
N

◂**Fig. 3** SctC secretin protein. **a** Topology diagram of SctC with domains/notable features highlighted. **b** Ribbon representations of open and closed conformations of SctC with shown surfaces and important residues highlighted with red spheres. Left: SctC monomer in the closed conformation, right: the open conformation. I–V: panels of notable residues forming molecular contacts, corresponding to the red spheres in the centre cartoon representation. VI–VIII: panels of conformational changes in the open SctC conformation, corresponding to the red spheres in the centre cartoon representation. Below, top views of opposing SctC monomers in the closed (IX) and open (X) conformation. Grey rings represent the SctC OM ring (PDB, accession numbers 6DV3, 5TCQ and 4G08; Hu et al. 2018; Worrall et al. 2016; Bergeron et al. 2013)

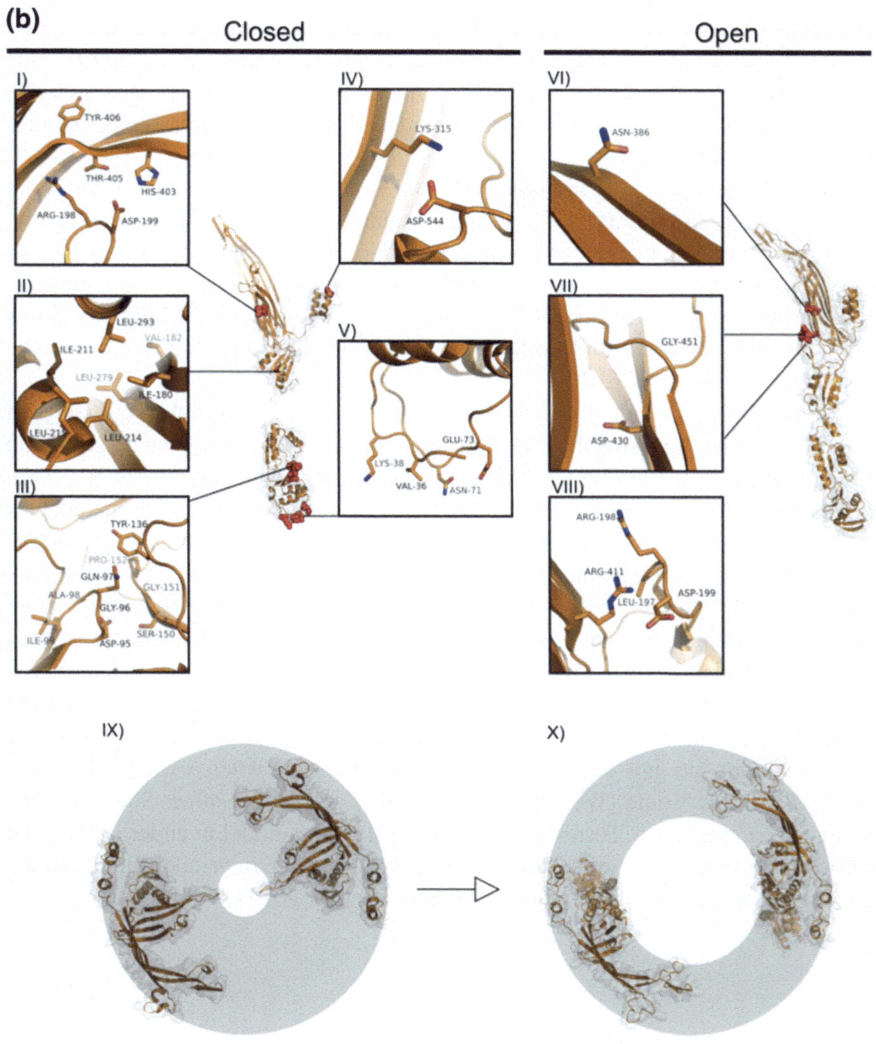

Fig. 3 (continued)

3.4 The Needle Adaptor Protein SctI

SctI (~11 kDa) forms a density connecting the export apparatus and the helical needle filament. It was originally named the inner rod of the needle complex (Marlovits et al. 2004) because it was first visualized as a tube-like density inside the SctC secretin attaching the needle filament from above to the socket-like export apparatus below (Marlovits et al. 2004). However, mass spectrometry strategies have revealed that the inner rod is smaller, with no more than six copies of SctI forming one helical turn (Zilkenat et al. 2016). Recent cryoEM data suggests that SctI assembles into a narrow density between the export apparatus and the base of the needle filament (Hu et al. 2018). Photo-crosslinking revealed that SctI tightly interacts with the export apparatus components (Torres-Vargas et al. 2019). This structure likely forms a connection between the needle and the export apparatus (Hu et al. 2018) and has been subsequently modelled as a hexameric needle adaptor (Torres-Vargas et al. 2019).

SctI is 101 residues long in *Salmonella* and is predicted to be a fold of two long α helices. It has a conserved C-terminus containing a seven residue region that causes immune activation of macrophages (Miao et al. 2010). Circular dichroism (CD) spectroscopy and three-dimensional NMR determined that SctI is a partially folded protein and with residues 65–82 containing a short α helix. Mutations in this region abolish secretion and some even abolish SctI incorporation into needle complexes (Zhong et al. 2012).

Unfortunately, the functional role of SctI in the needle complex is even less clear. Early work suggested that SctI is responsible for determining the length of the needle filament by anchoring the filament to the basal body (Marlovits et al. 2006). A model was proposed where SctI and SctF are secreted and assemble the adaptor and filament, respectively. Upon completion of the SctI complex, some conformation change occurs stopping the secreting needle filament proteins (Marlovits et al. 2006). In support of this model, mutations in SctI give rise to extra-long needle filaments suggesting these mutations abrogate SctI–SctI interactions, thus causing the SctI adaptor to assemble at a slower rate. This delays the observed switch from secretion of needle filament proteins to other substrates, increasing the length of the needle filaments (Lefebre and Galan 2014). However, because some of these mutations do not cause longer needle filaments when expressed in the *Salmonella* chromosome (Wee and Hughes 2015) and do not affect SctI assembly on the export apparatus (Torres-Vargas et al. 2019), the model is under debate. An additional protein, SctP, instead functions as a molecular ruler controlling needle filament length (Journet et al. 2003; Wee and Hughes 2015).

3.5 The Needle Filament Protein SctF

SctF (~8.9 kDa) forms the helical needle filament of the needle complex. The filament is one of the most recognizable components of the complex, however, structural studies on it have faced difficulty due to its inherent insolubility. Early work on *Shigella* T3SSs using negative-stain TEM determined the helical nature of the filament and that it is 70 Å wide and has a 25 Å-wide lumen (Cordes et al. 2003). The crystal and NMR solution structures of SctF from *Salmonella* (Wang et al. 2007; Poyraz et al. 2010), the crystal structure from *Shigella* (Deane et al. 2006) and NMR studies of the protomer in *Burkholderia mallei* (Zhang et al. 2006) gave structural insight on this protein family. Efforts using solid-state NMR and Rosetta modelling on in vitro reconstituted needles determined the atomic model of the filament from *Salmonella* (Loquet et al. 2012). This model was further supported by recent cryoEM analysis on isolated needle filaments (Hu et al. 2018).

SctF is eighty residues long in *Salmonella* and comprises of a fold of two anti-parallel helices. It consists of four structural elements: a highly ordered, N-terminal segment followed by an α helix, a loop and a C-terminal, kinked α helix (Loquet et al. 2012). In the cryoEM map, the kinked, N-terminal helix is more extended, and the C-terminal Arg80 has a flipped side chain and carboxylate, altering the interactions with surrounding monomers (Hu et al. 2018).

The overall structure of the filament protein is approximately similar among T3SSs of pathogens and the flagellar rod (Deane et al. 2006; Galán et al. 2014). However, structural variation between T3SS needle filaments is evident as protomers from *Salmonella* and *Shigella* are not structurally compatible (Poyraz et al. 2010). For instance, despite a conserved helical core, SctF homologues have different electrostatic surface arrangements which may cause variation in protein–protein interactions (Wang et al. 2007).

Monomers assemble into a hollow structure that has a width around 80 Å and which extends on average 800 Å as previously mentioned (Kubori et al. 1998; Loquet et al. 2012). Needle protomers assemble from the distal end in the absence of ATP, similar to flagellum elongation (Poyraz et al. 2010). Within each monomer, the N-terminal region of the first α helix interacts with the C-terminal portion of the second α helix. Contacts also occur between neighbouring monomers in the filament. There are approximately 11 subunits per two turns of the helical filament, and each subunit makes contacts with surrounding monomers laterally and axially. The helical rise is approximately 4.2 Å for the NMR model (Loquet et al. 2012) but slightly larger for the cryoEM model (Hu et al. 2018). Furthermore, in the cryoEM model, the lumen is approximately 15 Å, and it is a right-handed spiral groove which could suggest that substrates may rotate during translocation (Hu et al. 2018). Both models predict that the N-terminal domain of SctF is not α helical and is located on the exterior of the filament, extending outwards (Loquet et al. 2012; Hu et al. 2018). Conserved residues of the filament protein are located in the interior of the lumen, whereas the exterior residues are more divergent, reflecting a possible host invasion strategy by pathogens (Loquet et al. 2012).

Interestingly, SctF is likely involved in the regulation of protein secretion by the T3SS. It is believed that the distal tip complex senses host cells and transmits a signal via the SctF filament to the needle complex base, activating the system for protein translocation (Blocker et al. 2008). This transmission could be caused by changes in the needle filament architecture (Cordes et al. 2003). This hypothesis is supported by mutations along *Shigella* and *Yersinia* SctF causing defects in needle polymerization and protein secretion (Kenjale et al. 2005; Davis and Mecsas 2007).

Mutational analysis of SctF in *Salmonella* further supports its role in regulating T3SS secretion. Substitutions along the length of SctF caused different T3SS secretion phenotypes; secretion of only early substrates, enhanced secretion of middle/late substrates, or deficiency of SctE (SipB) secretion (Guo et al. 2019). Mutations causing an enhancement of secretion lie in the predicted tip–filament interface of SctF and could impede a signal originating from the tip complex to the needle filament (Guo et al. 2019). Mutations, causing the secretion of early substrates only, affect the stability or assembly of the needle filament, thus preventing substrate switching normally occurring upon full assembly of the needle filament (Guo et al. 2019). These secretion phenotypes could also be recapitulated by mutating residues specifically in the C-terminus of SctF, lining the lumen of the needle. Indeed, cryoEM and solid-state (SS)NMR data from these mutant strains revealed structural differences in the lumen, providing further evidence for the role of SctF in regulating protein secretion (Guo et al. 2019).

3.6 The Tip Complex Protein SctA

The needle filament is capped with SctA ($\sim$37.1 kDa) forming a complex which likely has multiple roles, including regulating protein secretion, sensing environmental signals and assembling and bridging to the pore complex in the host cell membrane (Sato et al. 2011). In *Yersinia*, SctA resembles a dumbbell of two globular domains connected by a coiled-coil domain (Derewenda et al. 2004; Chaudhury et al. 2013). This coiled-coil domain is found in other tip proteins and is responsible for oligomerization (Caroline et al. 2008) and has been modelled into a pentamer (Deane et al. 2006). In *Shigella* and *Burkholderia*, tip proteins also have a coiled-coil domain but have a different N-terminal domain, mostly a helical hairpin, and a C-terminal α–β domain (Johnson et al. 2007). In *Salmonella*, SctA has a similar structure to *Shigella* and *Burkholderia* and was recently visualized as a 5 nm density at the end of the needle filament in minicells (Park et al. 2018). It undergoes a conformational change when fused to SctF—the filament protein replaces a helix of the first domain of SctA providing insight on the tip complex assembly (Lunelli et al. 2011). The tip complexes found in EPEC are quite different from the others discussed so far. SctA forms a sheath-like, helical filament at the tip of the needle which extends over 600 nm (Sekiya et al. 2001; Wang et al. 2006) possibly reflecting the different infection strategy of EPEC (Hueck 1998).

3.7 The Export Apparatus Proteins SctV, SctR, SctS, SctT and SctU

The final component of the needle complex is an assembly of proteins SctV (~76 kDa), SctR (~25.2 kDa), SctS (~9.4 kDa), SctT (~28.5 kDa) and SctU (~40.1 kDa), which form the export apparatus in the needle complex. These proteins are required for needle complex assembly, and SctR and SctT can assemble into stable complexes independent of other components (Wagner et al. 2010). It is likely that SctR/S/T form a density in the centre of the needle complex base. Initial cryoEM maps showed a socket-like structure situated directly below the SctI adaptor and abutting to an upside-down cup-like structure in the centre of the SctD/J rings (Marlovits et al. 2004; Wagner et al. 2010). However, in cryoET maps of the in situ needle complex, the cup-like structure is less clear (Hu et al. 2017). Recently, a high-resolution cryoEM structure of the homologous FliP-FliQ-FliR complex in *Salmonella* was determined and placed into previous cryoEM maps of the needle complex basal body (Kuhlen et al. 2018).

SctR, SctS, SctT, SctU and SctV are found in a 5:4:1:1:9 stoichiometry in the needle complex (Zilkenat et al. 2016; Kuhlen et al. 2018). Earlier TEM data revealed that SctR forms a pentameric doughnut-shaped complex with an additional density on one side being SctT. It has a 15 Å-wide pore which might be the entrance to the secretion channel (Dietsche et al. 2016). These proteins were shown to interact with the needle adaptor SctI (Torres-Vargas et al. 2019), which would create a continuous secretion path in the needle complex. They also contact SctS and SctU, and together, this complex must assemble in order for the needle complex base to efficiently form (Wagner et al. 2010).

4 Assembly, Conformational Flexibility and Substrate Secretion

The needle complex adopts a variety of conformations during its lifetime: from the sequential assembly of individual components to the fully functioning effector-secreting machine. Assembly of the needle complex occurs in two main steps. The basal body is first assembled in the bacterial membranes via the Sec pathway, and then the final components, the adaptor and the needle filament, are assembled by being secreted through the system (Deng et al. 2017).

The polymerizing needle filament may induce a conformational change in the closed periplasmic gate of the secretin ring, leading to an opening of the basal body structure. Consequently, the periplasmic gate flattens against the outer β barrel (Marlovits et al. 2004; Hu et al. 2018) to render the needle complex active for secretion. The SctD/J ring IR1 appears to be the most stable part of the needle complex, while SctD IR2 is more flexible (Schraidt and Marlovits 2011). Recent work using cryoET on needle complexes in *Salmonella* and in *Shigella* minicells have revealed in situ structural differences in IR2 (Hu et al. 2015, 2017). The cytoplasmic ring in situ is separated into six patches and is located further down below the inner membrane. In the absence of the sorting platform structure, however, the SctD cytoplasmic ring was restored. It is hypothesized that the SctD N-terminus undergoes a conformational change upon sorting platform complex assembly, dividing into six patches which may serve as symmetry adaptors between the 24-fold basal body and the sixfold sorting platform complex (Hu et al. 2017). It is proposed that this ring structure collapses back closer to the N-terminus of SctD upon needle complex purification, and therefore, these structures were not seen via single-particle analysis (Hu et al. 2017). In other species such as in *Yersinia*, the basal body of the T3SS was found to vary by 20% in length in situ, denoting high flexibility. This could be due to high inter-domain flexibility of the SctD homologue and stretching of the SctC homologue (Kudryashev et al. 2013). As well, *Shigella* needle complexes are 40% longer when visualized by cryoET compared to single-particle approaches, further suggesting some conformational rearrangements occur upon needle complex purification (Kudryashev et al. 2013).

It is widely believed that the principal function of the needle complex is to translocate proteins from the bacterial cytoplasm into host cells. Structural studies revealed that the lumen of the needle is approximately 15 Å in diameter (Hu et al. 2018). In the base, the secretion channel starts with the gasket of the export apparatus, opening up to the atrium or lumen and then continuing on through the SctI adaptor and to the needle filament lumen (Fig. 1b; Radics et al. 2014; Hu et al. 2018; Kuhlen et al. 2018). This path is too small for a folded bacterial effector protein to pass through and efforts to trap proteins such as SptP in *Salmonella* and IpaB in *Shigella* demonstrated that effectors must be unfolded to traverse the needle complex (Dohlich et al. 2014; Radics et al. 2014). In these studies, effectors were fused C-terminally to proteins which resist T3SS-specific unfolding, therefore acting like a plug and trapping the unfolded effector into the secretion channel. This also demonstrated that protein export is a polar process where the N-terminus is secreted first.

Visualization of the trapped substrate inside the complex was even possible by cryoEM. The *Salmonella* effector SptP was visualized in the channel of the needle complex as a long, continuously unfolded density with a folded plug on the cytoplasmic side (Radics et al. 2014). It is proposed that the funnel shape or gasket at the beginning of the export apparatus acts as a checkpoint to ensure access of unfolded effectors prior to substrate secretion. Thus, the folded fusion tags are unable to enter the system and plug it. Surprising, few structural differences at resolutions of ~ 10 Å were seen between the substrate-trapped and non-trapped structures. There are some minor changes in the export apparatus densities indicating that the needle complex is largely rigid during secretion of at least the effector protein SptP (Radics et al. 2014). However, conformational changes before or after secretion, or with different effectors, or in other species cannot be ruled out. As such, cryoET on *Chlamydia* needle complexes in contact with HeLa cells revealed a 40–50 Å compaction of the basal body compared to non-contacted needle complexes. The in-contact complexes had a subtle widening of the export apparatus densities and the sorting platform components were more pronounced. These findings indicate that there is a conformational change of the needle complex upon contact with host cells, similar to a pumping-like action (Nans et al. 2015). However, in *Salmonella*, the needle complex/base does not undergo these observed structural changes upon contact with HeLa cells. These differences could represent species-specific differences or differences in image resolution and 3D reconstruction protocols (Park et al. 2018).

5 Future Directions

Over the many years of research on the T3SS, a lot of insight has been gained on the structural assembly of the needle complex. The overall structure of the complex is nearly known, but many of the more intricate structural details remain less clear and represent future research areas. Despite the unprecedented advances in cryoEM, there is still no complete, atomic structure of the needle complex. This is mostly due to structural heterogeneity, biochemical lability, structural flexibility and symmetry variations within the system. In particular, substructures such as the SctI adaptor or the cytoplasmic sorting platform complex are less resolved. Furthermore, to fully understand substrate transport, a high-resolution structure of active complexes is essential. Undoubtedly, technological advances in cryoEM, other complementary structural techniques and biological assays on the T3SS will enable scientists to push research in new directions and lead to these structures. Furthermore, recent studies have provided evidence for the needle complex being a dynamic molecular machine. CryoET will likely enable researchers to simultaneously determine different conformational states of the needle complex in situ, a previously impossible feat. Finally, it is generally assumed that the structure of the *Salmonella* needle complex, which has been primarily focused on in this chapter, is the similar across species. However and as mentioned, there are species-specific differences among T3SSs, and therefore, the field should continue to expand to other organisms, which will ultimately lead to a better understanding of bacterial pathogenesis overall.

References

Allaoui A, Sansonetti PJ, Parsot C (1992) MxiJ, a lipoprotein involved in secretion of *Shigella* Ipa invasins, is homologous to YscJ, a secretion factor of the *Yersinia* Yop proteins. J Bacteriol 174:7661–7669

Allaoui A, Woestyn S, Sluiters C, Cornelis GR (1994) YscU, a *Yersinia* enterocolitica inner membrane protein involved in Yop secretion. J Bacteriol 176:4534–4542

Bergeron JR (2016) Structural modeling of the flagellum MS ring protein FliF reveals similarities to the type III secretion system and sporulation complex. PeerJ 4:e1718. https://doi.org/10.7717/peerj.1718

Bergeron JRC, Worrall LJ, De S et al (2015) The modular structure of the inner-membrane ring component PrgK facilitates assembly of the Type III secretion system basal body. Structure 23:161–172. https://doi.org/10.1016/j.str.2014.10.021

Bergeron JRC, Worrall LJ, Sgourakis NG et al (2013) A refined model of the prototypical *Salmonella* SPI-1 T3SS basal body reveals the molecular basis for its assembly. PLoS Pathog. https://doi.org/10.1371/journal.ppat.1003307

Bitter W (2003) Secretins of *Pseudomonas aeruginosa*: large holes in the outer membrane. Arch Microbiol 179:307–314. https://doi.org/10.1007/s00203-003-0541-8

Blocker A, Jouihri N, Larquet E et al (2001) Structure and composition of the *Shigella flexneri* "needle complex", a part of its type III secreton. Mol Microbiol 39:652–663. https://doi.org/10.1046/j.1365-2958.2001.02200.x

Blocker AJ, Deane JE, Veenendaal AKJ et al (2008) What's the point of the type III secretion system needle? Proc Natl Acad Sci U S A 105:6507–6513. https://doi.org/10.1073/pnas.0708344105

Burghout P, van Boxtel R, Van Gelder P et al (2004) Structure and electrophysiological properties of the YscC secretin from the type III secretion system of *Yersinia enterocolitica*. J Bacteriol 186:4645–4654. https://doi.org/10.1128/JB.186.14.4645-4654.2004

Caroline G, Eric F, Bohn YST et al (2008) Oligomerization of PcrV and LcrV, protective antigens of *Pseudomonas aeruginosa* and *Yersinia pestis*. J Biol Chem 283:23940–23949. https://doi.org/10.1074/jbc.M803146200

Chaudhury S, Battaile KP, Lovell S et al (2013) Structure of the Yersinia pestis tip protein LcrV refined to 1.65 Å resolution. Acta Cryst 69:477–481. https://doi.org/10.1107/S1744309113008579

Collins RF, Ford RC, Kitmitto A et al (2003) Three-dimensional structure of the *Neisseria meningitidis* secretin PilQ determined from negative-stain transmission electron microscopy. J Bacteriol 185:2611–2617

Cordes FS, Komoriya K, Larquet E et al (2003) Helical structure of the needle of the Type III secretion system of *Shigella flexneri*. J Biol Chem 278:17103–17107. https://doi.org/10.1074/jbc.M300091200

Crago AM, Koronakis V (1998) *Salmonella* InvG forms a ring-like multimer that requires the InvH lipoprotein for outer membrane localization. Mol Microbiol 30:47–56

Davis AJ, Mecsas J (2007) Mutations in the *Yersinia pseudotuberculosis* type III secretion system needle protein, YscF, that specifically abrogate effector translocation into host cells. J Bacteriol 189:83–97. https://doi.org/10.1128/JB.01396-06

Deane JE, Roversi P, Cordes FS et al (2006) Molecular model of a type III secretion system needle: implications for host-cell sensing. PNAS 103:12529–12533

Deng W, Marshall NC, Rowland JL et al (2017) Assembly, structure, function and regulation of type III secretion systems. Nat Rev Microbiol 15:323–337. https://doi.org/10.1038/nrmicro.2017.20

Derewenda U, Mateja A, Devedjiev Y et al (2004) The structure of *Yersinia pestis* V-antigen, an essential virulence factor and mediator of immunity against plague. Structure 12:301–306. https://doi.org/10.1016/j.str.2004.01.010

Dietsche T, Tesfazgi Mebrhatu M, Brunner MJ et al (2016) Structural and functional characterization of the bacterial Type III secretion export apparatus. PLoS Pathog 12:e1006071. https://doi.org/10.1371/journal.ppat.1006071

Dohlich K, Zumsteg AB, Goosmann C, Kolbe M (2014) A substrate-fusion protein is trapped inside the type III secretion system channel in *Shigella flexneri*. PLoS Pathog. https://doi.org/10.1371/journal.ppat.1003881

Eichelberg K, Ginocchio CC, Galán JE (1994) Molecular and functional characterization of the *Salmonella typhimurium* invasion genes *invB* and *invC*: homology of InvC to the FoF1 ATPase family of proteins. J Bacteriol 176:4501–4510

Epler CR, Dickenson NE, Bullitt E, Picking WL (2012) Ultrastructural analysis of IpaD at the tip of the nascent MxiH type III secretion apparatus of *Shigella flexneri*. J Mol Biol 420:29–39. https://doi.org/10.1016/j.jmb.2012.03.025

Espina M, Olive AJ, Kenjale R et al (2006) IpaD localizes to the tip of the type III secretion system needle of *Shigella flexneri*. Infect Immun 74:4391–4400. https://doi.org/10.1128/IAI.00440-06

Fields KA, Plano GV, Straley SC (1994) A low-Ca2+ response (LCR) secretion (ysc) locus lies within the lcrB region of the LCR plasmid in *Yersinia pestis*. J Bacteriol 176:569–579

Francis NR, Sosinsky GE, Thomas D, DeRosier DJ (1994) Isolation, characterization and structure of bacterial flagellar motors containing the switch complex. J Mol Biol 235:1261–1270. https://doi.org/10.1006/jmbi.1994.1079

Galán JE, Ginocchio C, Costeas P (1992) Molecular and functional characterization of the *Salmonella* invasion gene *invA*: homology of InvA to members of a new protein family. J Bacteriol 174:4338–4349

Galán JE, Lara-Tejero M, Marlovits TC, Wagner S (2014) Bacterial type III secretion systems: specialized nanomachines for protein delivery into target cells. Annu Rev Microbiol 68:415–438. https://doi.org/10.1146/annurev-micro-092412-155725

Gough CL, Genin S, Zischek C, Boucher CA (1992) hrp genes of *Pseudomonas solanacearum* are homologous to pathogenicity determinants of animal pathogenic bacteria and are conserved among plant pathogenic bacteria. Mol Plant Microbe Interact 5:384–389

Groisman EA, Ochman H (1993) Cognate gene clusters govern invasion of host epithelial cells by *Salmonella typhimurium* and *Shigella flexneri*. EMBO J 12:3779–3787

Guo EZ, Desrosiers DC, Zalesak J et al (2019) A polymorphic helix of a *Salmonella* needle protein relays signals defining distinct steps in type III secretion. PLoS Biol 17:e3000351. https://doi.org/10.1371/journal.pbio.3000351

Hodgkinson JL, Horsley A, Stabat D et al (2009) Three-dimensional reconstruction of the *Shigella* T3SS transmembrane regions reveals 12-fold symmetry and novel features throughout. Nat Struct Mol Biol 16:477–485. https://doi.org/10.1038/nsmb.1599

Hu B, Lara-Tejero M, Kong Q et al (2017) In situ molecular architecture of the *Salmonella* type III secretion machine. Cell 168:1065–1074.e10. https://doi.org/10.1016/j.cell.2017.02.022

Hu B, Morado DR, Margolin W et al (2015) Visualization of the type III secretion sorting platform of *Shigella flexneri*. Proc Natl Acad Sci USA 112:1047–1052. https://doi.org/10.1073/pnas.1411610112

Hu J, Worrall LJ, Hong C et al (2018) Cryo-EM analysis of the T3S injectisome reveals the structure of the needle and open secretin. Nat Commun 9:3840. https://doi.org/10.1038/s41467-018-06298-8

Hueck CJ (1998) Type III protein secretion systems in bacterial pathogens of animals and plants. Microbiol Mol Biol Rev 62:379–433

Johnson S, Roversi P, Espina M et al (2007) Self-chaperoning of the type III secretion system needle tip proteins IpaD and BipD. J Biol Chem 282:4035–4044. https://doi.org/10.1074/jbc.M607945200

Journet L, Agrain C, Broz P, Cornelis GR (2003) The needle length of bacterial injectisomes is determined by a molecular ruler. Source Sci New Ser 302:1757–1760

Kaniga K, Bossio JC, Galán JE (1994) The *Salmonella typhimurium* invasion genes *invF* and *invG* encode homologues of the AraC and PulD family of proteins. Mol Microbiol 13:555–568

Kaur K, Chatterjee S, De Guzman RN (2016) Characterization of the *Shigella* and *Salmonella* type III secretion system tip-translocon protein-protein interaction by paramagnetic relaxation enhancement. ChemBioChem 17:745–752. https://doi.org/10.1002/cbic.201500556

Kenjale R, Wilson J, Zenk SF et al (2005) The needle component of the type III secreton of *Shigella* regulates the activity of the secretion apparatus. J Biol Chem 280:42929–42937. https://doi.org/10.1074/jbc.M508377200

Kimbrough TG, Miller SI (2000) Contribution of *Salmonella typhimurium* type III secretion components to needle complex formation. PNAS 97:11008–11013

Korotkov KV, Pardon E, Steyaert J, Hol WGJ (2009) Crystal structure of the N-terminal domain of the secretin GspD from ETEC determined with the assistance of a nanobody. Structure 17:255–265. https://doi.org/10.1016/J.STR.2008.11.011

Korotkov KV, Gonen T, Hol WGJ (2011) Secretins: dynamic channels for protein transport across membranes. Trends Biochem Sci 36:433–443. https://doi.org/10.1016/j.tibs.2011.04.002

Kubori T, Matsushima Y, Nakamura D et al (1998) Supramolecular structure of the *Salmonella typhimurium* type III protein secretion system. Science 280:602–605. https://doi.org/10.1126/science.280.5363.602

Kubori T, Sukhan A, Aizawa S-I, Galán JE (2000) Molecular characterization and assembly of the needle complex of the *Salmonella typhimurium* type III protein secretion system. Proc Natl Acad Sci 97:10225–10230. https://doi.org/10.1073/pnas.170128997

Kudryashev M, Stenta M, Schmelz S et al (2013) In situ structural analysis of the *Yersinia enterocolitica* injectisome. Elife 2:e00792. https://doi.org/10.7554/elife.00792

Kuhlen L, Abrusci P, Johnson S et al (2018) Structure of the core of the type III secretion system export apparatus. Nat Struct Mol Biol 25:583–590. https://doi.org/10.1038/s41594-018-0086-9

Lara-Tejero M, Kato J, Wagner S et al (2011) A sorting platform determines the order of protein secretion in bacterial type III systems. Science 331:1188–1191. https://doi.org/10.1126/science.1201476

Lefebre MD, Galan JE (2014) The inner rod protein controls substrate switching and needle length in a *Salmonella* type III secretion system. Proc Natl Acad Sci 111:817–822. https://doi.org/10.1073/pnas.1319698111

Loquet A, Sgourakis NG, Gupta R, Giller K (2012) Atomic model of the type III secretion system needle. Nature 486:276–279. https://doi.org/10.1038/nature11079.Atomic

Lunelli M, Hurwitz R, Lambers J, Kolbe M (2011) Crystal structure of PrgI-SipD: insight into a secretion competent state of the type three secretion system needle tip and its interaction with host ligands. PLoS Pathog 7:e1002163. https://doi.org/10.1371/journal.ppat.1002163

Makino K, Oshima K, Kurokawa K et al (2003) Genome sequence of *Vibrio parahaemolyticus*: a pathogenic mechanism distinct from that of *V cholerae*. Lancet 361:743–749. https://doi.org/10.1016/S0140-6736(03)12659-1

Marlovits TC, Kubori T, Lara-Tejero M et al (2006) Assembly of the inner rod determines needle length in the type III secretion injectisome. Nature 441:637–640. https://doi.org/10.1038/nature04822

Marlovits TC, Kubori T, Sukhan A et al (2004) Structural insights into the assembly of the type III secretion needle complex. Science 306:1040–1042. https://doi.org/10.1126/science.1102610

Miao EA, Mao DP, Yudkovsky N et al (2010) Innate immune detection of the type III secretion apparatus through the NLRC4 inflammasome. PNAS 1–7:3076–3080. https://doi.org/10.1073/pnas.0913087107

Michiels T, Wattiau P, Brasseur R, et al (1990) Secretion of Yop proteins by *Yersiniae*. Infect Immun 58:2840–2849

Mishima M, Shida T, Yabuki K et al (2005) Solution structure of the peptidoglycan binding domain of *Bacillus subtilis* cell wall lytic enzyme CwlC: characterization of the sporulation-related repeats by NMR. Biochemistry 44:10153–10163

Mueller CA, Broz P, Müller SA et al (2005) The V-Antigen of *Yersinia* forms a distinct structure at the tip of injectisome needles

Nans A, Kudryashev M, Saibil HR, Hayward RD (2015) Structure of a bacterial type III secretion system in contact with a host membrane *in situ*. Nat Commun 6:10114. https://doi.org/10.1038/ncomms10114

Park D, Lara-Tejero M, Waxham MN, et al (2018) Visualization of the type III secretion mediated *Salmonella*–host cell interface using cryo-electron tomography. Elife. https://doi.org/10.7554/elife.39514

Pegues DA, Hantman MJ, Behlau I, Miller SI (1995) PhoP/PhoQ transcriptional repression of *Salmonella typhimurium* invasion genes: evidence for a role in protein secretion. Mol Microbiol 17:169–181

Poyraz O, Schmidt H, Seidel K et al (2010) Protein refolding is required for assembly of the type three secretion needle. Nat Struct Mol Biol 17:788–792. https://doi.org/10.1038/nsmb.1822

Radics J, Königsmaier L, Marlovits TC (2014) Structure of a pathogenic type 3 secretion system in action. Nat Struct Mol Biol 21:82–87. https://doi.org/10.1038/nsmb.2722

Sanowar S, Singh P, Pfuetzner RA et al (2010) Interactions of the transmembrane polymeric rings of the *Salmonella enterica* Serovar Typhimurium type III secretion system. https://doi.org/10.1128/mbio.00158-10

Sato H, Frank DW, Schurr MJ, Yahr T (2011) Multi-functional characteristics of the *Pseudomonas aeruginosa*type III needle-tip protein, PcrV; comparison to orthologs in other gram-negative bacteria. Front Microbiol. https://doi.org/10.3389/fmicb.2011.00142

Schraidt O, Marlovits TC (2011) Three-dimensional model of *Salmonella*'s needle complex at subnanometer resolution. Science (80–) 331:1192–1195. https://doi.org/10.1126/science.1199358

Schraidt Oliver, Lefebre MD et al (2010) Topology and organization of the *Salmonella typhimurium* type III secretion needle complex components. PLoS Pathog 6:e1000824. https://doi.org/10.1371/journal.ppat.1000824

Sekiya K, Ohishi M, Ogino T et al (2001) Supermolecular structure of the enteropathogenic *Escherichia coli* type III secretion system and its direct interaction with the EspA-sheath-like structure. Proc Natl Acad Sci U S A 98:11638–11643. https://doi.org/10.1073/pnas.191378598

Spreter T, Yip CK, Sanowar S et al (2009) A conserved structural motif mediates formation of the periplasmic rings in the type III secretion system. Nat Struct Mol Biol 16:468–476. https://doi.org/10.1038/nsmb.1603

Sukhan A, Kubori T, Wilson J, Galán JE (2001) Genetic analysis of assembly of the *Salmonella enterica* serovar Typhimurium type III secretion-associated needle complex. J Bacteriol 183:1159–1167. https://doi.org/10.1128/JB.183.4.1159-1167.2001

Tamano K, Aizawa S, Katayama E et al (2000) Supramolecular structure of the *Shigella* type III secretion machinery: the needle part is changeable in length and essential for delivery of effectors. EMBO J 19:3876–3887. https://doi.org/10.1093/emboj/19.15.3876

Torres-Vargas CE, Kronenberger T, Roos N, et al (2019) The inner rod of virulence-associated type III secretion systems constitutes a needle adapter of one helical turn that is deeply integrated into the system's export apparatus. Mol Microbiol mmi.14327. https://doi.org/10.1111/mmi.14327

Valverde R, Edwards L, Regan L (2008) Structure and function of KH domains. FEBS J 275:2712–2726. https://doi.org/10.1111/j.1742-4658.2008.06411.x

Vander Broek CW, Stevens JM (2017) Type III secretion in the melioidosis pathogen *Burkholderia pseudomallei*. Front Cell Infect Microbiol 7:255. https://doi.org/10.3389/fcimb.2017.00255

Wagner S, Königsmaier L, Lara-Tejero M et al (2010) Organization and coordinated assembly of the type III secretion export apparatus. Proc Natl Acad Sci U S A 107:17745–17750. https://doi.org/10.1073/pnas.1008053107

Wang Y, Ouellette AN, Egan CW et al (2007) Differences in the electrostatic surfaces of the type III secretion needle proteins PrgI, BsaL, and MxiH. J Mol Biol 371:1304–1314. https://doi.org/10.1016/j.jmb.2007.06.034

Wang YA, Yu X, Yip C et al (2006) Structural polymorphism in bacterial EspA filaments revealed by cryo-EM and an improved approach to helical reconstruction. Structure 14:1189–1196. https://doi.org/10.1016/j.str.2006.05.018

Wee DH, Hughes KT (2015) Molecular ruler determines needle length for the *Salmonella* Spi-1 injectisome. Proc Natl Acad Sci USA 112:4098–4103. https://doi.org/10.1073/pnas.1423492112

Worrall LJ, Hong C, Vuckovic M et al (2016) Near-atomic-resolution cryo-EM analysis of the *Salmonella* T3S injectisome basal body. Nature 540:597–601. https://doi.org/10.1038/nature20576

Yip CK, Kimbrough TG, Felise HB et al (2005) Structural characterization of the molecular platform for type III secretion system assembly. Nature 435:702–707. https://doi.org/10.1038/nature03554

Zhang L, Wang Y, Picking WL et al (2006) Solution structure of monomeric BsaL, the type III secretion needle protein of *Burkholderia pseudomallei*. J Mol Biol 359:322–330. https://doi.org/10.1016/j.jmb.2006.03.028

Zhong D, Lefebre M, Kaur K et al (2012) The *Salmonella* type III secretion system inner rod protein PrgJ Is partially folded. J Biol Chem. https://doi.org/10.1074/jbc.M112.381574
Zilkenat S, Franz-Wachtel M, Stierhof Y-D et al (2016) Determination of the stoichiometry of the complete bacterial type III secretion needle complex using a combined quantitative proteomic approach. Mol Cell Proteomics. https://doi.org/10.1074/mcp.M115.056598

Molecular Organization and Assembly of the Export Apparatus of Flagellar Type III Secretion Systems

Tohru Minamino, Akihiro Kawamoto, Miki Kinoshita and Keiichi Namba

Contents

Abstract The bacterial flagellum is a supramolecular motility machine consisting of the basal body, the hook, and the filament. For construction of the flagellum beyond the cellular membranes, a type III protein export apparatus uses ATP and proton-motive force (PMF) across the cytoplasmic membrane as the energy sources to transport flagellar component proteins from the cytoplasm to the distal end of the growing flagellar structure. The protein export apparatus consists of a PMF-driven transmembrane export gate complex and a cytoplasmic ATPase complex. In

---

T. Minamino (✉) · A. Kawamoto · M. Kinoshita · K. Namba
Graduate School of Frontier Biosciences, Osaka University, 1-3 Yamadaoka, Suita, Osaka 565-0871, Japan
e-mail: tohru@fbs.osaka-u.ac.jp

A. Kawamoto
Institute for Protein Research, Osaka University, 3-2 Yamadaoka, Suita, Osaka 565-0871, Japan

K. Namba
Center for Biosystems Dynamics Research & SPring-8 Center, RIKEN, 1-3 Yamadaoka, Suita, Osaka 565-0871, Japan

Current Topics in Microbiology and Immunology (2020) 427: 91–107
https://doi.org/10.1007/82_2019_170
© Springer Nature Switzerland AG 2019
Published Online: 07 June 2019

addition, the basal body C ring acts as a sorting platform for the cytoplasmic ATPase complex that efficiently brings export substrates and type III export chaperone–substrate complexes from the cytoplasm to the export gate complex. In this book chapter, we will summarize our current understanding of molecular organization and assembly of the flagellar type III protein export apparatus.

1 Introduction

Bacteria swim in liquid environments and move on solid surfaces by rotating a very long filamentous assembly called the bacterial flagellum. The flagellum consists of at least three parts: the basal body as a bidirectional rotary motor, the hook as a universal joint, and the filament as a helical propeller. Flagellar assembly begins with the basal body, followed by the hook and finally the filament. Fourteen flagellar proteins are transported via a type III protein export apparatus into the central channel inside the growing structure and assemble at the distal end (Macnab 2004; Minamino et al. 2008; Minamino 2014).

The flagellar type III protein export apparatus is composed of a transmembrane export gate complex made of FlhA, FlhB, FliP, FliQ, and FliR and a cytoplasmic ATPase complex consisting of FliH, FliI, and FliJ. These proteins are evolutionarily related to components of the virulence-associated type III secretion system (T3SS) of pathogenic bacteria, also known as the injectisome (Fig. 1) (Galán et al. 2014; Wagner et al. 2018).

FliG, FliM, and FliN form the C ring on the cytoplasmic face of the basal body MS ring made of a transmembrane protein, FliF. The C ring acts not only as the rotor of the flagellar motor but also as the switch for bidirectional motor rotation, allowing the flagellar motor to rotate both in the counterclockwise and in the clockwise directions (Berg 2003; Morimoto and Minamino 2014). FliM and FliN, which are well conserved in virulence-associated T3SS families, provide the binding sites for the cytoplasmic ATPase complex in complex with export substrates and export chaperone–substrate complexes (González-Pedrajo et al. 2006; Minamino et al. 2009; Lara-Tejero et al. 2011). In this book chapter, we will describe the structure and assembly of the flagellar type III protein export apparatus in *Salmonella enterica*.

2 Structure of the Transmembrane Export Gate Complex

FlhA, FlhB, FliP, FliQ, and FliR form the transmembrane export gate complex inside the MS ring (Minamino and Macnab 1999; Fukumura et al. 2017). A transmembrane protein, FliO, which is not conserved in virulence-associated T3SS families of pathogenic bacteria, is required for efficient assembly of the export gate complex inside the MS ring although it is not essential for flagellar protein

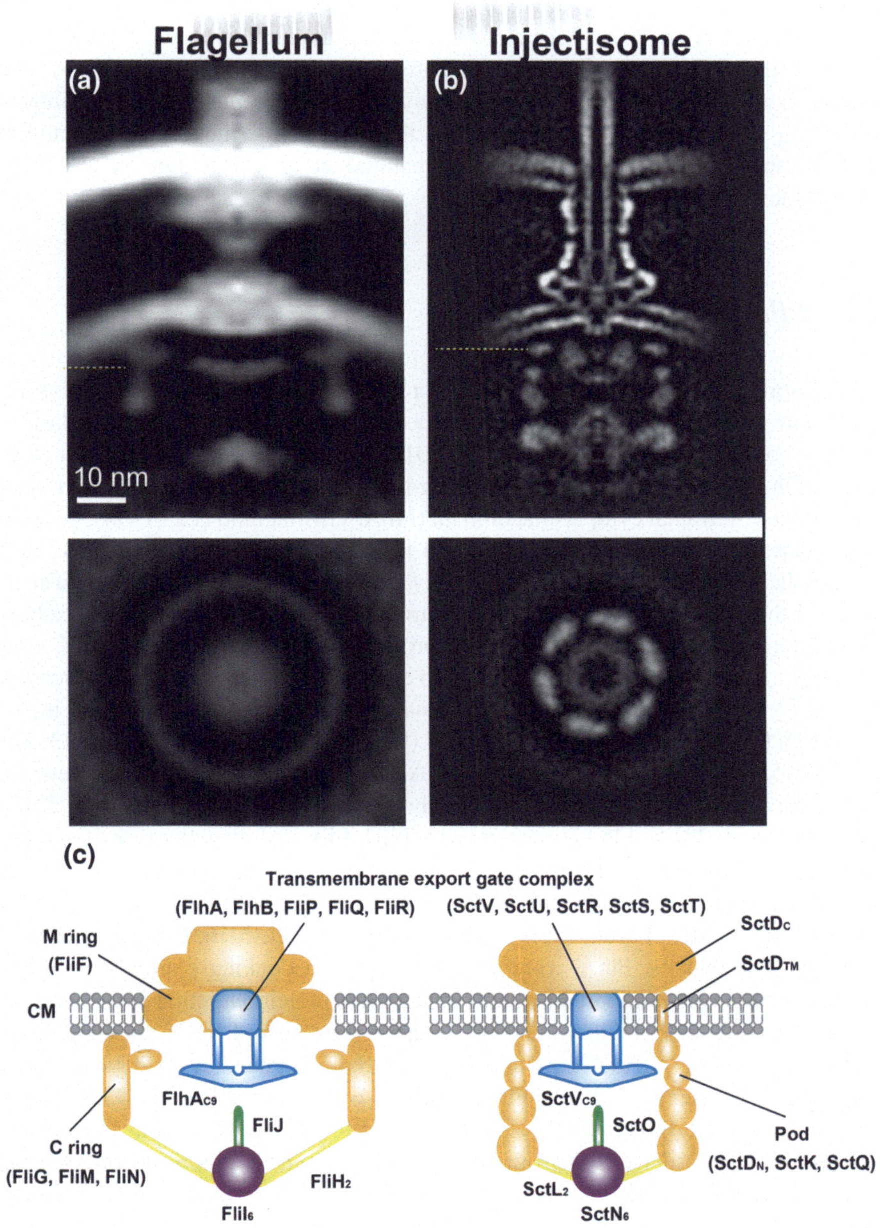

Fig. 1 Structural comparison between the flagellum and the injectisome. In situ structures of the basal bodies of the *Salmonella* flagellum (**a**) and injectisome (**b**) are visualized by electron cryotomography and subtomogram averaging. The central section maps of the flagellum (EMDB-2521) (**a**) and the injectisome (EMDB-8544) (**b**) after subtomogram averaging are shown. Upper panels, side view, lower panels, bottom view corresponding a cross section at height indicated by the dashed yellow line. **c** Schematic diagrams of cytoplasmic portions of the *Salmonella* flagellum (left panel) and injectisome (right panel). Name of each part of the basal body and component protein(s) are shown

export (Barker et al. 2010; Morimoto et al. 2014; Fukumura et al. 2017; Fabiani et al. 2017). The transmembrane export gate complex is powered by the proton-motive force (PMF) across the cytoplasmic membrane and facilitates unfolding and protein translocation across the cytoplasmic membrane (Minamino and Namba 2008; Paul et al. 2008; Minamino et al. 2011; Lee et al. 2014; Terashima et al. 2018).

2.1 FlhA Ring Structure

FlhA forms an ion channel to conduct protons and sodium ions and plays an important role in the energy coupling mechanism along with the cytoplasmic ATPase complex (Minamino et al. 2011, 2016; Morimoto et al. 2016; Erhardt et al. 2017). FlhA consists of a hydrophobic N-terminal transmembrane domain ($FlhA_{TM}$) and a large C-terminal cytoplasmic domain ($FlhA_C$) (Minamino et al. 1994). Genetic and biochemical analyses have shown that $FlhA_{TM}$ interacts with FlhB, FliF, and FliR (Kihara et al. 2001; Barker and Samatey 2012; Hara et al. 2011; Fukumura et al. 2017). $FlhA_C$ has been visualized by electron cryotomography (ECT) and subtomogram averaging to form a ring-shaped projection in the cavity within the C ring (Fig. 1a, c) (Abrusci et al. 2013; Kawamoto et al. 2013). Similar ring-like structures formed by the C-terminal cytoplasmic domain of a FlhA homologue of the injectisome, SctV, have been identified by ECT (Fig. 1b, c) (Kawamoto et al. 2013; Hu et al. 2015, 2017; Makino et al. 2016). $FlhA_C$ and $SctV_C$ form a homo-nonamer as part of the export gate complex (Fig. 2a) (Abrusci et al. 2013; Kawamoto et al. 2013; Morimoto et al. 2014). $FlhA_C$ interacts with FliH, FliI, FliJ, flagellar type III export chaperones, and export substrates and coordinates flagellar protein export with assembly (Minamino and Macnab 2000c; Bange et al. 2010; Minamino et al. 2012a; Kinoshita et al. 2013; Furukawa et al. 2016; Kinoshita et al. 2016; Inoue et al. 2018). This suggests that the $FlhA_C$ ring structure acts as a docking platform for these proteins and plays an important role in hierarchical protein targeting and export.

Crystal structures of $FlhA_C$ and $SctV_C$ have been solved by X-ray crystallography (Bange et al. 2010; Moore and Jia 2010; Saijo-Hamano et al. 2010; Worrall et al. 2010; Abrusci et al. 2013). $FlhA_C$ consists of four domains, D1, D2, D3, and D4, and a flexible linker ($FlhA_L$) connecting with $FlhA_{TM}$ (Fig. 2c). The crystal structure of $SctV_C$ derived from the *Shigella* injectisome forms a nonameric ring structure through D1–D3 and D3–D3 interactions (Abrusci et al. 2013). Because similar subunit interactions are observed in the crystal packing of the *Salmonella* $FlhA_C$ structure (Saijo-Hamano et al. 2010), D1–D3 and D3–D3 interactions are likely to be responsible for the $FlhA_C$ ring formation (Kawamoto et al. 2013). In addition to these interactions, interactions of $FlhA_L$ with domains D1 and D3 of its neighboring $FlhA_C$ subunit are involved in highly cooperative $FlhA_C$ ring formation in solution (Terahara et al. 2018). $FlhA_C$ adopts two distinct, open and closed conformations (Moore and Jia 2010; Saijo-Hamano et al. 2010; Worrall et al. 2010; Abrusci et al. 2013). A large open cleft between domains D2 and D4 is observed in

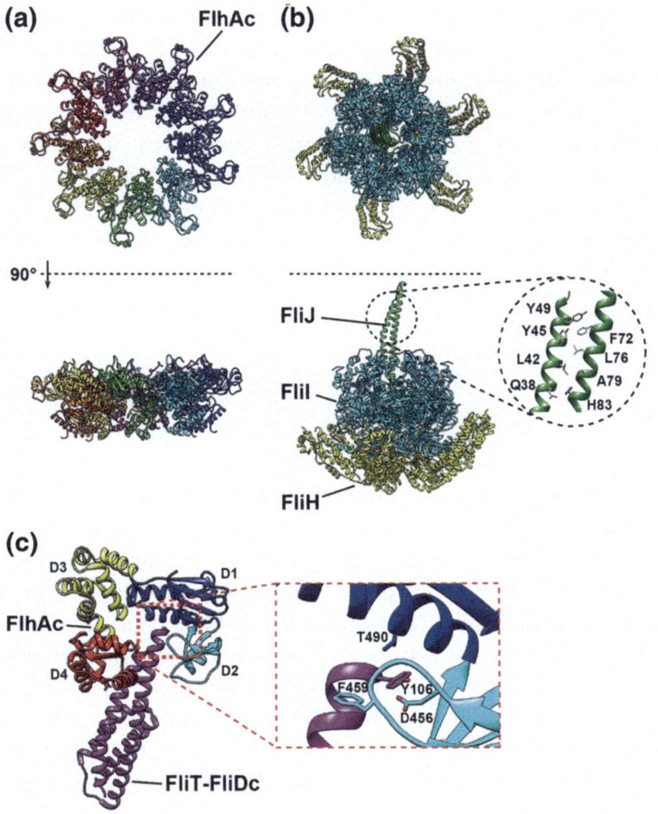

Fig. 2 Atomic models of the docking platform made of $FlhA_C$ and the cytoplasmic ATPase ring complex consisting of FliH, FliI and FliJ. **a** The $FlhA_C$ ring, which is involved in the interactions with FliH, FliI, FliJ, flagellar type III export chaperones and export substrates and **b** the $FliH_{12}FliI_6FliJ$ ring, which plays an important role in energy transduction. Cα ribbon representation of $FlhA_C$ (PDB ID: 3A5I), the $FliH_2FliI$ complex (PDB ID: 5B0O), and FliJ (PDB ID: 3AJW) is shown. Highly conserved Gln38, Leu42, Tyr45, Tyr49, Phe72, Leu76, Ala79, and His83 residues of FliJ (indicated as Q38, L42, Y45, Y49, F72, L76, A79, and H83, respectively) are responsible for the interaction with $FlhA_C$ and the FliJ–$FlhA_C$ interaction facilitates PMF-driven protein export by the transmembrane export gate complex **c** crystal structure of $FlhA_C$ in complex with a $FliD_C$–FliT fusion protein (PDB ID: 6CH2). $FlhA_C$ consists of four domains: D1, D2, D3, and D4. A highly conserved Tyr106 residue of FliT (magenta) binds to a conserved hydrophobic dimple at an interface between domains D1 and D2 of $FlhA_C$. Well conserved Asp456, Phe459 and Thr490 residues of $FlhA_C$ (indicated as D456, F459, and T490, respectively) are responsible for the interaction with Tyr108 of FliT

the open form, but the cleft is closed in the closed form. A conserved hydrophobic dimple containing Asp456, Phe459, and Thr490 resides is located at the interface between domains D1 and D2 and is directly involved in the interactions with the FlgN, FliS, and FliT chaperones in complex with their cognate substrates (Minamino et al. 2012a; Kinoshita et al. 2013). Crystal structures of $FlhA_C$ in complex with the chaperone–substrate complexes have shown that the chaperones bind to the hydrophobic dimple of the open form of $FlhA_C$ (Fig. 2c) (Xing et al. 2018). Mutations at residues involved in the interactions of $FlhA_L$ with its neighboring $FlhA_C$ subunits in the $FlhA_C$ ring structure significantly weaken the interaction of $FlhA_C$ with the chaperone–substrate complexes, thereby reducing the probability of filament assembly at the hook tip (Terahara et al. 2018). This leads to a plausible hypothesis that interactions of $FlhA_L$ with the D1 and D3 domains of its neighboring $FlhA_C$ subunits may convert the $FlhA_C$ ring structure from the closed conformation to the open one, allowing the chaperone–substrate complexes to bind to the $FlhA_C$ ring.

2.2 *FlhB*

The type III protein export apparatus undergoes substrate specificity switching upon completion of hook assembly, terminating hook assembly, and initiating filament assembly. FlhB is involved in the substrate specificity switching along with a secreted molecular ruler protein, FliK, which is also secreted via the type III protein export apparatus during hook assembly (Minamino 2018). FlhB consists of a hydrophobic N-terminal domain and a relatively large C-terminal cytoplasmic domain ($FlhB_C$) (Minamino et al. 1994). Crystal structures of $FlhB_C$ and its homologue of the injectisome, $SctU_C$, have been solved by X-ray crystallography. $FlhB_C$ and $SctU_C$ contain two distinct: CN and CC polypeptides (Zarivach et al. 2008; Meshcheryakov et al. 2013). The $FlhB_{CN}$ polypeptide consists of a long α helix ($\alpha 1$) and a β strand ($\beta 1$). The $FlhB_{CC}$ polypeptide is composed of three α helices ($\alpha 2$, $\alpha 3$, $\alpha 4$), three β strands ($\beta 2$, $\beta 3$, $\beta 4$), and a highly flexible C-terminal tail (residues 354–383, $FlhB_{CCT}$). Four α helices surround a four-stranded β sheet, forming a globular domain (Meshcheryakov et al. 2013). $FlhB_{CCT}$ is dispensable for FlhB function, but its truncation results in autonomous substrate specificity switching of the type III protein export apparatus in the absence of FliK (Kutsukake et al. 1994). This suggests that $FlhB_{CCT}$ may contribute to the well-regulated substrate specificity switching. A highly conserved NPTH sequence lies on a flexible loop located between $FlhB_{CN}$ and $FlhB_{CC}$ (Minamino and Macnab 2000a; Fraser et al. 2003). $FlhB_C$ undergoes autocatalytic cleavage between Asn269 and Pro270 within the NPTH sequence to be split into $FlhB_{CN}$ and $FlhB_{CC}$ by a mechanism involving cyclization of Asn269 (Ferris et al. 2005; Zarivach et al. 2008). A conserved hydrophobic patch formed by Ala286, Pro287, Ala341, and Leu344 in $FlhB_{CC}$ is directly involved in interactions with the N-terminal region of the hook protein, which contains an export signal recognized by the type III protein export apparatus (Evans et al. 2013). Photo-cross-linking experiments combined with mutational analyses have shown that the C-terminal domain of FliK binds to $FlhB_C$, thereby terminating the export of the rod and hook-type proteins and initiating the export of the filament-type proteins (Kinoshita et al. 2017).

2.3 Core Structure of the Export Gate Complex

FliP, FliQ, and FliR form a core structure of the transmembrane export gate complex (Fukumura et al. 2017). The structure of purified FliP–FliQ–FliR core complex has been determined at 4.2 Å resolution by electron cryomicroscopy (cryoEM) and single-particle image analysis. The cryoEM structure of the core complex adopts a right-handed helical assembly composed of five copies of FliP, four copies of FliQ, and one copy of FliR (Kuhlen et al. 2018) (Fig. 3). Although FliP, FliQ, and FliR are predicted to have four, two, and six transmembrane helices, respectively (Ohnishi et al. 1997), they do not adopt canonical integral membrane topologies in the core complex (Kuhlen et al. 2018). FliP and FliR form a $FliP_5FliR_1$ complex, and four FliQ subunits are associated with the $FliP_5FliR_1$ complex on its outside. FliR is a structural fusion of FliP and FliQ and so compensates for a helical rise between the first and the fifth FliP subunits to stabilize the helical structure. The assembled $FliP_5FliQ_4FliR_1$ core complex has a central pore with a diameter of 1.5 nm, which seems to be the protein translocation channel (Kuhlen et al. 2018). The most distal part of the core complex is likely to interact with the most proximal end of the rod inside the basal body MS ring (Dietsche et al. 2016; Kuhlen et al. 2018). Biochemical analyses have shown that SctR, SctS, and SctT, which are FliP, FliQ, and FliR homologues of the injectisome, respectively, form a core structure in the injectisome in a way similar to the assembly of the $FliP_5FliQ_4FliR_1$ complex (Wagner et al. 2010; Zilkenat et al. 2016; Dietsche et al. 2016).

FliO consists of an N-terminal periplasmic tail, a single transmembrane helix and a C-terminal cytoplasmic domain (Barker et al. 2010). FliO forms a 5 nm ring structure with three flexible clamp-like structures that bind to FliP to facilitate FliP oligomerization (Fukumura et al. 2017). Highly conserved Phe137, Phe150, and Glu178 residues of FliP, which are functionally important, contribute not only to FliO–FliP interaction but also for FliP–FliP interaction (Fukumura et al. 2017). The FliO ring complex protects FliP from proteolytic degradation and promotes stable $FliP_5FliR_1$ complex formation (Fabiani et al. 2017). Overexpression of FliP restores motility of a *Salmonella fliO* null mutant to the wild-type level, suggesting that the FliO ring complex acts as a structural scaffold to facilitate the helical assembly of the $FliP_5FliR_1$ complex (Barker et al. 2010; Fukumura et al. 2017; Fabiani et al. 2017).

3 Cytoplasmic ATPase Ring Complex

The cytoplasmic ATPase ring complex is composed of twelve copies of FliH, six copies of FliI and one copy of FliJ (Fig. 2b) (Imada et al. 2016). The ATPase ring structure has been visualized at the flagellar base by ECT (Fig. 1a, c) (Chen et al. 2011). ATP hydrolysis by the ATPase ring complex activates the transmembrane export gate complex through an interaction between FliJ and FlhA, allowing the export gate complex to utilize PMF across the cytoplasmic membrane to transport

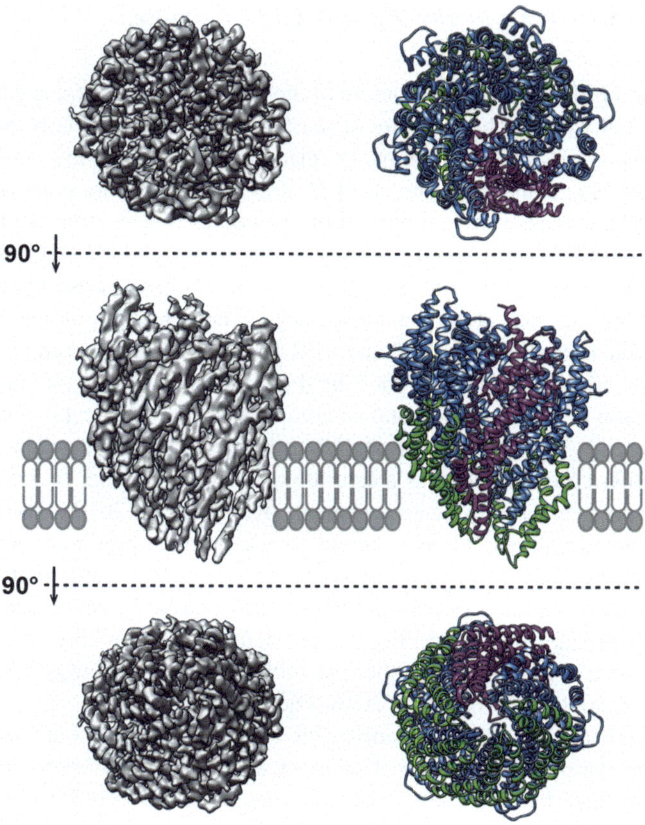

Fig. 3 CryoEM structure of the $FliP_5FliQ_4FliR_1$ complex. Left, cryoEM map of the $FliP_5FliQ_4FliR_1$ complex reconstructed from 98,000 particles with C1 symmetry (EMDB-4173). Right, Cα ribbon representation (PDB ID: 6F2D). Blue, FliP; green, FliQ; and magenta, FliR

flagellar proteins (Minamino et al. 2014). The $FliH_{12}FliI_6FliJ$ complex is structurally similar to F- and V-type rotary ATPases, suggesting that the flagellar ATPase ring complex is evolutionally related to these rotary ATPases (Imada et al. 2007; Ibuki et al. 2011; Imada et al. 2016).

FliI is a Walker-type ATPase and forms a homo-hexamer to fully exert its ATPase activity (Fig. 2b) (Fan et al. 1996; Claret et al. 2003). ATP binds to an ATP binding site located an interface between FliI subunits in a way similar to that found in F- and V-type rotary ATPases, and so FliI ring formation is required for ATP hydrolysis by the FliI ATPase (Kazetani et al. 2009). FliJ facilitates the assembly of FliI into the hexameric ring structure by binding to the center of the ring (Ibuki et al. 2011). Highly conserved, surface-exposed residues of FliJ, namely Gln38, Leu42, Tyr45, Tyr49, Phe72, Leu76, Ala79, and His83, are involved in the interaction with $FlhA_L$ (Fig. 2b) (Ibuki et al. 2013), and the interaction between FliJ and $FlhA_L$ coordinates ATP hydrolysis by the $FliI_6$ ring with proton-coupled flagellar protein export

(Minamino et al. 2014; Morimoto et al. 2016). FliH forms a homo-dimer (Minamino and Macnab 2000b) although the conformation of the two FliH monomers is different from each other (Minamino et al. 2002; Imada et al. 2016). The C-terminal domain of FliH binds to the extreme N-terminal region of FliI (Fig. 2b) (Minamino and Macnab 2000b; González-Pedrajo et al. 2002; Okabe et al. 2009; Imada et al. 2016). The FliH dimer binds to a C ring protein, FliN, and FlhA to allow the ATPase ring complex to efficiently localize to the flagellar base (González-Pedrajo et al. 2006; Minamino et al. 2009; Bai et al. 2014). Two conserved Trp7 and Trp10 residues of $FliH_N$ are directly involved in the interactions of FliH with FliN and FlhA (Minamino et al. 2009; Hara et al. 2012; Notti et al. 2015).

FliH and FliI also exist as a $FliH_2FliI_1$ complex in the cytoplasm (Minamino and Macnab 2000b; Minamino et al. 2001). Flagellar export chaperone–substrate complexes bind to the $FliH_2FliI_1$ complex through an interaction between the FliI ATPase and the chaperone (Thomas et al. 2004; Imada et al. 2010; Minamino et al. 2012b). More than six copies of FliI labeled with yellow fluorescent protein (FliI-YFP) are estimated to be associated with the basal body through the interactions of FliH with FliN and FlhA. Since FliI-YFP shows rapid exchanges between the flagellar basal body and the cytoplasmic pool, the $FliH_2FliI_1$ complex is thought to act as a dynamic carrier to bring export substrates and chaperone–substrate complexes from the cytoplasm to the $FlhA_C$–$FlhB_C$ docking platform of the transmembrane export gate complex (Bai et al. 2014; Terashima et al. 2018).

4 Sorting Platform

The C ring is composed of FliG, FliM, and FliN. The C ring acts not only as a rotor of the flagellar motor but also as a structural switch to change the direction of flagellar motor rotation (Fig. 1c) (Berg 2003; Morimoto and Minamino 2014). The C ring has 34-fold rotational symmetry (Thomas et al. 1999). FliG consists of three domains: N-terminal ($FliG_N$), middle ($FliG_M$), and C-terminal ($FliG_C$) domains (Lee et al. 2010). $FliG_N$ directly binds to the MS ring protein, FliF (Kihara et al. 2000). Crystal structures of $FliG_N$ in complex with the C-terminal portion of FliF ($FliF_C$) have been solved by X-ray crystallography (Lynch et al. 2017; Xue et al. 2018). Two α helices of $FliF_C$ are deeply inserted into a hydrophobic groove of $FliG_N$ formed by four α helices. Helix α4 of $FliG_N$ adopts two distinct conformations. One adopts an extended conformation whereas the other is divided into helices α4a and α4b. When the $FliF_C$–$FliG_N$ complex exists in a solution, the α4 helix adopts an extended conformation (Lynch et al. 2017). When FliG assembles into the FliG ring structure on the cytoplasmic face of the MS ring, the α4 helix induces a conformational change and so helix α4b interacts with helix α4a of its neighboring FliG subunit (Lynch et al. 2017; Xue et al. 2018). The $FliG_{MC}$ unit, which consists of $FliG_M$, $FliG_C$ and a helix linker connecting these two domains, adopts a compact conformation through an intramolecular interaction between $FliG_M$ and $FliG_C$, allowing FliG to exit as a monomer in solution (Baker et al. 2016;

Kinoshita et al. 2018a). In contrast, when the $FliG_{MC}$ unit adopts an extended conformation, intermolecular interactions between $FliG_M$ and $FliG_C$ promote the self-assembly of FliG into the ring structure (Baker et al. 2016; Kinoshita et al. 2018b). FliM and FliN form a stable $FliM_1FliN_3$ complex through an interaction between the C-terminal domain of FliM ($FliM_C$) and FliN (Notti et al. 2015; McDowell et al. 2016) and bind to the FliG ring through an interaction between the middle domain of FliM ($FliM_M$) and $FliG_M$ to form the C ring wall (Paul et al. 2011; Vartanian et al. 2012). Intermolecular interactions between $FliM_M$ domains are required for the formation of the continuous wall of the C ring (Park et al. 2006). $FliM_C$ and FliN together form a spiral structure at the bottom of the C ring (McDowell et al. 2016). It has been shown that surface-exposed hydrophobic residues of FliN, Val111, Val112 and Val113, are involved in the interaction with FliH (McMurry et al. 2006; Paul et al. 2006). Overexpression of FliI partially rescues the reduced ability of flagellar protein export by the *fliN* null mutant (McMurry et al. 2006), suggesting that the C ring is required for efficient recruitment of the $FliH_2FliI_1$ complex to the type III protein export apparatus for efficient flagellar protein export.

It has also been shown that the sorting platform of the injectisome contributes to a strict order of protein secretion by the type III protein export apparatus (Lara-Tejero et al. 2011). However, the structure and stoichiometry of the sorting platform of the injectisome are distinct from those of the flagellar C ring structure (Fig. 1) (Kawamoto et al. 2013; Hu et al. 2015; Makino et al. 2016; Hu et al. 2017). A FliM/FliN homologue of the injectisome, SctQ, forms six pod-like structures on the cytoplasmic face of the cytoplasmic membrane ring (Fig. 1b) (Hu et al. 2015; Makino et al. 2016; Hu et al. 2017). SctK, which is not conserved in the flagellar type III protein export system, associates with the pod-like structure (Hu et al. 2017). SctL, which is a FliH homologue of the injectisome, forms a linker connecting the pod and the ATPase ring complex made of a FliI homologue, SctN (Notti et al. 2015; Hu et al. 2017). It has been shown the sorting platform is highly dynamic structure during protein secretion (Diepold et al. 2015, 2017).

5 Assembly of the Type III Protein Export Apparatus

FliF assembles into the MS ring within the cytoplasmic membrane (Kubori et al. 1992; Ueno et al. 1992). Recently, it has been reported that FliG is required for efficient MS ring formation (Li and Sourjik 2011; Morimoto et al. 2014). This suggests that FliF and FliG together form the MS–FliG ring complex. FliP and FliR form a $FliP_5FliR_1$ complex in a FliO-dependent manner (Dietsche et al. 2016; Fukumura et al. 2017; Fabiani et al. 2017). FliQ is peripherally associated around the outside of the $FliP_5FliR_1$ complex and forms a helical $FliP_5FliQ_4FliR_1$ structure inside the MS ring (Kuhlen et al. 2018). FlhA and FlhB are associated with the $FliP_5FliQ_4FliR_1$ core complex (Fukumura et al. 2017). FlhA also binds to the MS ring directly (Fukumura et al. 2017). Since FlhA requires FliF, FliG, FliO, FliP,

FliQ, and FliR for its assembly to the flagellar basal body but not FlhB (Morimoto et al. 2014), it has been proposed that the assembly of the flagellar type III export gate complex begins with the formation of the $FliP_5FliR_1$ complex with the help of the FliO ring complex, followed by the assembly of FliQ and finally of FlhA and FlhB during MS ring formation (Fig. 4) (Wagner et al. 2010; Diepold et al. 2011; Dietsche et al. 2016; Fukumura et al. 2017).

The $FliM_1FliN_3$ complex forms the continuous wall of the C ring on the cytoplasmic face of the MS ring through the interactions between $FliG_M$ and $FliM_M$ (Paul et al. 2011; Vartanian et al. 2012). Finally, the cytoplasmic $FliH_{12}FliI_6FliJ_1$ ring complex is formed at the flagellar base through the interactions of FliH with FlhA and FliN (Fig. 4). Upon completion of the type III protein export apparatus, export substrates and flagellar export chaperone–substrate complexes are efficiently recruited via the cytoplasmic $FliH_2FliI_1$ complex to the type III protein export apparatus to be transported into the central channel of the growing flagellar structure.

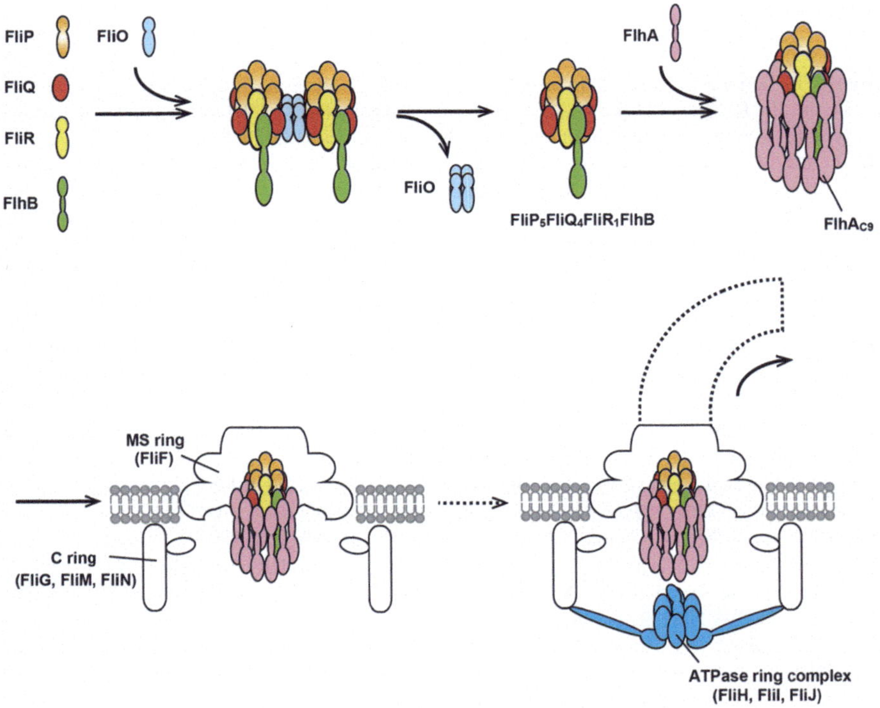

Fig. 4 Assembly mechanism of the type III protein export apparatus. FliP, FliQ, and FliR form a $FliP_5FliQ_4FliR_1$ complex with the help of the FliO complex, followed by the assembly of FlhB and finally of FlhA during MS ring formation in the cytoplasmic membrane. Then, the $FliM_1FliN_3$ complex binds to FliG to form the C ring on the cytoplasmic face of the MS ring. Finally, the cytoplasmic ATPase ring complex made of FliH, FliI, and FliJ is formed and is associated with the C ring through interactions between FliH and FliN

6 Conclusion

The type III protein export apparatus consists of a PMF-driven export gate complex made of FlhA, FlhB, FliP, FliQ, and FliR and a cytoplasmic ATPase complex consisting of FliH, FliI, and FliJ. The export apparatus utilizes ATP and PMF to efficiently couple the proton influx through the export gate complex with protein translocation into the central channel of the growing structure. Atomic structures of the C-terminal cytoplasmic domains of FlhA and FlhB, FliH, FliI, FliJ, and the $FliP_5FliQ_4FliR_1$ helical assembly have been solved. However, it still remains unknown how flagellar proteins are unfolded and transported by the PMF-driven export gate complex. High-resolution structural analysis of the entire protein export apparatus by cryoEM image analysis would be essential to advance our mechanistic understanding of the type III protein export process.

References

Abrusci P, Vergara-Irigaray M, Johnson S, Beeby MD, Hendrixson DR, Roversi P, Friede ME, Deane JE, Jensen GJ, Tang CM, Lea SM (2013) Architecture of the major component of the type III secretion system export apparatus. Nat Struct Mol Biol 20:99–104

Bai F, Morimoto YV, Yoshimura SDJ, Hara N, Kami-ike N, Namba K, Minamino T (2014) Assembly dynamics and the roles of FliI ATPase of the bacterial flagellar export apparatus. Sci Rep 4:6528

Baker MA, Hynson RM, Ganuelas LA, Mohammadi NS, Liew CW, Rey AA, Duff AP, Whitten AE, Jeffries CM, Delalez NJ, Morimoto YV, Stock D, Armitage JP, Turberfield AJ, Namba K, Berry RM, Lee LK (2016) Domain-swap polymerization drives the self-assembly of the bacterial flagellar motor. Nat Struct Mol Biol 23:197–203

Bange G, Kümmerer N, Engel C, Bozkurt G, Wild K, Sinning I (2010) FlhA provides the adaptor for coordinated delivery of late flagella building blocks to the type III secretion system. Proc Natl Acad Sci USA 107:11295–11300

Barker CS, Samatey FA (2012) Cross-complementation study of the flagellar type III export apparatus membrane protein FlhB. PLoS ONE 7:e44030

Barker CS, Meshcheryakova IV, Kostyukova AS, Samatey FA (2010) FliO regulation of FliP in the formation of the *Salmonella enterica* flagellum. PLoS Genet 6:e1001143

Berg HC (2003) The rotary motor of bacterial flagella. Ann Rev Biochem 72:19–54

Chen S, Beeby M, Murphy GE, Leadbetter JR, Hendrixson DR, Briegel A, Li Z, Shi J, Tocheva EI, Müller A, Dobro MJ, Jensen GJ (2011) Structural diversity of bacterial flagellar motors. EMBO J 30:2972–2981

Claret L, Susannah CR, Higgins M, Hughes C (2003) Oligomerization and activation of the FliI ATPase central to bacterial flagellum assembly. Mol Microbiol 48:1349–1355

Diepold A, Wiesand U, Cornelis GR (2011) The assembly of the export apparatus (YscR, S, T, U, V) of the Yersinia type III secretion apparatus occurs independently of other structural components and involves the formation of an YscV oligomer. Mol Microbiol 82:502–514

Diepold A, Kudryashev M, Delalez NJ, Berry RM, Armitage JP (2015) Composition, formation, and regulation of the cytosolic C-ring, a dynamic component of the type III secretion injectisome. PLoS Biol 13(1):e1002039

Diepold A, Sezgin E, Huseyin M, Mortimer T, Eggeling C, Armitage JP (2017) A dynamic and adaptive network of cytosolic interactions governs protein export by the T3SS injectisome. Nat Commun 8:15940

Dietsche T, Tesfazgi Mebrhatu M, Brunner MJ, Abrusci P, Yan J, Franz-Wachtel M, Schärfe C, Zilkenat S, Grin I, Galán JE, Kohlbacher O, Lea S, Macek B, Marlovits TC, Robinson CV, Wagner S (2016) Structural and functional characterization of the bacterial type III secretion export apparatus. PLoS Pathog 12:e1006071

Erhardt M, Wheatley P, Kim EA, Hirano T, Zhang Y, Sarkar MK, Hughes KT, Blair DF (2017) Mechanism of type-III protein secretion: regulation of FlhA conformation by a functionally critical charged-residue cluster. Mol Microbiol 104:234–249

Evans LD, Poulter S, Terentjev EM, Hughes C, Fraser GM (2013) A chain mechanism for flagellum growth. Nature 504:287–290

Fabiani FD, Renault TT, Peters B, Dietsche T, Gálvez EJC, Guse A, Freier K, Charpentier E, Strowig T, Franz-Wachtel M, Macek B, Wagner S, Hensel M, Erhardt M (2017) A flagellum-specific chaperone facilitates assembly of the core type III export apparatus of the bacterial flagellum. PLoS Biol 15:e2002267

Fan F, Macnab RM (1996) Enzymatic characterization of FliI: an ATPase involved in flagellar assembly in *Salmonella typhimurium*. J Biol Chem 271:31981–31988

Ferris HU, Furukawa Y, Minamino T, Kroetz MB, Kihara M, Namba K, Macnab RM (2005) FlhB regulates ordered export of flagellar components via autocleavage mechanism. J Biol Chem 280:41236–41242

Fraser GM, Hirano T, Ferris HU, Devgan LL, Kihara M, Macnab RM (2003) Substrate specificity of type III flagellar protein export in *Salmonella* is controlled by subdomain interactions in FlhB. Mol Microbiol 48:1043–1057

Fukumura T, Makino F, Dietsche T, Kinoshita M, Kato T, Wagner S, Namba K, Imada K, Minamino T (2017) Assembly and stoichiometry of the core structure of the bacterial flagellar type III export gate complex. PLoS Biol 15:e2002281

Furukawa Y, Inoue Y, Sakaguchi A, Mori Y, Fukumura T, Miyata T, Namba K, Minamino T (2016) Structural stability of flagellin subunits affects the rate of flagellin export in the absence of FliS chaperone. Mol Microbiol 102:405–416

Galán JE, Lara-Tejero M, Marlovits TC, Wagner S (2014) Bacterial type III secretion systems: specialized nanomachines for protein delivery into target cells. Annu Rev Microbiol 68: 415–438

González-Pedrajo B, Fraser GM, Minamino T, Macnab RM (2002) Molecular dissection of *Salmonella* FliH, a regulator of the ATPase FliI and the type III flagellar protein export pathway. Mol Microbiol 45:967–982

González-Pedrajo B, Minamino T, Kihara M, Namba K (2006) Interactions between C ring proteins and export apparatus components: a possible mechanism for facilitating type III protein export. Mol Microbiol 60:984–998

Hara N, Namba K, Minamino T (2011) Genetic characterization of conserved charged residues in the bacterial flagellar type III export protein FlhA. PLoS ONE 6:e22417

Hara N, Morimoto YV, Kawamoto A, Namba K, Minamino T (2012) Interaction of the extreme N-terminal region of FliH with FlhA is required for efficient bacterial flagellar protein export. J Bacteriol 194:5353–5360

Hu B, Morado DR, Margolin W, Rohde JR, Arizmendi O, Picking WL, Picking WD, Liu J (2015) Visualization of the type III secretion sorting platform of *Shigella flexneri*. Proc Natl Acad Sci USA 112:1047–1052

Hu B, Lara-Tejero M, Kong Q, Galán JE, Liu J (2017) In situ molecular architecture of the *Salmonella* type III secretion machine. Cell 168:1065–1074

Ibuki T, Imada K, Minamino T, Kato T, Miyata T, Namba K (2011) Common architecture between the flagellar protein export apparatus and F- and V-ATPases. Nat Struct Mol Biol 18:277–282

Ibuki T, Uchida Y, Hironaka Y, Namba K, Imada K, Minamino T (2013) Interaction between FliJ and FlhA, components of the bacterial flagellar type III export apparatus. J Bacteriol 195: 466–473

Imada K, Minamino T, Tahara A, Namba K (2007) Structural similarity between the flagellar type III ATPase FliI and F1-ATPase subunits. Proc Natl Acad Sci USA 104:485–490

Imada K, Minamino T, Kinoshita M, Furukawa Y, Namba K (2010) Structural insight into the regulatory mechanisms of interactions of the flagellar type III chaperone FliT with its binding partners. Proc Natl Acad Sci USA 107:8812–8817

Imada K, Minamino T, Uchida Y, Kinoshita M, Namba K (2016) Insight into the flagella type III export revealed by the complex structure of the type III ATPase and its regulator. Proc Natl Acad Sci USA 113:3633–3638

Inoue Y, Morimoto YV, Namba K, Minamino T (2018) Novel insights into the mechanism of well-ordered assembly of bacterial flagellar proteins in *Salmonella*. Sci Rep 8:1787

Kawamoto A, Morimoto YV, Miyata T, Minamino T, Hughes KT, Kato T, Namba K (2013) Common and distinct structural features of *Salmonella* injectisome and flagellar basal body. Sci Rep 3:3369

Kazetani K, Minamino T, Miyata T, Kato T, Namba K (2009) ATP-induced FliI hexamerization facilitates bacterial flagellar protein export. Biochem Biophys Res Commun 388:323–327

Kihara M, Miller GU, Macnab RM (2000) Deletion analysis of the flagellar switch protein FliG of Salmonella. J Bacteriol 182:3022–3028

Kihara M, Minamino T, Yamaguchi S, Macnab RM (2001) Intergenic suppression between the flagellar MS ring protein FliF of *Salmonella* and FlhA, a membrane component of its export apparatus. J Bacteriol 183:1655–1662

Kinoshita M, Hara N, Imada K, Namba K, Minamino T (2013) Interactions of bacterial flagellar chaperone-substrate complexes with FlhA contribute to co-ordinating assembly of the flagellar filament. Mol Microbiol 90:1249–1261

Kinoshita M, Nakanishi Y, Furukawa Y, Namba K, Imada K, Minamino T (2016) Rearrangements of α-helical structures of FlgN chaperone control the binding affinity for its cognate substrates during flagellar type III export. Mol Microbiol 101:656–670

Kinoshita M, Aizawa S, Inoue Y, Namba K, Minamino T (2017) The role of intrinsically disordered C-terminal region of FliK in substrate specificity switching of the bacterial flagellar type III export. Mol Microbiol 105:572–588

Kinoshita M, Furukawa Y, Uchiyama S, Imada K, Namba K, Minamino T (2018a) Insight into adaptive remodeling of the rotor ring complex of the bacterial flagellar motor. Biochem Biophys Res Commun 496:12–17

Kinoshita M, Namba K, Minamino T. (2018) Effect of a clockwise-locked deletion in FliG on the FliG ring structure of the bacterial flagellar motor. Genes Cells 23:241–247 (2018)

Kubori T, Shimamoto N, Yamaguchi S, Namba K, Aizawa S (1992) Morphological pathway of flagellar assembly in *Salmonella typhimurium*. J Mol Biol 226:433–446

Kuhlen L, Abrusci P, Johnson S, Gault J, Deme J, Caesar J, Dietsche T, Mebrhatu MT, Ganief T, Macek B, Wagner S, Robinson CV, Lea SM (2018) Structure of the core of the type III secretion system export apparatus. Nat Struct Mol Biol 25:583–590

Kutsukake K, Minamino T, Yokoseki T (1994) Isolation and characterization of FliK-independent flagellation mutants from *Salmonella typhimurium*. J Bacteriol 176:7625–7629

Lara-Tejero M, Kato J, Wagner S, Liu X, Galán JE (2011) A sorting platform determines the order of protein secretion in bacterial type III systems. Science 331:1188–1191

Lee KL, Ginsburg MA, Crovace C, Donohoe M, Stock D (2010) Structure of the torque ring of the flagellar motor and the molecular basis for rotational switching. Nature 466:996–1000

Lee PC, Zmina SE, Stopford CM, Toska J, Rietsch A (2014) Control of type III secretion activity and substrate specificity by the cytoplasmic regulator PcrG. Proc Natl Acad Sci USA 111:2027–2036

Li H, Sourjik V (2011) Assembly and stability of flagellar motor in *Escherichia coli*. Mol Microbiol 80:886–899

Lynch MJ, Levenson R, Kim EA, Sircar R, Blair DF, Dahlquist FW, Crane BR (2017) Co-folding of a FliF-FliG split domain forms the basis of the MS: C ring interface within the bacterial flagellar motor. Structure 25:317–328

Macnab RM (2004) Type III flagellar protein export and flagellar assembly. Biochim Biophys Acta 1694:207–217

Makino F, Shen D, Kajimura N, Kawamoto A, Pissaridou P, Oswin H, Pain M, Murillo I, Namba K, Blocker AJ (2016) The architecture of the cytoplasmic region of type III secretion systems. Sci Rep 6:33341

McDowell MA, Marcoux J, McVicker G, Johnson S, Fong YH, Stevens R, Bowman LA, Degiacomi MT, Yan J, Wise A, Friede ME, Benesch JL, Deane JE, Tang CM, Robinson CV, Lea SM (2016) Characterisation of *Shigella* Spa33 and *Thermotoga* FliM/N reveals a new model for C-ring assembly in T3SS. Mol Microbiol 99:749–766

McMurry JL, Murphy JW, González-Pedrajo B (2006) The FliN-FliH interaction mediates localization of flagellar export ATPase FliI to the C ring complex. Biochemistry 45:11790–11798

Meshcheryakov VA, Kitao A, Matsunami H, Samatey FA (2013) Inhibition of a type III secretion system by the deletion of a short loop in one of its membrane proteins. Acta Crystallogr D Biol Crystallogr 69:812–820

Minamino T (2014) Protein export through the bacterial flagellar type III export pathway. Biochim Biophys Acta 1843:1642–1648

Minamino T (2018) Hierarchical protein export mechanism of the bacterial flagellar type III protein export apparatus. FEMS Microbiol Lett 365:fny117

Minamino T, Macnab RM (1999) Components of the *Salmonella* flagellar export apparatus and classification of export substrates. J Bacteriol 181:1388–1394

Minamino T, Macnab RM (2000a) Domain structure of *Salmonella* FlhB, a flagellar export component responsible for substrate specificity switching. J Bacteriol 182:4906–4919

Minamino T, Macnab RM (2000b) FliH, a soluble component of the type III flagellar export apparatus of *Salmonella*, forms a complex with FliI and inhibits its ATPase activity. Mol Microbiol 37:1494–1503

Minamino T, Macnab RM (2000c) Interactions among components of the *Salmonella* flagellar export apparatus and its substrates. Mol Microbiol 35:1052–1064

Minamino T, Namba K (2008) Distinct roles of the FliI ATPase and proton motive force in bacterial flagellar protein export. Nature 451:485–488

Minamino T, Iino T, Kutuskake K (1994) Molecular characterization of the *Salmonella typhimurium flhB* operon and its protein products. J Bacteriol 176:7630–7637

Minamino T, Tame JRH, Namba K, Macnab RM (2001) Proteolytic analysis of the FliH/FliI complex, the ATPase component of the type III flagellar export apparatus of *Salmonella*. J Mol Biol 312:1027–1036

Minamino T, González-Pedrajo B, Oosawa K, Namba K, Macnab RM (2002) Structural properties of FliH, an ATPase regulatory component of the *Salmonella* type III flagellar export apparatus. J Mol Biol 322:281–290

Minamino T, Imada K, Namba K (2008) Mechanisms of type III protein export for bacterial flagellar assembly. Mol BioSyst 4:1105–1111

Minamino T, Yoshimura SDJ, Morimoto YV, González-Pedrajo B, Kami-ike N, Namba K (2009) Roles of the extreme N-terminal region of FliH for efficient localization of the FliH-FliI complex to the bacterial flagellar type III export apparatus. Mol Microbiol 74:1471–1483

Minamino T, Morimoto YV, Hara N, Namba K (2011) An energy transduction mechanism used in bacterial type III protein export. Nat Commun 2:475

Minamino T, Kinoshita M, Hara N, Takeuchi S, Hida A, Koya S, Glenwright H, Imada K, Aldridge PD, Namba K (2012a) Interaction of a bacterial flagellar chaperone FlgN with FlhA is required for efficient export of its cognate substrates. Mol Microbiol 83:775–788

Minamino T, Kinoshita M, Imada K, Namba K (2012b) Interaction between FliI ATPase and a flagellar chaperone FliT during bacterial flagellar export. Mol Microbiol 83:168–178

Minamino T, Morimoto YV, Kinoshita M, Aldridge PD, Namba K (2014) The bacterial flagellar protein export apparatus processively transports flagellar proteins even with extremely infrequent ATP hydrolysis. Sci Rep 4:7579

Minamino T, Morimoto YV, Hara N, Aldridge PD, Namba K (2016) The Bacterial flagellar type III export gate complex is a dual fuel engine that can use both H^+ and Na^+ for flagellar protein export. PLoS Pathog 12:e1005495

Moore SA, Jia Y (2010) Structure of the cytoplasmic domain of the flagellar secretion apparatus component FlhA from *Helicobacter pylori*. J Biol Chem 285:21060–21069

Morimoto YV, Minamino T (2014) Structure and function of the bi-directional bacterial flagellar motor. Biomolecules 4:217–234

Morimoto YV, Ito M, Hiraoka KD, Che Y-S, Bai F, Kami-ike N, Namba K, Minamino T (2014) Assembly and stoichiometry of FliF and FlhA in *Salmonella* flagellar basal body. Mol Microbiol 91:1214–1226

Morimoto YV, Kami-ike N, Miyata T, Kawamoto A, Kato T, Namba K, Minamino T (2016) High-resolution pH imaging of living bacterial cell to detect local pH differences. mBio 7: e01911-16

Notti RQ, Bhattacharya S, Lilic M, Stebbins CE (2015) A common assembly module in injectisome and flagellar type III secretion sorting platforms. Nat Commun 6:7125

Ohnishi K, Fan F, Schoenhals GJ, Kihara M, Macnab RM (1997) The FliO, FliP, FliQ, and FliR proteins of *Salmonella typhimurium*: putative components for flagellar assembly. J Bacteriol 179:6092–6099

Okabe M, Minamino T, Imada K, Namba K, Kihara M (2009) Role of the N-terminal domain of FliI ATPase in bacterial flagellar protein export. FEBS Lett 583:743–748

Park SY, Lowder B, Bilwes AM, Blair DF, Crane BR (2006) Structure of FliM provides insight into assembly of the switch complex in the bacterial flagella motor. Proc Natl Acad Sci USA 103:11886–11891

Paul K, Harmon JG, Blair DF (2006) Mutational analysis of the flagellar rotor protein FliN: identification of surfaces important for flagellar assembly and switching. J Bacteriol 188:5240–5248

Paul K, Erhardt M, Hirano T, Blair DF, Hughes KT (2008) Energy source of the flagellar type III secretion. Nature 451:489–492

Paul K, Gonzalez-Bonet G, Bilwes AM, Crane BR, Blair D (2011) Architecture of the flagellar rotor. EMBO J 30:2962–2971

Saijo-Hamano Y, Imada K, Minamino T, Kihara M, Shimada M, Kitao A, Namba K (2010) Structure of the cytoplasmic domain of FlhA and implication for flagellar type III protein export. Mol Microbiol 76:260–268

Terahara N, Inoue Y, Kodera N, Morimoto YV, Uchihashi T, Imada K, Ando T, Namba K, Minamino T (2018) Insight into structural remodeling of the FlhA ring responsible for bacterial flagellar type III protein export. Sci Adv 4:eaao7054

Terashima H, Kawamoto A, Tastumi C, Namba K, Minamino T, Imada K (2018) In vitro reconstitution of functional type III protein export and insights into flagellar assembly. mBio 9: e00988-18

Thomas DR, Morgan DG, DeRosier DJ (1999) Rotational symmetry of the C ring and a mechanism for the flagellar rotary motor. Proc Natl Acad Sci USA 96:10134–10139

Thomas J, Stafford GP, Hughes C (2004) Docking of cytosolic chaperone-substrate complexes at the membrane ATPase during flagellar type III protein export. Proc Natl Acad Sci USA 101:3945–3950

Ueno T, Oosawa K, Aizawa S (1992) M ring, S ring and proximal rod of the flagellar basal body of *Salmonella typhimurium* are composed of subunits of a single protein, FliF. J Mol Biol 227:672–677

Vartanian AS, Paz A, Fortgang EA, Abramson J, Dahlquist FW (2012) Structure of flagellar motor proteins in complex allows for insights into motor structure and switching. J Biol Chem 287:35779–35783

Wagner S, Königsmaier L, Lara-Tejero M, Lefebre M, Marlovits TC, Galán JE (2010) Organization and coordinated assembly of the type III secretion export apparatus. Proc Natl Acad Sci USA 107:17745–17750

Wagner S, Grin I, Grin I, Malmsheimer S, Singh N, Torres-Vargas CE, Westerhausen S (2018) Bacterial type III secretion systems: a complex device for the delivery of bacterial effector proteins into eukaryotic host cells. FEMS Microbiol Lett 365:fny201

Worrall LJ, Vuckovic M, Strynadka NCJ (2010) Crystal structure of the C-terminal domain of the *Salmonella* type III secretion system export apparatus protein InvA. Protein Sci 19:1091–1096

Xing Q, Shi K, Portaliou A, Rossi P, Economou A, Kalodimos CG (2018) Structure of chaperone-substrate complexes docked onto the export gate in a type III secretion system. Nat Commun 9:1773

Xue C, Lam KH, Zhang H, Sun K, Lee SH, Chen X, Au SWN (2018) Crystal structure of the FliF-FliG complex from *Helicobacter pylori* yields insight into the assembly the motor MS–C ring in the bacterial flagellum. J Biol Chem 293:2066–2078

Zarivach R, Deng W, Vuckovic M, Felise HB, Nguyen HV, Miller SI, Finlay BB, Strynadka NC (2008) Structural analysis of the essential self-cleaving type III secretion proteins EscU and SpaS. Nature 453:124–127

Zilkenat S, Franz-Wachtel M, Stierhof YD, Galán JE, Macek B, Wagner S (2016) Determination of the stoichiometry of the complete bacterial type III secretion needle complex using a combined quantitative proteomic approach. Mol Cell Proteomics 15:1598–1609

Structures of Type III Secretion System Needle Filaments

Birgit Habenstein, Nadia El Mammeri, James Tolchard, Gaëlle Lamon, Arpita Tawani, Mélanie Berbon and Antoine Loquet

Contents

Abstract Among the Gram-negative bacterial secretion systems, type III secretion systems (T3SS) possess a unique extracellular molecular apparatus called the needle. This macromolecular protein assembly is a nanometre-size filament formed by the helical arrangement of hundreds of copies of a single, small protein, which is highly conserved between T3SSs from animal to plant bacterial pathogens. The needle filament forms a hollow tube with a channel ~20 Å in diameter that serves as a conduit for proteins secreted into the targeted host cell. In the past ten years, technical breakthroughs in biophysical techniques such as cryo-electron microscopy (cryo-EM) and solid-state NMR (SSNMR) spectroscopy have uncovered atomic resolution details about the T3SS needle assembly. Several high-resolution

B. Habenstein (✉) · N. El Mammeri · J. Tolchard · G. Lamon · A. Tawani · M. Berbon · A. Loquet (✉)
University of Bordeaux, CNRS, UMR 5248, European Institute of Chemistry and Biology, 2 rue Robert Escarpit, Pessac 33607, France
e-mail: b.habenstein@iecb.u-bordeaux.fr

A. Loquet
e-mail: a.loquet@iecb.u-bordeaux.fr

Current Topics in Microbiology and Immunology (2020) 427: 109–131
https://doi.org/10.1007/82_2019_192
© Springer Nature Switzerland AG 2019
Published Online: 24 January 2020

structures of *Salmonella typhimurium* and *Shigella flexneri* T3SS needles have been reported demonstrating a common structural fold. These structural models have been used to explain the active role of the needle in transmitting the host-cell contact signal from the tip to the base of the T3SS through conformational changes as well as during the injection of effector proteins. In this chapter, we summarize the current knowledge about the structure and the role of the T3SS needle during T3SS assembly and effector secretion.

1 Introduction

The type III secretion system (T3SS) is a complex virulence mechanism that has been reported for more than 20 Gram-negative bacteria, including *Yersinia*, *Shigella* and *Salmonella* species. These nanomachines lie at the heart of Gram-negative bacteria's ability to directly deliver effector protein from their cytoplasm to that of eukaryotic cells and do so via a conserved battery of homologous genes (Galan and Wolf-Watz 2006). Termed "injectisome", this appendage traverses the biological barrier composed of plasma membranes, the peptidoglycan layer and the extracellular space. T3SSs are composed of several highly conserved substructures: a cytoplasmic complex, a basal body that spans both the inner and the outer membrane, and an extracellular segment comprising the needle filament and a translocator pore. The T3SS needle is a filamentous protein multimer that extends to the external milieu and is composed of numerous copies of a single subunit protein. The needle filament reaches a length of tens of nanometres and displays an outer diameter of $\sim$8 nm. It is a hollow tube serving as a conduit to export effectors to the host cells. Several structural models described an inner diameter of $\sim$2.0–2.5 nm (Galkin et al. 2010; Fujii et al. 2012; Loquet et al. 2012); however, a recent cryo-EM model proposed by Strynadka and co-workers showed a more restrained inner lumen of $\sim$1.5 nm (Hu et al. 2018). T3SS injectisomes were first observed at high resolution in 1998 by Kubori and co-workers (Kubori et al. 1998), whom described a supramolecular assembly spanning the inner and outer membranes of flagellated and non-flagellated Gram-negative strains of *Salmonella typhimurium* and revealed an extracellular part forming a thin filament termed the needle. This study, based on electron microscopy, reported the first observation of the T3SS needle filament and opened the way towards understanding the structure and functioning of such bacterial appendages. Considering the needle filament, bridging the extracellular space towards the host cell, this first study engendered scientific interest that has led to low-resolution structures and finally to the elucidation of the needle's high-resolution three-dimensional structures of increasing resolution: in 2012 by solid-state nuclear magnetic resonance (NMR) (Loquet et al. 2012), in 2014 by combining solid-state NMR with limited resolution cryo-electron microscopy (EM) (Demers et al. 2014) and in 2018 by high-resolution cryo-EM microscopy (Hu et al. 2018).

The structural conservation of the T3SS across the Gram-negative bacterial sub-group suggests that an understanding of the T3SSs components, their structural composition, arrangements and interactions will provide the mechanistic basis to describe the T3SS complex assembly, the biological activities and effector delivery pathways. Although most structural components of the injectisome are well-explored, many open questions still remain as the complexity of the system (including various components and potentially several conformational states) so far impedes an all-encompassing comprehension of its function. The inter-dependent interactions within the nanomachine during protein secretion into the host cytosol, the mechanistic aspects of the sorting platform, and how the overall secretory hierarchy can be tightly maintained are some of the persisting challenges. The chapter hereafter focuses on the mentioned T3SS filamentous needles. Structure, assembly and prospects will be discussed in detail.

2 The T3SS Needle: Biological Functions

T3SSs avail the interaction of Gram-negative bacteria to host cells by transporting its proteins from the bacterial cytoplasm to the host cell through the periplasmic space and across the extracellular environment. Electron microscopy studies of T3SSs have revealed a wide base embedded in the membrane of the bacteria with a protruding thin extracellular filament that resembles a syringe, first observed in *Salmonella* (Galan and Wolf-Watz 2006) and *Shigella* (Tamano et al. 2000; Blocker et al. 2001). During functioning of the T3SS, a direct interaction between the extracellular complex and the host cell is required for inducing the transportation of bacterial effector proteins into the host-cell cytoplasm (Fig. 1a) (Michiels et al. 1991; Salmond and Reeves 1993). A key feature of the T3SS is the assembly of an extracellular appendage termed the needle. The needle filament is assembled by numerous copies of a single protein subunit forming a tube that serves as a conduit to secrete effectors to the host cell (Tamano et al. 2000; Kubori et al. 2000; Kimbrough and Miller 2000). High concentrations of recombinant needle subunit protein have been shown to promote in vivo self-assembly of the needle filaments (Fig. 1b), wherein they reassemble the structural features of sheared needles (Loquet et al. 2012; Demers et al. 2014; Verasdonck et al. 2015). The filament is $\sim$50 nm long with an 8 nm external diameter formed by the helical assembly of needle protomers (Fig. 1c–d). The structural proteins constituting the T3SS injectisome are highly conserved in primary sequence, which is also observed for the needle protein specifically (Fig. 2). The sequences are poorly conserved at the N-terminal extension (amino acids 1 to $\sim$20), the central region is moderately conserved, and the C-terminal region is highly conserved. In the case of *Pseudomonas aeruginosa*, two molecular chaperones (PscG and PscE) have been identified by Attree, Dessen and colleagues (Quinaud et al. 2005). PscG and PscE impede the polymerization of the needle protein (PscF in *P. aeruginosa* and SctF following the unified Sct nomenclature) by forming a stable ternary complex

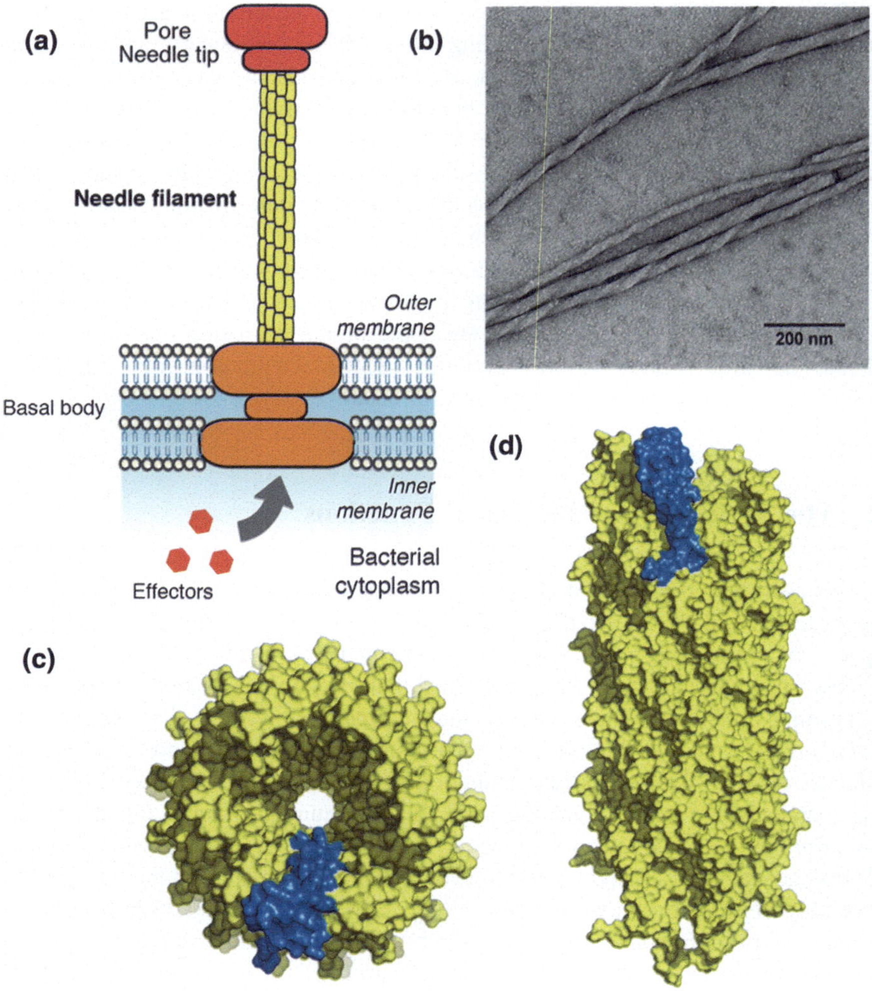

Fig. 1 Needle filament of the type III secretion system. **a** Schematic diagram of the T3SS and its needle filament. **b** Electron micrograph of the needle filaments of *Pseudomonas aeruginosa*. **c–d** High-resolution structure of the *Salmonella typhimurium* needle structure (Hu et al. 2018), **c** Top view. **d** Front view. The needle subunit PrgI is coloured in blue

(Quinaud et al. 2007) in the cytoplasmic milieu. Similar observations have been made for the YscF-YscG-YscE complex in *Yersinia* (Sun et al. 2008). Interestingly, such chaperones have not been observed for *Salmonella* or *Shigella* species, implying an important difference in the needle subunit stability and polymerization efficiency despite high sequence homology at the inter-species level.

The assembly of the injectisome is a coordinated process, starting with the self-organization of the basal body. The proteins forming the needle adapter (Torres-Vargas et al. 2019) (previously called the inner rod) and the needle filament

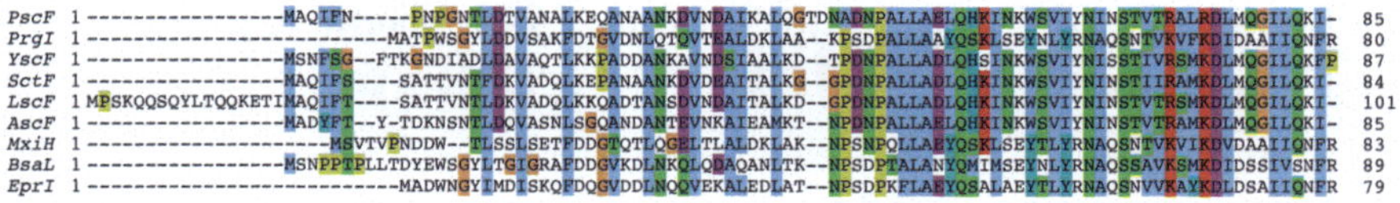

Fig. 2 Inter-species comparison of T3SS needle subunit proteins. As shown for PscF (*Pseudomonas aeruginosa*), PrgI (*Salmonella typhimurium*), YscF (*Yersinia enterocolitica*), SctF (*Photorhabdus luminescens subsp. laumondii*), LscF (*Photorhabdus luminescens*), AscF (*Aeromonas salmonicida subsp. salmonicida*), MxiH (*Shigella flexneri*), BsaL (*Burkholderia pseudomallei*) and EprI (*Escherichia coli*). Generated using Jalview, with the Clustal-Omega alignment program (default parameters), displayed using Clustal residue-type coloration

are considered to be early substrates, in opposition to middle and late substrates (tip protein, translocases, effectors). The needle adapter forms the channel inside the basal body and has historically been proposed to act as a conduit on which the needle can extend (Lefebre and Galan 2014). It has been demonstrated that it is formed by a single helical turn (Torres-Vargas et al. 2019) that helps to anchor the needle filament (Torres-Vargas et al. 2019; Hu et al. 2019) It has been proposed by Cornelis and co-workers that the length of the needle is crucial to efficiently trigger the secretion (Mota and Cornelis 2005), raising the question of how the needle length is controlled by the apparatus. One hypothesis [the "molecular ruler" (Journet et al. 2003)] from studies on *Yersinia spp.* implies that an accessory protein of the T3SS (YscP in *Yersinia*, InvJ in *Salmonella*, SctP using the Sct nomenclature) acts as the needle length regulator. It has been proposed that SctP can probe the degree of elongation of the needle and once it reaches the suitable length, it triggers substrate switching. This suggestion is supported by sequence modifications in SctP that result in a change of the needle length (Journet et al. 2003; Wagner et al. 2009, 2010). In contrast to this model of needle length regulation, work on *Salmonella* by Galán and colleagues (Marlovits et al. 2006) has led to an alternative model. In turn, a step-wise mechanism implied a conformational change after the needle adapter assembly at the level of the base that triggers the substrate switching. Because SctP is required for a proper assembly of the needle adapter (Lefebre and Galan 2014; Marlovits et al. 2006), the regulation of the needle length would indirectly involve SctP through its interaction with the needle adapter. Recent studies (Torres-Vargas et al. 2019; Hu et al. 2019) have demonstrated that the needle adapter (SctI) does not play a specific role in substrate switching or needle length control, although its proper assembly might provide the suitable structural scaffold for the needle polymerization. It is worth mentioning that in vitro sample preparation of recombinant needle filaments lacks a biochemical cap (the tip protein) and any length regulators (the needle adapter SctI or YscP/InvJ). As a result, in vitro needle filaments of *Salmonella* (Loquet et al. 2011, 2012) and *Shigella* (Demers et al. 2014) are considerably longer compared to native needles observed at the bacterial surfaces. This aspect makes structural characterization of the molecular mechanisms involved in needle length regulation substantially difficult when using techniques such as crystallography or NMR spectroscopy.

3 Structures of Crystal and Soluble T3SS Needle Subunit Fragments

The size and complexity of T3SS needles make them particularly challenging targets for the common high-resolution structural techniques such as solution nuclear magnetic resonance spectroscopy (solution NMR) and X-ray crystallography. Therefore, many structural investigations directed towards T3SSs (and indeed other bacterial macromolecular assemblies) often approach the problem from the level of the monomeric subunits in order to simplify sample preparation and facilitate both experimentation and analysis. A selection of such monomeric protein structures is presented in Fig. 3, and Table 1 summarizes the high-resolution structures of the needle in its monomeric or filamentous states, as deposited at the PDB.

The first reported high-resolution structure of a T3SS needle subunit protein was that of the BsaL (SctF using the Sct nomenclature) protein from *Burkholderia pseudomallei* by the De Guzman group in 2006 (Zhang et al. 2006). Here, analysis of the monomeric BsaL protein necessitated a C-terminal truncation of five residues to block polymerization, as had previously been shown for the needle protein of *Shigella* (MxiH) (Kenjale et al. 2005). Preliminary analysis by circular dichroism suggested the protein was 70% α–helical, in good agreement with analyses of other needle protein homologues. The calculated structural ensemble of BsaL, determined by solution NMR, described a core tertiary structure with an α–helical hairpin separated by a short four-residue linker (Fig. 3). The presence and location of the two α–helices were confirmed by both chemical shift analysis and through-space internuclear NMR distance restraints and indicated they comprised 17 and 18 residues, from the N-terminus, respectively. The helical hairpin was revealed to be stabilized by inter-helix hydrophobic interactions from residues well conserved

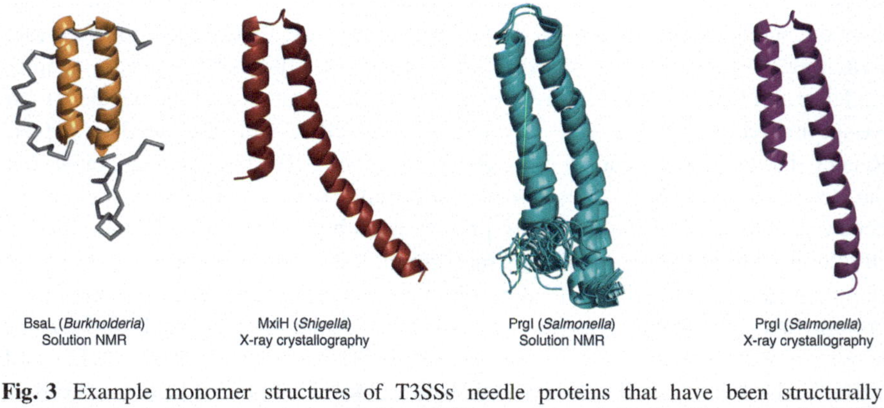

Fig. 3 Example monomer structures of T3SSs needle proteins that have been structurally characterized at high resolution: solution NMR structure of BsaL (Zhang et al. 2006) (*B. pseudomallei*, orange; core residues 30–68), X-ray structure of MxiH (Deane et al. 2006) (*S. flexneri*, red), solution NMR structure of PrgI (Poyraz et al. 2010) (*S. typhimurium*, teal) and X-ray structure of PrgI (Poyraz et al. 2010) (*S. typhimurium*, purple)

Table 1 Structures of T3SS needles deposited at the Protein Data Bank

Name	Organism	Structural completeness	PDB code	Primary technique (resolution)	Date	References
MxiH	*Shigella*	Monomer	2CA5	X-ray (2.1 Å)	2006	Deane et al. (2006a, b)
		Filament	2V6L	EM (16 Å)	2007	Deane et al. (2006a, b)
		Filament	3J0R	EM (7.7 Å)	2011	Fujii et al. (2012)
		Filament	2MME	SSNMR & EM	2014	Demers et al. (2014)
PrgI	*Salmonella*	Monomer	2X9C	X-ray (2.4 Å)	2010	Poyraz et al. (2010)
		Monomer	2KV7	Solution NMR	2010	Poyraz et al. (2010)
		Filament	2LPZ	SSNMR	2012	Loquet et al. (2012)
		Tetramer	2MEX	SSNMR	2013	Loquet et al. (2013)
		Filament	6DWB	EM (3.3 Å)	2018	Hu et al. (2018)
PscF	*Pseudomonas*	Complexed monomer	2UWJ	X-ray (2.0 Å)	2007	Quinaud et al. (2007)
YscF	*Yersinia*	Complexed monomer	2P58	X-ray (1.8 Å)	2008	Sun et al. (2008)
BsaL	*Burkholderia*	Monomer	2G0U	Solution NMR	2006	Zhang et al. (2006)

among homologous needle subunits. Similarly, the identified *trans*-conformation for proline 47 and proline 50, residues also well conserved among needle proteins (Fig. 2), was found to be crucial for the rigidity of the four-residue linker region and the subsequent positioning of the two α–helices. Aside from the hairpin region, the extreme N-terminal seven residues of BsaL exhibited a lack of secondary structure. However, although the ~30 residues stretch leading up to the first α–helix and the extreme C-terminal ~20 residues reflected some degree of α–helical content, this was hypothesized to be only partially formed due to the lack of conclusive through-space NMR restraints.

Shortly after the publication of the BsaL structure, Deane and co-workers reported the X-ray structure of the MxiH needle protein from *Shigella flexneri* (Deane et al. 2006a, b). MxiH is highly similar in sequence to BsaL (61.8%) and the group exploited the same five-residue C-terminal truncation to inhibit polymerization for crystallization (Deane et al. 2006a, b). The resulting structure was resolved to 2.1 Å and yielded a conformation in general agreement with that of BsaL, wherein the needle protein subunit was comprised of two α-helices with a hairpin fold. The existence of a partial α–helix at the N-terminus, as observed for

BsaL, was not evident from the electron density map, contradicting secondary structure predictions from the MxiH sequence. The resolvable C-terminal α-helix however was found to be six residues longer than BsaL with a bent hinge midway along. The start of the angled region of the C-terminal helix interestingly corresponds to the region of partial helicity from the BsaL solution structure.

In 2009, work by Kolbe, Lange and co-workers further contributed to the description of the T3SS needle, this time with regard to the PrgI needle subunit from *Salmonella typhimurium* (Poyraz et al. 2010). Unique to this work was the identification of a double mutation of the full-length PrgI monomer (V65A/V67A) which allowed the study of the monomeric form while, unlike the C-terminal truncations of BsaL or MxiH, maintaining partial wild type functionality in the presence of other T3SS components. The structure of the PrgI monomer was explicitly determined by both solution NMR and X-ray crystallography, with highly comparable structures both portraying a two α-helix hairpin fold (Fig. 3). Structures were highly similar to those of BsaL and MxiH, although the hinge in the second α–helix was less pronounced than the latter. More interesting however were the observations at the protomers' termini. Firstly, although the orientation was different (this time contiguous with the first hairpin helix), the N-terminal helix of PrgI was eight residues longer in the NMR model than in the X-ray structure of the same construct. This discrepancy between the observable secondary structures would agree with the previous reports of BsaL and MxiH with regard to the structural techniques used in their study. In both models of PrgI, the C-terminal hairpin α-helix was nevertheless present in both models, similar to that of MxiH, albeit without such a pronounced bend. The second intriguing aspect arose from the observation by Fourier transform infrared experiments that protomer polymerization was concomitant with a significant (~20%) α–helix to β-strand transition in secondary structure. Magic-angle spinning solid-state NMR was then employed to study the intact needle assembly and, in stark comparison to the monomeric models, identified distinct chemical shifts for two β-strand elements within the C-terminal 18 residues. The significance of this structural transition observed for this particular double mutant is still to be fully understood. However, β-strands are not observed in the most recent structures of whole T3SS needle complexes (Loquet et al. 2012; Hu et al. 2018).

Apart from the *Salmonella*, *Shigella* and *Burkholderia* T3SS needles, Attree, Dessen and co-workers reported the high-resolution structure of the needle subunit of *Pseudomonas aeruginosa* (PscF) in complex with its two chaperones PscE and PscG. A truncated version of the needle subunit (amino acids 55–85) was used to crystallize the complex and avoid PscF polymerization. PscF in the complex consists of an extended coil (amino acids 55–67) and a C-terminal helix (amino acids 68–85). The PscF construct used in this study would virtually match $PrgI^{49-79}$, this stretch corresponding to the C-terminal part of the second helix, exhibiting an α–helical conformation in the context of the needle filament assembly (Loquet et al. 2012). Studies on *Yersinia* have revealed a similar behaviour with the formation of a complex between YscF, YscG and YscE (Sun et al. 2008). So far, no specific chaperones of the needle protein have been reported in *Salmonella* and *Shigella*.

4 Atomic Resolution Models of T3SS Needle Filaments

The T3SS needle filament is a homopolymeric system made by the self-organization of ~100–200 copies of a single protein subunit. Such high-order protein complexes are ubiquitous in cells, e.g. molecular motors, actin filaments, bacterial secretion systems, viral capsids, etc. Conventional methods in structural biology are traditionally used to solve high-resolution structure of protein systems, namely X-ray diffraction techniques and solution NMR spectroscopy. Such large high-order assemblies however pose important technical challenges for these conventional approaches (Cuniasse et al. 2017; Joseph et al. 2017), that usually rely on in vitro self-assembled samples. Firstly, due to the inherent tendency of needle proteins to rapidly self-assemble after purification, tremendous efforts are typically required to find suitable crystallization conditions. As an example, the T3SS needle subunit of *S. typhimurium* readily self-assembles into filaments at room temperature at a protein concentration of 0.1 mM (Loquet et al. 2012), hampering its crystallization. A biochemical approach to circumvent this self-assembly involves truncating small protein segments that are crucial for its aggregation. As it was demonstrated for the T3SS needle subunit, the removal of the C-terminal five residues prevents its self-assembly and leads to a truncated form that can be handled more easily (Zhang et al. 2006; Kenjale et al. 2005). When assembled into filamentous objects, the lack of long-range non-crystallographic symmetries usually restricts the use of conventional X-ray crystallography to derive high-resolution information. Moreover, the high-molecular weight of self-assembled T3SS needle filaments (~1 MDa) drastically reduces their molecular tumbling in solution, a requirement to perform high-resolution solution NMR. As a consequence, high-resolution studies of T3SS needle filaments have been carried out using two approaches: (i) the use of integrative methods to combine structural information of different sources and (ii) the use of high-resolution cryo-EM. The intensive investigation of the needle filamentous structures has exploited the tremendous integrative set of tools composed of cryo-EM, solid-state and solution NMR, X-ray crystallography, X-ray diffraction and scanning transmission electron microscopy. Such endeavours have led to the deposition of a small number of high-resolution T3SS needle filament structures at the Protein Data Bank (PDB). These studies have been built upon the outcomes of countless other reports concerning recombinant expression of needle proteins, needle reconstitution, filament stabilization and helical symmetry assessment. Hereafter are discussed the most momentous reports, now considered as milestones in the structural elucidation of the T3SS needle.

In 2006, Egelman, Blocker and Lea (Cordes et al. 2003) reported a ~16 Å structural model of the T3SS needle filament of *Shigella flexneri* based on X-ray fibre diffraction and electron microscopy. X-ray fibre diffraction measurements of oriented needles sheared from *Shigella flexneri* cultures, combined with low-resolution data obtained on negatively stained needles, prompted the determination of the global needle symmetry with a number of subunits per turn of ~5.6 MxiH subunits and a 24 Å-pitch helix. It corresponds to an axial rise per subunit of ~4.3 Å. Although the helical handedness was not explicitly determined in this study, it was assumed to be right-handed in analogy to the right-handed flagellar hook (Mimori et al. 1995), based on strong homology between the two systems (Blocker et al. 2003). The same year, the combination of the 16 Å density map, the helical parameters and the crystal structure of a truncated form of MxiH (residues 20–81) led to a structural model of the *S. flexneri* needle filament (Deane et al. 2006a, b). The N-terminus was missing in the crystal structure and assumed to be a helical extension to the helix-turn-helix fold of the monomeric MxiH subunit. The best energetic fit of the rigid crystal subunit structure inside the density map resulted in an overall architecture characterized by the N-terminus pointing inside the needle lumen. The best fit also suggested the helical arrangement to be right-handed.

In 2010, Egelman and Marlovitz (Galkin et al. 2010) used negative staining and cryo-EM to derive the helical arrangements of sheared *Salmonella typhimurium* T3SS needles, leading to a ~6.3 subunits/turn symmetry. Scanning transmission electron microscopy (STEM) offered a powerful approach to obtain the mass-per-unit length in the needle filaments, giving a ~2 kDa/Å for *S. typhimurium* T3SS needles. Considering the molecular weight of the *S. typhimurium* subunit PrgI of 8 kDa, the axial rise per subunit could be estimated to 4.2 Å. The authors derived an ~18 Å resolution three-dimensional reconstruction of the T3SS needle filament. Although this resolution was not enough to determine the helical handedness, it was assumed to be similar to the *S. flexneri* needle model, supported by the homology to the *S. typhimurium* flagellar filament. The truncated form of PrgI (residues 18–60) could be docked to the present reconstruction, based on the homology to the crystal structure of the N-terminal truncated MxiH (*S. flexneri*). Although no high-resolution structure was disclosed in this study, the symmetry parameters and overall supramolecular morphology laid the groundwork for the following helical atomic model of the T3SS needle.

While integrative approaches based on the docking of the subunit crystal structure inside density maps obtained from cryo-EM are offering tremendous insights into macromolecular protein assemblies (Karaca and Bonvin 2013; Alber et al. 2008), these approaches are inherently limited by the resolution of the density map. Moreover, putative structural rearrangements of the subunit structure between the monomeric and the assembled state might lead to inaccuracies during the fitting procedure. In the case of the T3SS needle, the truncation of several N- or C-terminal residues is required to obtain a soluble and monomeric form of the subunit to perform crystallographic studies. These residues are crucial for the needle assembly, and the determination of their native conformation in the context of the filamentous assembly has been an enduring limitation towards the structure determination of the needle filament. In 2011, Lange and co-workers reported high-resolution solid-state NMR (SSNMR) data of in vitro self-assembled T3SS needles of *S. typhimurium* (Loquet et al. 2011). The recombinant wild type protein PrgI was used during the sample preparation, and the in vitro preparation (*i.e.* assembly) led to a single polymorphic filament conformation, as observed by very sharp SSNMR ^{13}C line-width comparable to microcrystalline samples. It contrasts with SSNMR data reported on a double mutant (V65A/V67A) of PrgI

(Poyraz et al. 2010) that exhibited more pronounced structural polymorphism due to structural changes induced by the two mutations at the C-terminal domain. The very high spectral quality of the *S. typhimurium* needle SSNMR data was the basis to solve its atomic structure in 2012 (Loquet et al. 2012) (Fig. 4a). SSNMR chemical shifts revealed an unexpected conformation for the N-terminal residues, with the presence of a rigid and extended stretch for residues 3–8. The collection of SSNMR distance restraints defined three inter-subunit interfaces: two lateral interfaces and one axial interface. SSNMR is a powerful method to derive secondary structure and local structural information; however, the technique suffers from an inability to measure long-distance (>10 Å) structural restraints. The SSNMR restraints were used to define the axial interface and derive an axial translation of ~24 Å between subunit *i* and *i* + 11. SSNMR-derived symmetry and distance restraints were combined with STEM data (Galkin et al. 2010) (i.e. an axial rise/subunit of ~4.2 Å) to solve an atomic resolution model of the *S. typhimurium* needle (Loquet et al. 2012) (Fig. 4a). Experimental restraints were consistent with an architecture presenting the N-terminus located on the outside face of the needle, in contrast to previous models based on a 16 Å (Deane et al. 2006a, b) and 7 Å (Fujii et al. 2012) cryo-EM density maps. Immunogold labelling experiments performed in vitro and in vivo corroborate this finding for *S. typhimurium* (Loquet et al. 2012) and for *S. flexneri* needles (Demers et al. 2013; Verasdonck et al. 2015). Further methodological developments in SSNMR approaches by the Lange group demonstrated that SSNMR data in combination with cryo-EM density map (Demers et al. 2014) provided a powerful integrative approach to solve the *S. flexneri* needle structure (Fig. 4b).

In both SSNMR studies, the helical handedness could not be explicitly determined because both right- and left-handed filament would satisfy the structural restraints and energy minimization. An approach (Loquet et al. 2013) was developed based on SSNMR experiments to determine the structure of the non-symmetric building block of the T3SS needle at high precision. It corresponds to the smallest protein network that includes the three inter-subunit interfaces, *i.e.* a tetramer formed by the subunit *i*, *i* + 5, *i* + 6 and *i* + 11 (Fig. 5). The high resolution of the needle building block structure determined by SSNMR allowed for the extraction of the helical handedness (right-handed, see Fig. 5d) and the axial rise/subunit (4.16 Å) without the use of electron microscopy on in vitro polymerized *S. typhimurium* needles. These results were corroborated with STEM analysis (Loquet et al. 2013) and were consistent with measurements performed on sheared *S. typhimurium* needles (Galkin et al. 2010).

The 7 Å resolution cryo-EM density map reported by Fujii et al. (Fujii et al. 2012) on sheared *S. flexneri* needles in 2012 showed a sufficient quality to observe several secondary structure elements, but not enough to perform an accurate de novo structure reconstruction. It has been unclear if in vitro polymerized needle filaments (as used in the SSNMR studies) have the same molecular structure as sheared needles from cell cultures (used by cryo-EM). In 2018, Strynadka and co-workers (Hu et al. 2018) reported a high-resolution structure of the *S. typhimurium* T3SS injectisome. Although the structure of the needle filament was not

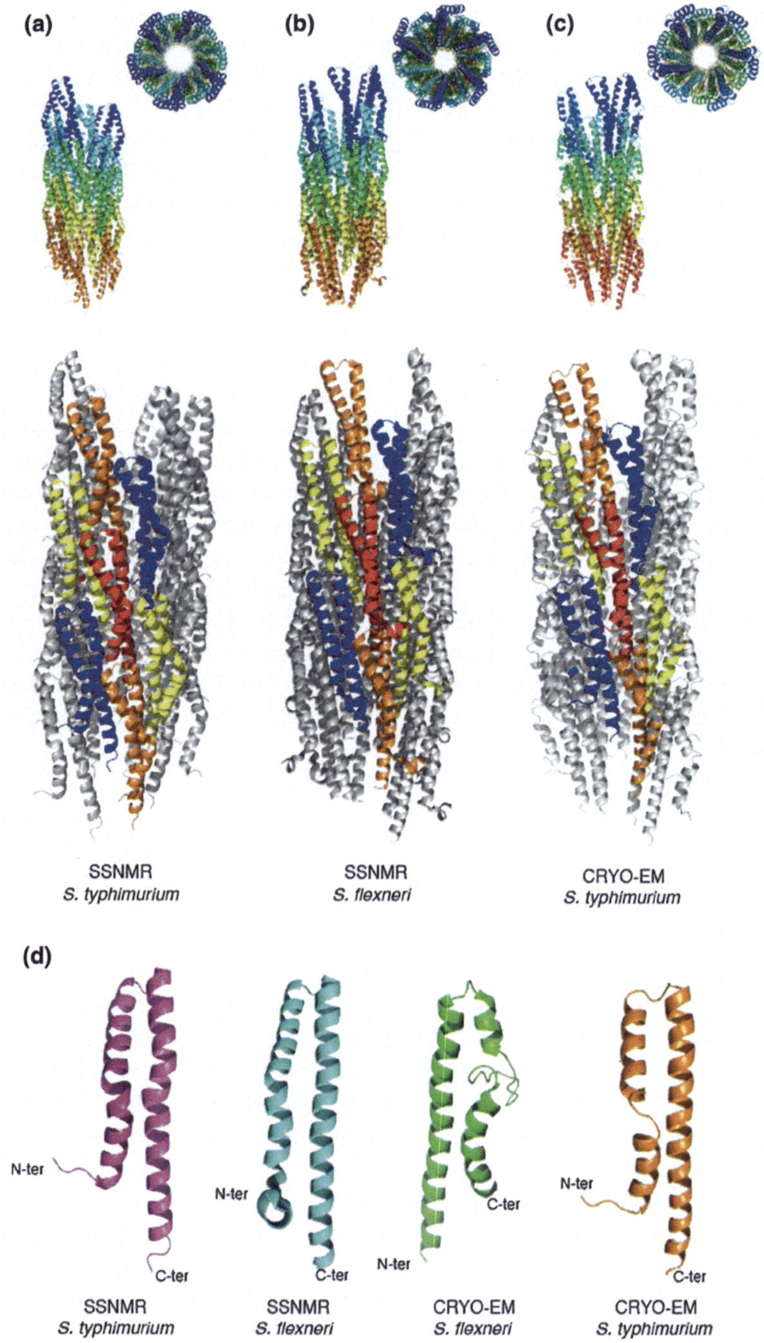
(a)
(b)
(c)
SSNMR
S. typhimurium
SSNMR
S. flexneri
CRYO-EM
S. typhimurium
(d)
N-ter
C-ter
N-ter
C-ter
C-ter
N-ter
N-ter
C-ter
SSNMR
S. typhimurium
SSNMR
S. flexneri
CRYO-EM
S. flexneri
CRYO-EM
S. typhimurium

◂**Fig. 4** Top view, side view and helical symmetry (subunit *i* in red, *i* + 6 in dark blue, *i* + 5 in yellow, *i* + 11 in orange) of the T3SS needle structure of **a** *Salmonella* by solid-state NMR (Loquet et al. 2012). **b** *Shigella* by solid-state NMR (Demers et al. 2014). **c** *Salmonella* by cryo-EM (Hu et al. 2018). **d** T3SS needle subunits extracted from the filament structure, as observed in different bacterial strains and using different biophysical tools

explicitly solved on the complete apparatus, Hu et al. sheared the needles and solved the high-resolution structure using single-particle cryo-EM reconstruction at a resolution of 3.3 Å (Fig. 4c), which represents the highest resolution model of the T3SS needle to date. The cryo-EM structure supports the SSNMR model with the presence of an extended N-terminal domain crucial for the inter-subunit interfaces and faces the external side of the needle. Several subtle differences were observed between the two models: (i) the precise side-chain conformations for several residues differ and (ii) the axial rise per subunit is slightly different (~4.2 Å for SSNMR vs. 4.33 Å for cryo-EM). The most striking difference is the inner diameter of the filament, considerably thinner in the cryo-EM model (~15 Å) compared to previous studies (~25 Å). This observation substantially questions the mechanism

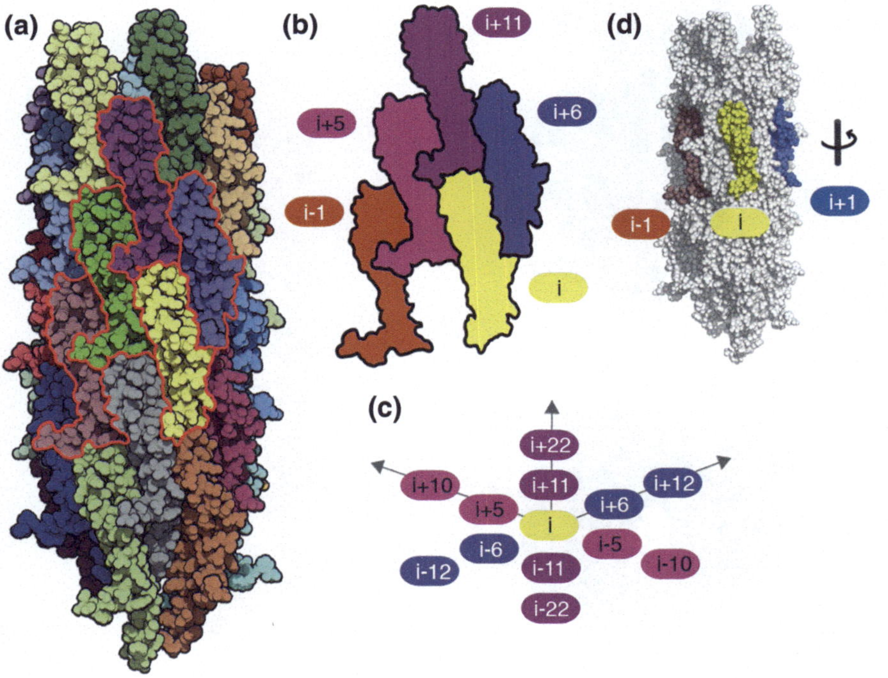

Fig. 5 Helical architecture of the *S. typhimurium* T3SS needle filament. **a** The needle structure. **b** The different subunit–subunit interfaces are shown, corresponding to the axial interface (between subunit *i* and *i* + 11), and the two lateral interfaces (between subunit *i* and *i* + 5/*i* + 6). Note that the subunits *i* and *i* + 1 are not in close contact in the needle assembly. **c** The helical arrangement is depicted. **d** The T3SS needle has a right-handed helical symmetry

used by the injectisome to secrete biomolecules inside the needle tube. Previous models exhibiting a 25 Å inner diameter would advocate for a possible passive diffusion of secreted molecules from the basal body to the end of the needle. The discovery of a much thinner diameter in the cryo-EM model proposed by Strynadka reinforces the hypothesis that subtle residue–residue interaction, possibly between charged amino acids, might impose a particular path of secretion along the needle. Indeed, most conserved residues are observed at the inner surface (Fig. 6a), suggesting a conserved secretion mechanism in the channel. Moreover, the internal surface of *S. typhimurium* needles exposes an intriguing symmetrical helical pattern of charged residues, namely K66, K69, D70 and R80 in PrgI (Fig. 6b). All these residues are highly conserved among needle subunits (Fig. 1), and it has been shown that their mutations to alanine cause defective strains, with these residues

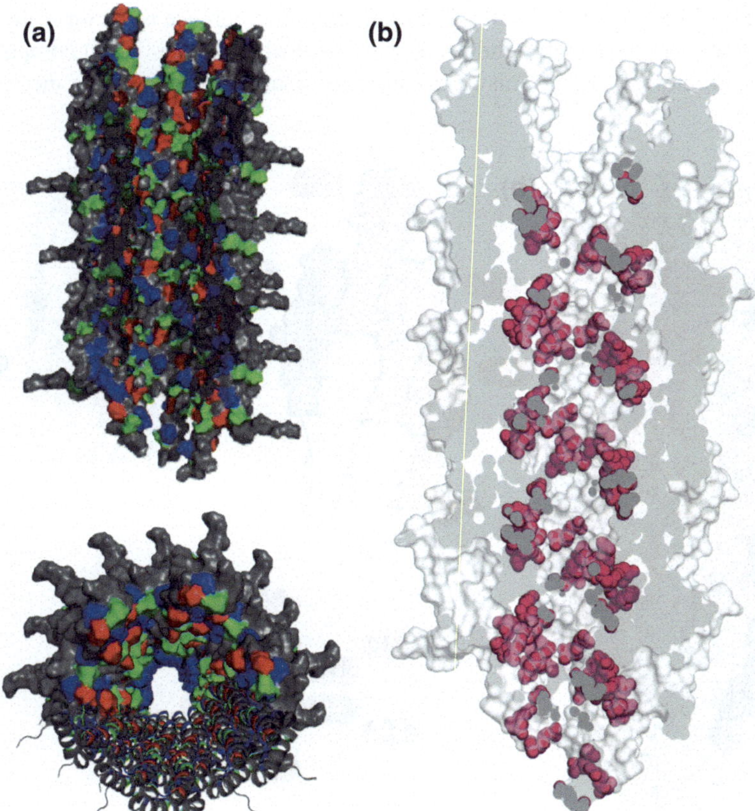

Fig. 6 **a** Surface representation of the residue conservation on the *Salmonella* T3SS needle structure. Conservation varies from full (red) to strong (blue) and weak (green). Front view and top view are presented. **b** The inner channel of the *Salmonella* T3SS needle exhibits a helical pattern of charged residues, coloured in magenta (K66, K69, D70 and R80)

affecting the needle length, stability and secretion (Kenjale et al. 2005). Considering this puzzling helical pattern observed at the luminal surface of the needle and the thin diameter (~ 15 Å), a complex and not yet understood mechanism of secretion involving side-chain/side-chain interactions between the charged residues of the needle subunit and the secreted molecules appears plausible.

Further experiments will be required to elaborate a mechanistic model of secretion in the needle filament. It is important to note that the cryo-EM (Hu et al. 2018) and SSNMR (Loquet et al. 2012) models differ by the sample preparation (using, respectively, sheared needles vs. in vitro polymerized needles) and it still remains unclear if the differences observed between the two structures arise from the sample preparation or by the accuracy of the reconstruction protocols.

5 Structural Variability Between *Salmonella* and *Shigella* Needle Filaments

The high-resolution structures from *Salmonella* (Loquet et al. 2012; Hu et al. 2018) and *Shigella* (Demers et al. 2014) needles, respectively, currently represent our best atomistic descriptions of the T3SS needle assembly. Both models were achieved using a fundamentally similar methodology, combining SSNMR distance restraints, high-resolution cryo-EM density maps and computational modelling with symmetric restraints. The needle complexes are depicted as 11-start helical arrangements of subunits, wherein the protomer N-termini locate to the external faces and the C-terminal helices form the wall of the needle lumen. Overall, the common architecture of the two structures can of course be considered a reflection of their high degree of sequence similarity, and this is evident to varying extents for distinct regions (Fig. 4d). For example, both the loop region and the residues forming the terminal stretch of the second helix are highly conserved regions. These correspond to the structurally important hairpin loop, which contain the *trans*-proline stitch, and the residues closest to the lumen presumably involved with substrate interactions, respectively. The first, N-terminal, external facing, α-helix shows an intermediate level of conservation between all needle proteins. Yet, conversely, the extreme N-terminal 10 residues of the proteins are the least conserved. Intriguingly, the PrgI N-terminus, as compared to that of MxiH, has been shown to have a negative impact upon host-cell recognition of the needle (NF-KB/AP-1 activation), implicating the variability of the N-terminal as a potential defensive mechanism (Osei-Owusu et al. 2015). Nonetheless, in spite of such strong sequence conservation, agreement between the different needle structures was not simply to be expected. Two earlier structures had proposed slightly different models for the MxiH needle based on low-resolution cryo-EM density maps. The first model combined an atomic description of the protomer provided by X-ray crystallography with a previously published low-resolution (16 Å) electron-microscopy map to propose an atomic model for the complete needle structure (Deane et al. 2006a, b). The optimal arrangement of monomers in this macromolecular needle suggested the monomers stacked in a symmetric helical fashion, with the major axis of the monomer aligning to the major axis of the needle. But most importantly, the disordered N-termini were incorrectly shown to line the interior of the needle lumen while the C-terminal helices formed the needle surface. The diameter of the assembled needle lumen was however found to be sufficient for the secretion of unfolded substrates. Yet, as is still the case, most of the helical surface area was shown to form inter-subunit interfaces and helped to provide rational explanations for deleterious phenotypes previously observed in some MxiH mutants. The second MxiH structure was determined from EM data alone (albeit to the higher resolution of 7.7 Å) and was less controversial, whereby greater resolution enabled determination of the correct subunit orientation (C-termini lining the lumen) (Fujii et al. 2012). The precise EM map used was actually the one to be later combined with SSNMR data (Demers et al. 2014) and, perhaps unsurprisingly, the model also proposed an 11-start helical subunit organization. However, one difference between this structure and its later refinement was that the terminal-half of the exterior section (the section described as a partial or dynamic α-helix in other models) occupied a β-hairpin conformation. The rectification of this error in the most recent investigation (Loquet et al. 2012; Hu et al. 2018) highlights the atomic-level insights that SSNMR and high-resolution cryo-EM reconstructions are able to provide.

6 Towards an Assembly Mechanism of T3SS Needle Filaments

The building of a functional T3SS requires that early substrates, such as the proteins that assemble the needle adapter SctI and the needle filament SctF, must be secreted by the injectisome after the basal body is formed (Sukhan et al. 2001). The non-polymerized recombinant forms of the needle proteins have a strong tendency to aggregate in solution (Kenjale et al. 2005), suggesting a similar issue for the bacteria: they must also prevent inappropriate and intracellular polymerization of SctF protomers. To face this problem, several bacterial species have developed a chaperone-assisted protomer assembly as shown for *Pseudomonas* (Quinaud et al. 2005) and *Yersinia* (Sun et al. 2008). In this case, the needle subunit was seen in complex with its specific intracellular chaperones: the face of the C-terminal helix that forms the majority of the subunit–chaperone contacts is also the face of the helix that forms the majority of the subunit–subunit contacts in the assembled needle. Therefore, by protecting this polypeptide stretch from protein–protein interaction, a premature polymerization is greatly limited. Because no chaperone homologues have been identified in *Salmonella* and *Shigella* T3SS, it still remains unclear how the subunits MxiH and PrgI can efficiently avoid unwanted cytoplasmic polymerization.

Moreover, the needle model exhibits a pronounced homology concerning its structural features to the packing of the flagellar filaments (Yonekura et al. 2003), questioning if the filament assembly mechanism is comparable between the two supramolecular complexes. In the case of the flagellum, the correct insertion of the protomers at the capping end of the filament is controlled by a cap protein (Yonekura et al. 2000). Because non-flagellar T3SSs lack such cap protein during the needle protomers secretion, it has been proposed that the relatively small size of the needle protomers, compared to the flagellar ones, could explain that the termini of the needle subunits remain close enough to the top of the assembling filament to directly polymerize at its end without a capping mechanism (Blocker et al. 2008). It is worth mentioning that the recent discovery of OrgC in *S. typhimurium* (Kato et al. 2018), a protein able to physically interact and speed up the polymerization of SscF from the outside, suggests an evolved capping-like mechanism, although the protein is not detected at the tip of SctF filaments. The needles subunits are stabilized by multiple intra- and inter-subunit contacts (Loquet et al. 2012) resulting in a very rigid structure, showing almost no structural polymorphism as seen independently by SSNMR and cryo-EM (Loquet et al. 2012; Hu et al. 2018). Minor perturbations of the needle sequence, as demonstrated for the double mutant V65A/V67A of PrgI (Poyraz et al. 2010), can lead to structural polymorphism and a change of secondary structure of the C-terminal helix as observed by SSNMR chemical shift perturbation. However, the MxiH mutants D673A and Q51A generate nearly identical cryo-EM density maps (Fujii et al. 2012). A recent study by Galan and co-workers combining solid-state NMR and cryo-EM identified the C-terminal region of SctF as a protein segment prone to conformational plasticity (Guo et al. 2019), enabling subtle side-chain rearrangements that could modify the electrostatic properties of the needle lumen surface. Several SctF mutations located at the C-terminal [a detailed list of previously known mutants is presented here (Torres-Vargas et al. 2019)] have profound impact on the needle assembly and generation of defined secretion phenotypes (Guo et al. 2019).

The high-resolution cryo-EM density map of *Salmonella* needles showed a ~15 Å inner diameter (Hu et al. 2018). The diameter of a single polypeptide helix is ~9–13 Å, while the maximum diameter of the helix-turn-helix motif of the needle subunit is ~16–21 Å. It suggests that a well-folded needle protomer could not travel through the needle channel and that the secretion mechanism might take into account (i) a partial folding or (ii) a complete unfolding of the needle protomers. In the case of the secretion of substrates such as effectors, that have a much larger molecular size than the needle subunit, a fully unfolded conformation might be required. Cryo-EM experiments conducted by the Marlovits group (Radics et al. 2014) and biochemical results from Kolbe and co-workers (Dohlich et al. 2014) support this hypothesis. Additionally, the size of the portal, as seen by cryo-EM, at the cytoplasmic site showed an opening of ~10–15 Å (Radics et al. 2014) which is even narrower than the needle channel diameter. Such dynamic processes are still very challenging to probe by high-resolution techniques such as electron microscopy or NMR spectroscopy, because these approaches mostly capture a "frozen" and rigid structural state. Alternative biophysical methods will be required to decipher the structural state of dynamic needle protomers travelling through the filament.

7 Towards Functionalized T3SS Needles as Delivery Biotechnological Tools

Several successful attempts at exploiting the fascinating T3SS machine's properties have been reported in a wide range of applications, from biomedicine to bioengineering. Both mechanistic and structural insights on such a remarkable machine set in motion numerous approaches in governing the type III secretion system. Many studies have first focused on the formulation of T3SS protein vaccines as prevention against infectious diseases related to Gram-negative bacteria. *Salmonella enterica* alone counts thousands of serotypes and causes numerous diseases such as gastroenteritis and enteric fever. With $\sim$1.3 billion cases of non-typhoid salmonellosis and $\sim$17 million occurrences of typhoid fever infection each year, *Salmonella enterica* is annually responsible for hundreds of thousands of deaths in developing countries and several billions of dollars in health care costs (Martinez-Becerra et al. 2018). Many protein-based formulations have been developed and tested with the aim of inducing protective immunity against *Chlamydia muridarum*, *Salmonella enterica*, *Shigella flexneri*, *Salmonella typhimurium* and *Burkholderia pseudomallei*, using the T3SS needle proteins as antigens (Martinez-Becerra et al. 2018; Koroleva et al. 2017; Barrett et al. 2010; O'Meara et al. 2017), and might lead the way for valuable vaccine distribution and ultimate prevention against frequent infections. In particular, Barrett et al. reported a formulation based on the C-terminal truncated versions of the needle subunits from *S. flexneri* and *S. typhimurium*. Interestingly, the polymerized form of MxiH displays much higher immunogenic response compared to its soluble, monomeric counterpart, highlighting the potential of needle filaments in future formulations.

In addition to potential medical applications, the field of biotechnology has been inspired in utilizing this bacterial weapon for research purposes. One excellent example would be the T3SS engineering for sensing analyses and energy transduction experiments in vivo, as suggested in a compelling essay of Azam and Tullman-Ercek, whom designed a PrgI-based filamentous scaffold for nanostructured materials (Azam and Tullman-Ercek 2016). Other reports have promoted the use of the T3SS export/delivery function in several interesting and inspired schemes, in which a protein, peptide, transcription factor, nuclease, antigen or antibody for example directly delivered into a target cell or compartment, while fused to T3SS effectors or secretion signals. Dozens of proteins have indeed been delivered by bacterial T3SSs in different fields (Bai et al. 2018; Walker et al. 2017): export of spider silk monomers (Widmaier et al. 2009), selective purification of recombinant peptides (Singer et al. 2012), recombinant tumour-associated antigens delivery (Xu et al. 2014), antigen/epitope delivery for antiviral/bacterial vaccines (Russmann et al. 1998), antigen/epitope delivery for anti-tumour immunotherapy (Panthel et al. 2006; Epaulard et al. 2006), delivery of transcription factors (Bichsel et al. 2013, 2011), protein delivery for research purposes (Polack et al. 2000; Chamekh et al. 2008; Wolke et al. 2011; Ittig et al. 2015) or even genome editing (Jia et al. 2014, 2015).

Such tremendous advances have long been encouraged and facilitated through many efforts made in the purpose of controlling and regulating the injectisome, both genetically and mechanistically. The use of engineered bacteria for heterologous protein delivery presupposes an attenuation of cell toxicity and pathogenic capacity with regard to potential eukaryotic target cells, thus avoiding host-cell signal transduction tampering or apoptosis inducement (Deslandes and Rivas 2012, Hilbi et al. 1998; Gong et al. 2010). A recent review on T3SS-based protein delivery tools thoroughly discusses such genetic alterations in the context of *Salmonella*, *Yesinia* and *Pseudomonas* species (Bai et al. 2018). The design of a simplified and minimal type III secretion system which would aid and enhance our control over its use (Song et al. 2017) should be mentioned as well. Considering the extent of these existing biotechnological applications, one can assume that additional structural or mechanistic data regarding the T3SS injectisome will only serve to further their development, such as by functionalizing the T3SS needle or designing more efficient antigens/epitopes for vaccine formulation immunotherapies.

8 Concluding Remarks

Since its first visualization by electron microscopy, the structural aspects of the needle filament of the T3SS have considerably evolved, leading in the past years to several high-resolution structural models. Combining recent technological advances in structural biology using solid-state and solution NMR, cryo-EM and X-ray crystallography, the past ten years of research in the field have led to fundamental insights into the structural features of the T3SS needle filaments. Several high-resolution structural models of different needle filaments have uncovered a rather common structural fold, both at the level of the needle subunit and at the level of the filament supramolecular architecture. Numerous questions remain to be addressed based on these structures. In particular, the tight space inside the needle tube and the specific electrostatic pattern observed at the internal surface directly question the nature of the secretion process inside the needle: Is passive diffusion of secreted molecules possible? Might subtle, local, protein–protein interactions trigger a secretion force to export the molecules along the needle? The folding process of secreted needle protomers at the end of the growing filament is also still poorly understood. Further biophysical characterization of the T3SS needle filament, in the context of its native conformation at the cell surface, will be welcome to answer these questions.

Acknowledgements This work was funded by the ERC StG (639020 to A.L.), IdEx Bordeaux (Chaire d'Installation to B.H., ANR-10-IDEX-03-02), the ANR (ANR-14-CE09-0020-01 to A.L. and ANR-16-CE11-0020 to A.L. and G.L.) and the CNRS (MOMENTUM 2019 to B.H.).

References

Alber F, Forster F, Korkin D, Topf M, Sali A (2008) Integrating diverse data for structure determination of macromolecular assemblies. Annu Rev Biochem 77:443–477

Azam A, Tullman-Ercek D (2016) Type-III secretion filaments as scaffolds for inorganic nanostructures. J R Soc Interface 13(114):20150938

Bai F, Li Z, Umezawa A, Terada N, Jin S (2018) Bacterial type III secretion system as a protein delivery tool for a broad range of biomedical applications. Biotechnol Adv 36(2):482–493

Barrett BS, Markham AP, Esfandiary R, Picking WL, Picking WD, Joshi SB, Middaugh CR (2010) Formulation and immunogenicity studies of type III secretion system needle antigens as vaccine candidates. J Pharm Sci 99(11):4488–4496

Bichsel C, Neeld DK, Hamazaki T, Wu D, Chang LJ, Yang L, Terada N, Jin S (2011) Bacterial delivery of nuclear proteins into pluripotent and differentiated cells. PLoS ONE 6(1):e16465

Bichsel C, Neeld D, Hamazaki T, Chang LJ, Yang LJ, Terada N, Jin S (2013) Direct reprogramming of fibroblasts to myocytes via bacterial injection of MyoD protein. Cell Reprogram 15(2):117–125

Blocker A, Jouihri N, Larquet E, Gounon P, Ebel F, Parsot C, Sansonetti P, Allaoui A (2001) Structure and composition of the *Shigella flexneri* "needle complex", a part of its type III secreton. Mol Microbiol 39(3):652–663

Blocker A, Komoriya K, Aizawa S (2003) Type III secretion systems and bacterial flagella: insights into their function from structural similarities. Proc Natl Acad Sci USA 100(6):3027–3030

Blocker AJ, Deane JE, Veenendaal AK, Roversi P, Hodgkinson JL, Johnson S, Lea SM (2008) What's the point of the type III secretion system needle? Proc Natl Acad Sci USA 105 (18):6507–6513

Chamekh M, Phalipon A, Quertainmont R, Salmon I, Sansonetti P, Allaoui A (2008) Delivery of biologically active anti-inflammatory cytokines IL-10 and IL-1ra in vivo by the *Shigella* type III secretion apparatus. J Immunol 180(6):4292–4298

Cordes FS, Komoriya K, Larquet E, Yang S, Egelman EH, Blocker A, Lea SM (2003) Helical structure of the needle of the type III secretion system of Shigella flexneri. J Biol Chem 278 (19):17103–17107

Cuniasse P, Tavares P, Orlova EV, Zinn-Justin S (2017) Structures of biomolecular complexes by combination of NMR and cryoEM methods. Curr Opin Struct Biol 43:104–113

Deane JE, Roversi P, Cordes FS, Johnson S, Kenjale R, Daniell S, Booy F, Picking WD, Picking WL, Blocker AJ et al (2006a) Molecular model of a type III secretion system needle: implications for host-cell sensing. Proc Natl Acad Sci USA 103(33):12529–12533

Deane JE, Cordes FS, Roversi P, Johnson S, Kenjale R, Picking WD, Picking WL, Lea SM, Blocker A (2006b) Expression, purification, crystallization and preliminary crystallographic analysis of MxiH, a subunit of the Shigella flexneri type III secretion system needle. Acta Crystallogr, Sect F: Struct Biol Cryst Commun 62(Pt 3):302–305

Demers JP, Sgourakis NG, Gupta R, Loquet A, Giller K, Riedel D, Laube B, Kolbe M, Baker D, Becker S et al (2013) The common structural architecture of *Shigella flexneri* and *Salmonella typhimurium* type three secretion system needles. PLoS Pathog, accepted

Demers JP, Habenstein B, Loquet A, Kumar Vasa S, Giller K, Becker S, Baker D, Lange A, Sgourakis NG (2014) High-resolution structure of the Shigella type-III secretion needle by solid-state NMR and cryo-electron microscopy. Nat Commun 5:4976

Deslandes L, Rivas S (2012) Catch me if you can: bacterial effectors and plant targets. Trends Plant Sci 17(11):644–655

Dohlich K, Zumsteg AB, Goosmann C, Kolbe M (2014) A substrate-fusion protein is trapped inside the Type III secretion system channel in *Shigella flexneri*. PLoS Pathog 10(1):e1003881

Epaulard O, Toussaint B, Quenee L, Derouazi M, Bosco N, Villiers C, Le Berre R, Guery B, Filopon D, Crombez L et al (2006) Anti-tumor immunotherapy via antigen delivery from a live attenuated genetically engineered *Pseudomonas aeruginosa* type III secretion system-based vector. Mol Ther 14(5):656–661

Fujii T, Cheung M, Blanco A, Kato T, Blocker AJ, Namba K (2012) Structure of a type III secretion needle at 7-A resolution provides insights into its assembly and signaling mechanisms. Proc Natl Acad Sci USA 109(12):4461–4466

Galan JE, Wolf-Watz H (2006) Protein delivery into eukaryotic cells by type III secretion machines. Nature 444(7119):567–573

Galkin VE, Schmied WH, Schraidt O, Marlovits TC, Egelman EH (2010) The structure of the *Salmonella typhimurium* type III secretion system needle shows divergence from the flagellar system. J Mol Biol 396(5):1392–1397

Gong H, Vu GP, Bai Y, Yang E, Liu F, Lu S (2010) Differential expression of *Salmonella* type III secretion system factors InvJ, PrgJ, SipC, SipD, SopA and SopB in cultures and in mice. Microbiology 156(Pt 1):116–127

Guo EZ, Desrosiers DC, Zalesak J, Tolchard J, Berbon M, Habenstein B, Marlovits T, Loquet A, Galan JE (2019) A polymorphic helix of a *Salmonella* needle protein relays signals defining distinct steps in type III secretion. PLoS Biol 17(7):e3000351

Hilbi H, Moss JE, Hersh D, Chen Y, Arondel J, Banerjee S, Flavell RA, Yuan J, Sansonetti PJ, Zychlinsky A (1998) Shigella-induced apoptosis is dependent on caspase-1 which binds to IpaB. J Biol Chem 273(49):32895–32900

Hu J, Worrall LJ, Hong C, Vuckovic M, Atkinson CE, Caveney N, Yu Z, Strynadka NCJ (2018) Cryo-EM analysis of the T3S injectisome reveals the structure of the needle and open secretin. Nat Commun 9(1):3840

Hu J, Worrall LJ, Vuckovic M, Hong C, Deng W, Atkinson CE, Brett Finlay B, Yu Z, Strynadka NCJ (2019) T3S injectisome needle complex structures in four distinct states reveal the basis of membrane coupling and assembly. Nat Microbiol 4(11):2010–2019

Ittig SJ, Schmutz C, Kasper CA, Amstutz M, Schmidt A, Sauteur L, Vigano MA, Low SH, Affolter M, Cornelis GR et al (2015) A bacterial type III secretion-based protein delivery tool for broad applications in cell biology. J Cell Biol 211(4):913–931

Jia J, Jin Y, Bian T, Wu D, Yang L, Terada N, Wu W, Jin S (2014) Bacterial delivery of TALEN proteins for human genome editing. PLoS ONE 9(3):e91547

Jia J, Bai F, Jin Y, Santostefano KE, Ha UH, Wu D, Wu W, Terada N, Jin S (2015) Efficient gene editing in pluripotent stem cells by bacterial injection of transcription activator-like effector nuclease proteins. Stem Cells Transl Med 4(8):913–926

Joseph AP, Polles G, Alber F, Topf M (2017) Integrative modelling of cellular assemblies. Curr Opin Struct Biol 46:102–109

Journet L, Agrain C, Broz P, Cornelis GR (2003) The needle length of bacterial injectisomes is determined by a molecular ruler. Science 302(5651):1757–1760

Karaca E, Bonvin AM (2013) Advances in integrative modeling of biomolecular complexes. Methods 59(3):372–381

Kato J, Dey S, Soto JE, Butan C, Wilkinson MC, De Guzman RN, Galan JE (2018) A protein secreted by the *Salmonella* type III secretion system controls needle filament assembly. Elife 7

Kenjale R, Wilson J, Zenk SF, Saurya S, Picking WL, Picking WD, Blocker A (2005) The needle component of the type III secreton of Shigella regulates the activity of the secretion apparatus. J Biol Chem 280(52):42929–42937

Kimbrough TG, Miller SI (2000) Contribution of *Salmonella typhimurium* type III secretion components to needle complex formation. Proc Natl Acad Sci USA 97(20):11008–11013

Koroleva EA, Kobets NV, Shcherbinin DN, Zigangirova NA, Shmarov MM, Tukhvatulin AI, Logunov DY, Naroditsky BS, Gintsburg AL (2017) Chlamydial type III secretion system needle protein induces protective immunity against *Chlamydia muridarum* intravaginal infection. Biomed Res Int 2017:3865802

Kubori T, Matsushima Y, Nakamura D, Uralil J, Lara-Tejero M, Sukhan A, Galan JE, Aizawa SI (1998) Supramolecular structure of the *Salmonella typhimurium* type III protein secretion system. Science 280(5363):602–605

Kubori T, Sukhan A, Aizawa SI, Galan JE (2000) Molecular characterization and assembly of the needle complex of the *Salmonella typhimurium* type III protein secretion system. Proc Natl Acad Sci USA 97(18):10225–10230

Lefebre MD, Galan JE (2014) The inner rod protein controls substrate switching and needle length in a *Salmonella* type III secretion system. Proc Natl Acad Sci USA 111(2):817–822

Loquet A, Lv G, Giller K, Becker S, Lange A (2011) (13)C spin dilution for simplified and complete solid-state NMR resonance assignment of insoluble biological assemblies. J Am Chem Soc 133(13):4725–4727

Loquet A, Sgourakis NG, Gupta R, Giller K, Riedel D, Goosmann C, Griesinger C, Kolbe M, Baker D, Becker S et al (2012) Atomic model of the type III secretion system needle. Nature 486(7402):276–279

Loquet A, Habenstein B, Chevelkov V, Vasa SK, Giller K, Becker S, Lange A (2013) Atomic structure and handedness of the building block of a biological assembly. J Am Chem Soc 135 (51):19135–19138

Marlovits TC, Kubori T, Lara-Tejero M, Thomas D, Unger VM, Galan JE (2006) Assembly of the inner rod determines needle length in the type III secretion injectisome. Nature 441(7093): 637–640

Martinez-Becerra FJ, Kumar P, Vishwakarma V, Kim JH, Arizmendi O, Middaugh CR, Picking WD, Picking WL (2018) Characterization and protective efficacy of type III secretion proteins as a broadly protective subunit vaccine against *Salmonella enterica* Serotypes. Infect Immun 86(3)

Michiels T, Vanooteghem JC, Lambert de Rouvroit C, China B, Gustin A, Boudry P, Cornelis GR (1991) Analysis of virC, an operon involved in the secretion of Yop proteins by *Yersinia enterocolitica*. J Bacteriol 173(16):4994–5009

Mimori Y, Yamashita I, Murata K, Fujiyoshi Y, Yonekura K, Toyoshima C, Namba K (1995) The structure of the R-type straight flagellar filament of *Salmonella* at 9 A resolution by electron cryomicroscopy. J Mol Biol 249(1):69–87

Mota LJ, Cornelis GR (2005) The bacterial injection kit: type III secretion systems. Ann Med 37 (4):234–249

O'Meara CP, Armitage CW, Andrew DW, Kollipara A, Lycke NY, Potter AA, Gerdts V, Petrovsky N, Beagley KW (2017) Multistage vaccines containing outer membrane, type III secretion system and inclusion membrane proteins protects against a *Chlamydia* genital tract infection and pathology. Vaccine 35(31):3883–3888

Osei-Owusu P, Jessen Condry DL, Toosky M, Roughead W, Bradley DS, Nilles ML (2015) The *N* terminus of type III secretion needle protein YscF from *Yersinia pestis* functions to modulate innate immune responses. Infect Immun 83(4):1507–1522

Panthel K, Meinel KM, Sevil Domenech VE, Geginat G, Linkemann K, Busch DH, Russmann H (2006) Prophylactic anti-tumor immunity against a murine fibrosarcoma triggered by the Salmonella type III secretion system. Microbes Infect 8(9–10):2539–2546

Polack B, Vergnaud S, Paclet MH, Lamotte D, Toussaint B, Morel F (2000) Protein delivery by *Pseudomonas* type III secretion system: ex vivo complementation of p67(phox)-deficient chronic granulomatous disease. Biochem Biophys Res Commun 275(3):854–858

Poyraz O, Schmidt H, Seidel K, Delissen F, Ader C, Tenenboim H, Goosmann C, Laube B, Thunemann AF, Zychlinsky A et al (2010) Protein refolding is required for assembly of the type three secretion needle. Nat Struct Mol Biol 17(7):788–792

Quinaud M, Chabert J, Faudry E, Neumann E, Lemaire D, Pastor A, Elsen S, Dessen A, Attree I (2005) The PscE-PscF-PscG complex controls type III secretion needle biogenesis in *Pseudomonas aeruginosa*. J Biol Chem 280(43):36293–36300

Quinaud M, Ple S, Job V, Contreras-Martel C, Simorre JP, Attree I, Dessen A (2007) Structure of the heterotrimeric complex that regulates type III secretion needle formation. Proc Natl Acad Sci USA 104(19):7803–7808

Radics J, Konigsmaier L, Marlovits TC (2014) Structure of a pathogenic type 3 secretion system in action. Nat Struct Mol Biol 21(1):82–87

Russmann H, Shams H, Poblete F, Fu Y, Galan JE, Donis RO (1998) Delivery of epitopes by the *Salmonella* type III secretion system for vaccine development. Science 281(5376):565–568

Salmond GP, Reeves PJ (1993) Membrane traffic wardens and protein secretion in gram-negative bacteria. Trends Biochem Sci 18(1):7–12

Singer HM, Erhardt M, Steiner AM, Zhang MM, Yoshikami D, Bulaj G, Olivera BM, Hughes KT (2012) Selective purification of recombinant neuroactive peptides using the flagellar type III secretion system. MBio 3(3)

Song M, Sukovich DJ, Ciccarelli L, Mayr J, Fernandez-Rodriguez J, Mirsky EA, Tucker AC, Gordon DB, Marlovits TC, Voigt CA (2017) Control of type III protein secretion using a minimal genetic system. Nat Commun 8:14737

Sukhan A, Kubori T, Wilson J, Galan JE (2001) Genetic analysis of assembly of the *Salmonella enterica* serovar Typhimurium type III secretion-associated needle complex. J Bacteriol 183 (4):1159–1167

Sun P, Tropea JE, Austin BP, Cherry S, Waugh DS (2008) Structural characterization of the *Yersinia pestis* type III secretion system needle protein YscF in complex with its heterodimeric chaperone YscE/YscG. J Mol Biol 377(3):819–830

Tamano K, Aizawa S, Katayama E, Nonaka T, Imajoh-Ohmi S, Kuwae A, Nagai S, Sasakawa C (2000) Supramolecular structure of the Shigella type III secretion machinery: the needle part is changeable in length and essential for delivery of effectors. EMBO J 19(15):3876–3887

Torres-Vargas CE, Kronenberger T, Roos N, Dietsche T, Poso A, Wagner S (2019) The inner rod of virulence-associated type III secretion systems constitutes a needle adapter of one helical turn that is deeply integrated into the system's export apparatus. Mol Microbiol 112(3):918–931

Verasdonck J, Shen DK, Treadgold A, Arthur C, Bockmann A, Meier BH, Blocker AJ (2015) Reassessment of MxiH subunit orientation and fold within native Shigella T3SS needles using surface labelling and solid-state NMR. J Struct Biol 192(3):441–448

Wagner S, Sorg I, Degiacomi M, Journet L, Dal Peraro M, Cornelis GR (2009) The helical content of the YscP molecular ruler determines the length of the Yersinia injectisome. Mol Microbiol 71(3):692–701

Wagner S, Stenta M, Metzger LC, Dal Peraro M, Cornelis GR (2010) Length control of the injectisome needle requires only one molecule of Yop secretion protein P (YscP). Proc Natl Acad Sci USA 107(31):13860–13865

Walker BJ, Stan GV, Polizzi KM (2017) Intracellular delivery of biologic therapeutics by bacterial secretion systems. Expert Rev Mol Med 19:e6

Widmaier DM, Tullman-Ercek D, Mirsky EA, Hill R, Govindarajan S, Minshull J, Voigt CA (2009) Engineering the *Salmonella* type III secretion system to export spider silk monomers. Mol Syst Biol 5:309

Wolke S, Ackermann N, Heesemann J (2011) The *Yersinia enterocolitica* type 3 secretion system (T3SS) as toolbox for studying the cell biological effects of bacterial Rho GTPase modulating T3SS effector proteins. Cell Microbiol 13(9):1339–1357

Xu X, Hegazy WA, Guo L, Gao X, Courtney AN, Kurbanov S, Liu D, Tian G, Manuel ER, Diamond DJ et al (2014) Effective cancer vaccine platform based on attenuated salmonella and a type III secretion system. Cancer Res 74(21):6260–6270

Yonekura K, Maki S, Morgan DG, DeRosier DJ, Vonderviszt F, Imada K, Namba K (2000) The bacterial flagellar cap as the rotary promoter of flagellin self-assembly. Science 290 (5499):2148–2152

Yonekura K, Maki-Yonekura S, Namba K (2003) Complete atomic model of the bacterial flagellar filament by electron cryomicroscopy. Nature 424(6949):643–650

Zhang L, Wang Y, Picking WL, Picking WD, De Guzman RN (2006) Solution structure of monomeric BsaL, the type III secretion needle protein of *Burkholderia pseudomallei*. J Mol Biol 359(2):322–330

The Type III Secretion System Sorting Platform

María Lara-Tejero

Contents

Abstract A central feature of type III protein secretion machines is their ability to engage their substrates in a hierarchical and organized fashion. The hierarchy in the secretion process is first observed during the assembly of the type III secretion injectisome when the secretion machine exclusively engages proteins required for building the needle complex substructure (early substrates). After completion of the needle complex, the secretion system loads the proteins that will form the needle tip substructure as well as the protein translocases (middle substrates), which upon contact with host cells will mediate the passage of effectors (late substrates) through the host plasma membrane. The hierarchy of the secretion process is orchestrated by a very large cytoplasmic complex known as the sorting platform, which selects and initiates the substrates into the secretion pathway.

Bacterial type III secretion systems (T3SSs) have specifically evolved to deliver effector proteins directly into the cytosol of eukaryotic cells to modulate cellular functions (Galan et al. 2014). A fundamental aspect of type III protein secretion systems is that they engage their substrates in a hierarchical manner (Kubori et al. 2000; Lara-Tejero et al. 2011) This hierarchy is first observed during the assembly of the needle complex (NC) when the base substructure exclusively engages early

M. Lara-Tejero (✉)
Department of Microbial Pathogenesis, Yale University School of Medicine, New Haven 06536, CT, USA
e-mail: maria.lara-tejero@yale.edu

Current Topics in Microbiology and Immunology (2020) 427: 133–142
https://doi.org/10.1007/82_2019_167
© Springer Nature Switzerland AG 2019
Published Online: 11 June 2019

substrates necessary for the assembly of the inner rod (SctI) and needle filament (SctF) substructures as well as the regulatory protein SctP (Kubori et al. 2000; Lara-Tejero et al. 2011). However, the hierarchy in the secretion process also operates after the complete assembly of the needle complex (NC) to ensure the placement of the tip protein (SctA) at the end of the polymerized needle and the subsequent secretion of the protein translocases (middle substrates) so that the effector proteins (late substrates) can be productively delivered into target eukaryotic cells. This hierarchical process is orchestrated by a very large (several MDa) and highly conserved cytoplasmic complex known as the sorting platform (Lara-Tejero et al. 2011). Here, I will review what is known about this fundamental element of all T3SSs.

1 Composition

Earlier studies showed that a significant proportion of SctQ, a conserved component of T3SSs that shares limited amino acid sequence similarity to a component of the flagella known as the C-ring, was organized in a high-molecular-weight complex that co-localized with the NC substructure (Lara-Tejero et al. 2011). Analysis of the composition of these SctQ complexes by LC-MS/MS after 1-D SDS-PAGE and 2-D BN-PAGE showed the presence of several T3SS proteins including SctK, SctL, and SctN. Notably, although the protein translocases were readily detected in this complex, the effector proteins and their chaperones were largely absent or were present in very low amounts. The abundance of the translocases (middle substrates) coupled to the absence or low abundance of the effectors (late substrates) in these complexes isolated from bacteria that had not been signaled for activation of secretion, suggested that this complex may act as a sorting platform to direct the orderly secretion of substrates. Consistent with this hypothesis, in the absence of the translocases, a much larger proportion of effector proteins were detected in the SctQ high MW complex. Furthermore, no translocases were detected in SctQ complexes obtained from a strain lacking the regulatory protein SctP, which is “locked” in a secretion mode that is only competent for the secretion of the needle filament and inner rod proteins, SctF and SctI (Kubori et al. 2000; Tamano et al. 2002; Magdalena et al. 2002). Taken together, these observations indicated that a complex made up of the soluble cytosolic proteins SctQ, SctK, SctL, and SctN constitutes a sorting platform that may help establish the order of secretion (Lara-Tejero et al. 2011). SctO was not detected in these studies because of its small size and low stoichiometry, but it is also a component of the sorting platform that presumably provides a functional linkage to the export apparatus (see below).

The estimated stoichiometry of the injectisome-associated sorting platform components has been calculated from fluorescently labeled components using single-molecule super-resolution nanoscopy (SMSN) (Zhang et al. 2017) and conventional confocal microscopy (Diepold et al. 2017). Since SctQ exhibits a stoichiometric relationship with SctD, it is predicted to be present in 24 copies per sorting platform. The other components of the sorting platform, SctK ($\sim$6 copies), SctL ($\sim$12 copies), and SctN (6 copies), are estimated to be in substoichiometric amounts relative to SctQ. Because no clusters were detected for SctO when observed by super-resolution microscopy, the stoichiometry could not be determined but the absence of SctO clusters suggests that this protein is present in much lower stoichiometry (one or two copies per sorting platform), which is consistent with previous reports on a related component of the flagellar apparatus (Ibuki et al. 2011).

2 Structure

Because the NC substructure can be isolated in a manner that is suitable for single-particle cryo-electron microscopy (cryo-EM), its structure has been resolved to near atomic resolution (Marlovits et al. 2004; Worrall et al. 2016; Loquet et al. 2012; Hu et al. 2018). The sorting platform, however, disassembles during this type of purification, and therefore, its structure cannot be solved using this technique. Recent advances in cryo-electron tomography (cryo-ET) coupled with the use of bacterial mini-cells have allowed the in situ visualization of the complete injectisome of *Shigella* (Hu et al. 2015; Makino et al. 2016) and *Salmonella* (Hu et al. 2017) including the sorting platform substructure. In keeping with the high degree of conservation of the sorting platform components, both structures present a similar architecture consisting of a six-pod structure that is different from the related C-ring structure in the flagellar apparatus (Thomas et al. 2001; Kawamoto et al. 2013) (see Figs. 1 and 2 for details). The flagellar structure appears as a continuous ring beneath the flagellar basal body, an organization that may be necessary for its function in flagellar rotation. Thus, the different organizations of these related structures may reflect the different functions of these evolutionarily related bacterial nanomachines. The six-pod core structure of the sorting platform is linked to the inner rings of the NC base on its membrane-proximal side and capped on its cytoplasmic side by a six-spoke wheel structure with a central nave-like hub. The entire scaffold encloses a cage-like space where substrates are likely to be engaged and targeted to the export apparatus for secretion. The high-resolution structural details obtained in the *Salmonella* cryo-ET average maps allowed the precise localization of the individual components of the cytoplasmic sorting platform by adding traceable densities to each of its components (Hu et al. 2017). SctK localizes to the most proximal region of the pods close to the cytoplasmic domain of the inner ring protein SctD and therefore most likely serves as the link between the pods and the inner ring 2 (IR2) of the NC base. SctL maps to the spokes of the wheels that caps the structure on its cytoplasmic side. SctQ makes up the bulk of the central segment of the pods with its N terminus pointing toward the NC and its C terminus merging with the spokes of the wheel, which is in keeping with studies demonstrating the interaction between the C-terminal region of SctQ and the N terminus of SctL (Notti et al. 2015). The ATPase SctN seats on top of the hexameric nave of the wheel with its C terminus, the predicted substrate-binding domain (see below), oriented toward the toroidal shape structure formed by the

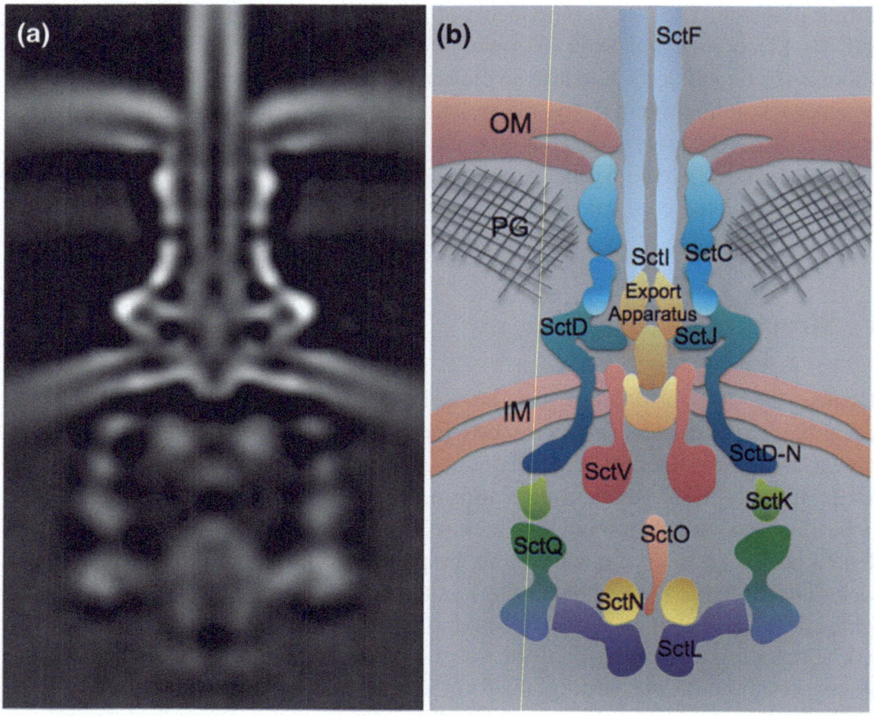

Fig. 1 Molecular architecture of the sorting platform in the intact injectisome a. A central section of the subtomogram average of the *Salmonella* injectisome as observed by cryo-ET. **b** Schematic representation of the type III secretion system with the different components labeled

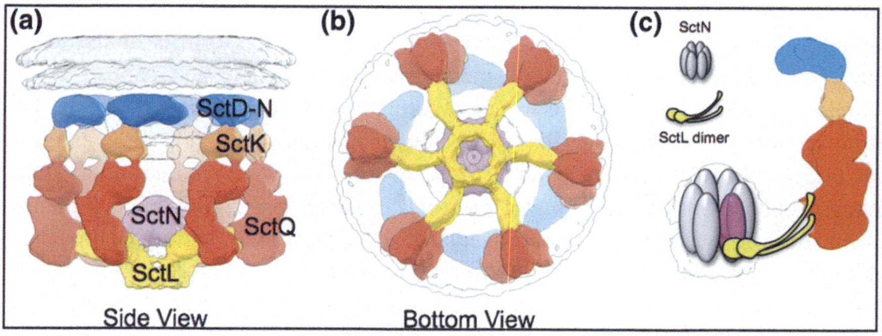

Fig. 2 3D representation of the cage-like structure of the type III secretion system sorting platform. A side view (**a**) and bottom view (**b**) of the sorting platform. **c** Schematic representation of one of the six pods that integrate the sorting platform

cytoplasmic C terminus of the export apparatus component SctV. This orientation may be critical for the coordination of the SctN and SctV activities in the initiation of the T3S substrates through the secretion channel (see below). SctO localizes between SctN and SctV, suggesting a potential role in coordinating their activities.

Through an internal translation start site, *SctQ* encodes a small C-terminal protein ($SctQ_C$, which in the T3SS of some plant pathogens is encoded by a separate gene and is equivalent to the flagellar protein FliN (Bzymek et al. 2012; Yu et al. 2011; Fadouloglou et al. 2004). This internally translated product has been reported to be essential for T3SS function in *Yersinia* (Bzymek et al. 2012) and *Shigella* (McDowell et al. 2016), and it has therefore been suggested to be a structural component of the sorting platform. In the case of *Salmonella*, however, the absence of $SctQ_C$ (from both of its T3SSs) results in a mild secretion phenotype and non-detectable functional phenotype, observations that are not consistent with an essential structural role for this component of the sorting platform. For these reasons, a regulatory role during assembly and disassembly of the sorting platform was proposed for $SctQ_C$, perhaps by acting as a chaperone for the full-length protein, because in its absence, SctQ becomes less stable (Yu et al. 2011; Lara-Tejero et al. 2019). More studies will be necessary to completely understand the role of this conserved element of the sorting platform.

3 Assembly

Several experimental approaches have provided insights into the assembly of the sorting platform and have helped define its core structural components. Insight into the assembly of this complex structure can be gleaned by observing the effect of the removal of some of the components of the sorting platform on the overall integrity of the structure. Cryo-ET injectisome structures from strains lacking SctK, SctL, or SctQ showed the complete absence of the sorting platform (Hu et al. 2017) indicating that the removal of any of these components prevents its assembly. Similarly, the removal of SctK or SctL resulted in the disappearance of the sorting platform-associated clusters identified by super-resolution microscopy of fluorescently labeled SctQ (Zhang et al. 2017). On the other hand, the removal of the other components, SctO or SctN, had little impact on the assembly of the sorting platform as observed both by microscopy (Zhang et al. 2017) and cryo-ET (Hu et al. 2017), suggesting a minor structural role for these components despite their essential role in type III protein secretion (Dreyfus et al. 1993; Eichelberg et al. 1994; Collazo et al. 1995; Payne and Straley 1998).

Important insight into the assembly of the sorting platform has also been obtained by comparing the in situ structures of the needle complex in the presence or absence of the sorting platform. In the absence of the sorting platform, the inner ring 2 of the needle complex, which is located within the bacterial cytoplasm and is formed by the cytoplasmic domain of SctD, adopted a 24-fold configuration similar to that observed in isolated needle complexes. In contrast, in the presence of the sorting platform, this structure re-organized very significantly, appearing as six discrete patches arranged in circular fashion and linked to each one of the pods of the sorting platform most likely through SctK (Hu et al. 2017). This observation indicates that assembly of the sorting platform results in a significant rearrangement of the IR2 to accommodate the 24-fold symmetry of the NC to the 6-fold symmetry of the sorting platform. These findings are also consistent with the observed flexibility of the IR2 (Schraidt and Marlovits 2011; Worrall et al. 2016) and explain why the N terminus of SctD does not oligomerize in solution (Spreter et al. 2009). Thus, the NC substructure must be in place prior to the incorporation of the sorting platform. In fact, the NC most likely templates the assembly of the sorting platform. This hypothesis is supported by the observation that in the absence of the NC, the sorting platform components are seen in live bacteria as large aggregates that localize to one of the bacterial poles (Zhang et al. 2017). It is therefore likely that in the absence of the NC, the sorting platform intermediates coalesce into misassembled aggregates. Furthermore, the sorting platform was assembled in the absence of SctV (Hu et al. 2017; Zhang et al. 2017), a component of the export apparatus, indicating that an active type III secretion system is not required for sorting platform assembly. Nevertheless, SctQ aggregates were observed in mutants lacking other components of the export apparatus (SctR, SctS, SctT, and SctU). However, these export apparatus components are required for templating the assembly of the NC (Wagner et al. 2010); therefore, the aggregation of sorting platforms in these mutant strains is likely due to the reduced number of NC structures present in these mutants.

Imaging of sorting platform components in live bacteria has shown that a significant proportion of complexes of these components can be observed within the bacterial cytosol not in association with the NC (Zhang et al. 2017; Diepold et al. 2017). This observation is intriguing as it suggests that preassembled sorting platforms, poised to be engaged by the NC, may be present in the bacterial cytoplasm. As discussed above, in the absence of the NC, the sorting platform components aggregate at the poles of the bacteria. Therefore, the observation of apparently fully or partially assembled sorting platforms that are not in association with the NC suggests that the cytoplasmic complexes of sorting platform components must have interacted with the NC prior to their relocation to the cytoplasmic pool. Therefore, the sorting platform may undergo cycles of assembly and partial or total disassembly, which may be central to its function. Consistent with this hypothesis, experiments in live bacteria have shown that cytoplasmic and NC-associated components of the sorting platform are dynamic and that the cytoplasmic and needle complex associated pools of sorting platform components exchange with one another (Diepold et al. 2017).

4 Substrate Engagement

The mechanisms by which the sorting platform engages its substrates are poorly understood. Studies have shown that the recruitment of the translocases to the sorting platform requires their cognate chaperones. Thus, in the absence of SctW, the translocases SctE and SctB were not detected within the SctQ sorting platform complex (Lara-Tejero et al. 2011). Likewise, in a mutant lacking the translocases, which allows the recruitment of effectors to the sorting platform (see above), such recruitment could be prevented by the removal of their respective cognate chaperones. Furthermore, no translocases were detected in the SctQ complex in a strain lacking the regulatory protein SctP, in which substrate switching does not occur and the T3SS is locked in secretion mode that is only able to secrete the filament and inner rod proteins SctF and SctI. Taken together, these findings suggest a mechanism by which the hierarchy in the secretion may be established by the different affinities of the chaperone/substrate complexes with the sorting platform (Lara-Tejero et al. 2011; Portaliou et al. 2017) (Fig. 3). The actual component of the sorting platform that acts as a receptor for the type III secreted substrates has not been formally identified. However, the best candidate is the conserved ATPase SctN. The previous studies have shown that SctN can recognize the chaperone/substrate complex, triggering the release and unfolding of the substrate in an ATP-dependent manner (Akeda and Galan 2005). It is therefore possible that SctN, scaffolded by the structural components SctK, SctL, and SctQ, acts as a receptor for

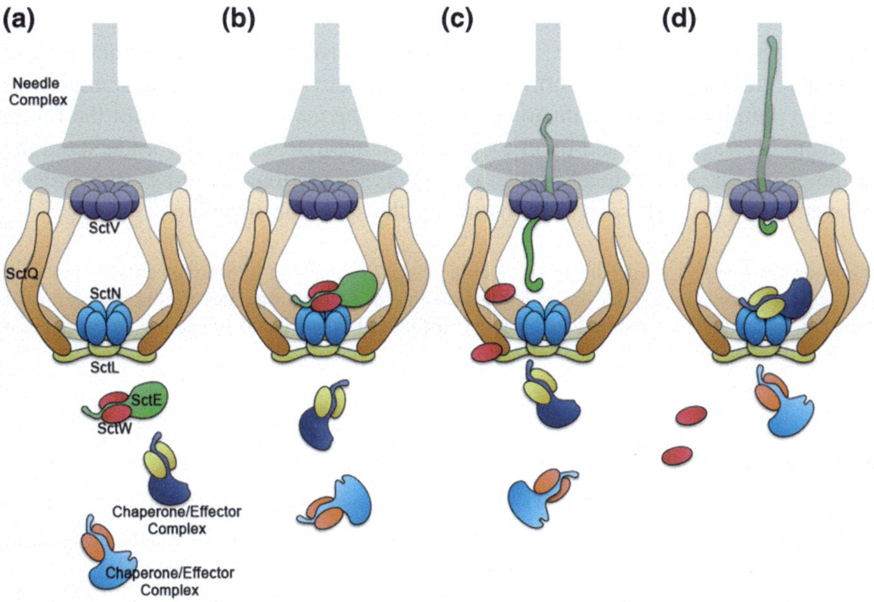

Fig. 3 Model for substrate engagement by the sorting platform. a Substrates are recruited to the sorting platform by their cognate chaperone. Different chaperone/substrate complexes have different affinities for the sorting platform, and the order in which they are brought to the ATPase component (SctN) of the sorting platform is based on this affinity. **b** Thus, upon completion of the needle complex substructure, the translocases, such as SctE, are brought to the sorting platform by their cognate chaperone SctW. **c** SctN then triggers the release and unfolding of the substrate that can now go through the T3SS channel, while the chaperones are recycled. **d** After the translocases are secreted, another effector/chaperone complex is recruited to the sorting platform based on the affinity of the chaperone for the sorting platform and the process is repeated to engage additional effectors

the chaperone/substrate complexes and may play a central role in determining the hierarchy of secretion. The architecture of the in situ structure of the sorting platform observed by cryo-ET would be compatible with such model. SctN seats at the center of the wheel-like structure that caps the sorting platform's cage on its cytoplasmic side (Fig. 2). Importantly, its C terminus, the predicted substrate-binding domain (Akeda and Galan 2004), is oriented toward the toroidal shape structure formed by the cytoplasmic C terminus of SctV (Hu et al. 2017; Majewski et al. 2019). Therefore, the chaperone/substrate complexes could enter the chamber-like space limited by the sorting platform pods and engage the substrate-binding domain of the ATPase SctN. Binding would trigger the release of the chaperone and the unfolding of the effector that would be adequately positioned to enter the channel formed by SctV and the other components of the export apparatus (Fig. 3). More experiments will be required to substantiate this model.

5 Concluding Remarks

Recent cryo-ET studies have provided major insights into the structure and molecular architecture of the sorting platform, and in vivo live imaging studies have begun to illuminate its mechanism of assembly. The major challenges for the future will be to leverage these and other experimental approaches to fully understand the mechanisms of hierarchical substrate engagement that ensures the orderly engagement of type III secreted proteins by the secretion machine.

Acknowledgements I would like to thank Jorge E. Galán and José E. Soto for critical review of this manuscript and Jun Liu for the generous donation of graphic materials for the elaboration of figures. This work was supported by National Institute of Allergy and Infectious Diseases grant R21AI126158 (to M.L.T.)

References

Akeda Y, Galan JE (2004) Genetic analysis of the Salmonella enterica type III secretion-associated ATPase InvC defines discrete functional domains. J Bacteriol 186:2402–2412

Akeda Y, Galan JE (2005) Chaperone release and unfolding of substrates in type III secretion. Nature 437:911–915

Bzymek KP, Hamaoka BY, Ghosh P (2012) Two translation products of Yersinia yscQ assemble to form a complex essential to type III secretion. Biochemistry 51:1669–1677

Collazo CM, Zierler MK, Galan JE (1995) Functional analysis of the Salmonella typhimurium invasion genes invl and invJ and identification of a target of the protein secretion apparatus encoded in the inv locus. Mol Microbiol 15:25–38

Diepold A, Sezgin E, Huseyin M, Mortimer T, Eggeling C, Armitage JP (2017) A dynamic and adaptive network of cytosolic interactions governs protein export by the T3SS injectisome. Nat Commun 8:15940

Dreyfus G, Williams AW, Kawagishi I, Macnab RM (1993) Genetic and biochemical analysis of Salmonella typhimurium FliI, a flagellar protein related to the catalytic subunit of the F0F1

ATPase and to virulence proteins of mammalian and plant pathogens. J Bacteriol 175:3131–3138

Eichelberg K, Ginocchio CC, Galan JE (1994) Molecular and functional characterization of the Salmonella typhimurium invasion genes invB and invC: homology of InvC to the F0F1 ATPase family of proteins. J Bacteriol 176:4501–4510

Fadouloglou VE, Tampakaki AP, Glykos NM, Bastaki MN, Hadden JM, Phillips SE, Panopoulos NJ, Kokkinidis M (2004) Structure of HrcQB-C, a conserved component of the bacterial type III secretion systems. Proc Natl Acad Sci U S A 101:70–75

Galan JE, Lara-Tejero M, Marlovits TC, Wagner S (2014) Bacterial type III secretion systems: specialized nanomachines for protein delivery into target cells. Annu Rev Microbiol 68:415–438

Hu B, Lara-Tejero M, Kong Q, Galan JE, Liu J (2017) In situ molecular architecture of the Salmonella type III secretion machine. Cell 168(1065–1074):e10

Hu B, Morado DR, Margolin W, Rohde JR, Arizmendi O, Picking WL, Picking WD, Liu J (2015) Visualization of the type III secretion sorting platform of Shigella flexneri. Proc Natl Acad Sci U S A 112:1047–1052

Hu J, Worrall LJ, Hong C, Vuckovic M, Atkinson CE, Caveney N, Yu Z, Strynadka NCJ (2018) Cryo-EM analysis of the T3S injectisome reveals the structure of the needle and open secretin. Nat Commun 9:3840

Ibuki T, Imada K, Minamino T, Kato T, Miyata T, Namba K (2011) Common architecture of the flagellar type III protein export apparatus and F- and V-type ATPases. Nat Struct Mol Biol 18:277–282

Kawamoto A, Morimoto YV, Miyata T, Minamino T, Hughes KT, Kato T, Namba K (2013) Common and distinct structural features of Salmonella injectisome and flagellar basal body. Sci Rep 3:3369

Kubori T, Sukhan A, Aizawa SI, Galan JE (2000) Molecular characterization and assembly of the needle complex of the Salmonella typhimurium type III protein secretion system. Proc Natl Acad Sci U S A 97:10225–10230

Lara-Tejero M, Kato J, Wagner S, Liu X, Galan JE (2011) A sorting platform determines the order of protein secretion in bacterial type III systems. Science 331:1188–1191

Lara-Tejero M, Qin Z, Hu B, Butan C, Liu J, Galan JE (2019) Role of SpaO in the assembly of the sorting platform of a Salmonella type III secretion system. PLoS Pathog 15:e1007565

Loquet A, Sgourakis NG, Gupta R, Giller K, Riedel D, Goosmann C, Griesinger C, Kolbe M, Baker D, Becker S, Lange A (2012) Atomic model of the type III secretion system needle. Nature 486:276–279

Magdalena J, Hachani A, Chamekh M, Jouihri N, Gounon P, Blocker A, Allaoui A (2002) Spa32 regulates a switch in substrate specificity of the type III secreton of Shigella flexneri from needle components to Ipa proteins. J Bacteriol 184:3433–3441

Majewski DD, Worrall LJ, Hong C, Atkinson CE, Vuckovic M, Watanabe N, Yu Z, Strynadka NCJ (2019) Cryo-EM structure of the homohexameric T3SS ATPase-central stalk complex reveals rotary ATPase-like asymmetry. Nat Commun 10:626

Makino F, Shen D, Kajimura N, Kawamoto A, Pissaridou P, Oswin H, Pain M, Murillo I, Namba K, Blocker AJ (2016) The architecture of the cytoplasmic region of type III secretion systems. Sci Rep 6:33341

Marlovits TC, Kubori T, Sukhan A, Thomas DR, Galan JE, Unger VM (2004) Structural insights into the assembly of the type III secretion needle complex. Science 306:1040–1042

McDowell MA, Marcoux J, McVicker G, Johnson S, Fong YH, Stevens R, Bowman LA, Degiacomi MT, Yan J, Wise A, Friede ME, Benesch JL, Deane JE, Tang CM, Robinson CV, Lea SM (2016) Characterisation of Shigella Spa33 and Thermotoga FliM/N reveals a new model for C-ring assembly in T3SS. Mol Microbiol 99:749–766

Notti RQ, Bhattacharya S, Lilic M, Stebbins CE (2015) A common assembly module in injectisome and flagellar type III secretion sorting platforms. Nat Commun 6:7125

Payne PL, Straley SC (1998) YscO of Yersinia pestis is a mobile core component of the Yop secretion system. J Bacteriol 180:3882–3890

Portaliou AG, Tsolis KC, Loos MS, Balabanidou V, Rayo J, Tsirigotaki A, Crepin VF, Frankel G, Kalodimos CG, Karamanou S, Economou A (2017) Hierarchical protein targeting and secretion is controlled by an affinity switch in the type III secretion system of enteropathogenic Escherichia coli. EMBO J 36:3517–3531

Schraidt O, Marlovits TC (2011) Three-dimensional model of Salmonella's needle complex at subnanometer resolution. Science 331:1192–1195

Spreter T, Yip CK, Sanowar S, Andre I, Kimbrough TG, Vuckovic M, Pfuetzner RA, Deng W, Yu AC, Finlay BB, Baker D, Miller SI, Strynadka NC (2009) A conserved structural motif mediates formation of the periplasmic rings in the type III secretion system. Nat Struct Mol Biol 16:468–476

Tamano K, Katayama E, Toyotome T, Sasakawa C (2002) Shigella Spa32 is an essential secretory protein for functional type III secretion machinery and uniformity of its needle length. J Bacteriol 184:1244–1252

Thomas D, Morgan DG, Derosier DJ (2001) Structures of bacterial flagellar motors from two FliF-FliG gene fusion mutants. J Bacteriol 183:6404–6412

Wagner S, Konigsmaier L, Lara-Tejero M, Lefebre M, Marlovits TC, Galan JE (2010) Organization and coordinated assembly of the type III secretion export apparatus. Proc Natl Acad Sci U S A 107:17745–17750

Worrall LJ, Hong C, Vuckovic M, Deng W, Bergeron JR, Majewski DD, Huang RK, Spreter T, Finlay BB, Yu Z, Strynadka NC (2016) Near-atomic-resolution cryo-EM analysis of the Salmonella T3S injectisome basal body. Nature

Yu XJ, Liu M, Matthews S, Holden DW (2011) Tandem translation generates a chaperone for the Salmonella type III secretion system protein SsaQ. J Biol Chem 286:36098–36107

Zhang Y, Lara-Tejero M, Bewersdorf J, Galan JE (2017) Visualization and characterization of individual type III protein secretion machines in live bacteria. Proc Natl Acad Sci U S A 114:6098–6103

Export Mechanisms and Energy Transduction in Type-III Secretion Machines

Thibaud T. Renault, Alina Guse and Marc Erhardt

Contents

Abstract The remarkably complex architecture and organization of bacterial nanomachines originally raised the enigma to how they are assembled in a coordinated manner. Over the years, the assembly processes of the flagellum and evolutionary-related injectisome complexes have been deciphered and were shown to rely on a conserved protein secretion machine: the type-III secretion system.

Authors Thibaud T. Renault and Alina Guse equally contributed to this chapter.

T. T. Renault · A. Guse · M. Erhardt (✉)
Institute for Biology—Bacterial Physiology, Humboldt-Universität zu Berlin, Berlin, Germany
e-mail: marc.erhardt@hu-berlin.de

T. T. Renault
Max Planck Institute for Infection Biology, Berlin, Germany

T. T. Renault
CNRS UMR 5234 Microbiologie Fondamentale et Pathogénicité, Bordeaux, France

T. T. Renault
Institut Européen de Chimie et Biologie, University of Bordeaux, Pessac, France

Current Topics in Microbiology and Immunology (2020) 427: 143–159
https://doi.org/10.1007/82_2019_166
© Springer Nature Switzerland AG 2019
Published Online: 20 June 2019

In this book chapter, we demonstrate how individually evolved mechanisms cooperate in highly versatile and robust secretion machinery to export and assemble the building blocks of those nanomachines.

1 Introduction

Type-III secretion systems (T3SS) are multi-component protein export complexes responsible for recognizing and exporting at a high-speed the building blocks of self-assembling bacterial nanomachines. The most ancestral T3SS, the flagellar T3SS (fT3SS), assembles the bacterial flagellum, a macromolecular structure found both in gram-positive (e.g. *Bacillus subtilis*) and gram-negative (e.g. *Salmonella* spp.) bacteria (Hueck 1998). This long extracellular appendage enables motility on surfaces and in liquid environments (Berg and Anderson 1973; Kubori et al. 1998). Another evolutionarily related nanomachine is assembled with the help of a T3SS and is found in many gram-negative pathogens. The T3SS of this needle-like nanomachine, the injectisome, has evolved an additional function and does not only enable assembly of the injectisome, but additionally energizes the translocation of effectors into host cells during invasion and colonization. Given its involvement in bacterial virulence by the additional secretion of those non-structural proteins, this machinery is termed the virulence-associated T3SS (vT3SS) (Blocker et al. 2003; Cornelis 2006; Galan and Wolf-Watz 2006).

The flagellum and injectisome share several structural homologies; in particular, their respective T3SS consist of about ten cytoplasmic and inner membrane proteins, which display similarity in either sequence or membrane topology (Kubori et al. 1998; Blocker et al. 2003; Cornelis 2006).

The nanomachines are composed of three main structural parts (Blocker et al. 2003; Macnab 2004; Erhardt et al. 2010): a basal body with a highly conserved T3SS embedded within the MS-ring (FliF/SctDJ) in the inner membrane (IM) and a rod that traverses the periplasmic space. The external structures differ, however (Blocker et al. 2003; Macnab 2004; Erhardt et al. 2010). The flagellar basal body extends by a hook and a filament, and the injectisome features a needle with a translocon pore at the distal end. Assembly of these nanomachines is highly regulated and occurs in a specific order. Flagellar assembly is based on a transcriptional hierarchy of three promoter classes, while injectisome assembly depends rather on varying affinities and kinetics of protein–protein interactions (Macnab 2003; Chevance and Hughes 2008; Diepold and Wagner 2014).

Over the years, several models have been proposed to explain how protein transport via the T3SS is energized. The current understanding of protein secretion via T3SS proposes a mechanism that involves multi-simultaneous processes (Fig. 1). Export and assembly through the T3SS can be subdivided into three major steps: (1) substrate recognition and association at the cytosolic side of the export gate; (2) translocation across the inner membrane; and (3) further translocation and assembly beyond the inner membrane. Substrate protein translocation across the

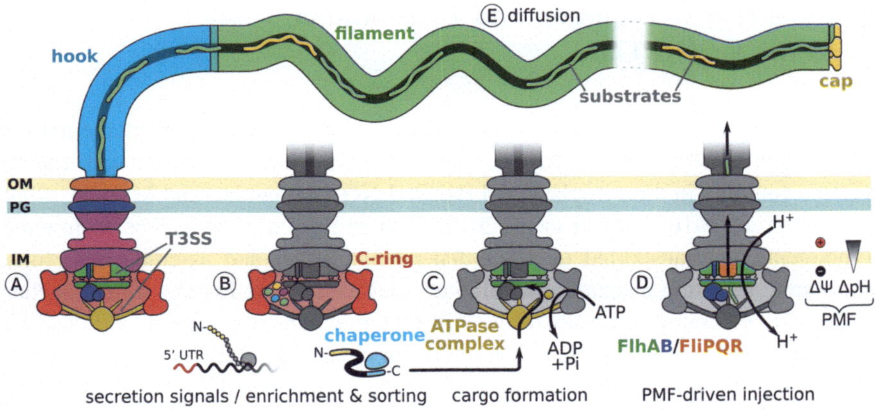

Fig. 1 Molecular mechanisms of substrate export during flagellar assembly. **a** Structural model of the flagellum featuring its major components. The T3SS, composed of FliPQR (orange), FlhA (green) and FlhB (blue), is located at the base of the basal body, embedded within the MS-ring (magenta) in the inner membrane (IM). Underneath the MS-ring assembles the C-ring (red). The flagellar rod (purple/blue shades) spans periplasmic space until the outer membrane (OM) and traverses the peptidoglycan layer (PG). Outside of the cell, the flagellum extends by the hook (blue), the hook–filament junction (turquoise) and the filament (green) which polymerizes underneath the filament cap (yellow). Substrate recognition, delivery to the export gate and injection across the IM are a multi-step process involving complementary mechanisms. **b** Multiple secretion signals for fT3SS substrates and binding to the C-ring participate in the localized enrichment of cargo near the export gate. **c** Chaperone-bound substrates are delivered by the ATPase complex (dark yellow) to the FlhA. **d** The proton motive force (PMF) is the primary energy source driving the secretion process across the IM. **e** Further translocation of substrates beyond the IM after injection by the T3SS does not require additional energization. It occurs through diffusion in the narrow secretion channel towards the distal end of the growing flagellar structure

inner membrane relies mainly on the proton motive force (PMF) as the major energy providing source (Paul et al. 2008; Minamino and Namba 2008), although the associated ATPase plays an important role in the recruitment and unfolding of substrates (Terashima et al. 2018).

This chapter highlights the mechanisms underlying the substrate recognition and energization processes, which enable efficient protein export and self-assembly by the T3SS and its accessory components. In the following, the term T3SS will be used for general descriptions that apply to both, the fT3SS and the vT3SS. General information will be given using the flagellar system as a reference, yet major differences in the vT3SS will be emphasized. The nomenclature refers to the flagellum or the unified Sct (secretion and cellular translocation) nomenclature in case of the injectisome (Hueck 1998).

2 Substrate Recognition and Presentation to the Export Gate

The T3SS is composed of an export gate organized in five integral membrane proteins (FliPQR/SctRST, FlhA/SctV and FlhB/SctU) and of an associated soluble ATPase complex in the cytoplasm (FliHIJ/SctLNO) (Minamino and Macnab 1999; Zilkenat et al. 2016; Fabiani et al. 2017; Fukumura et al. 2017). The membrane components are transported in a Sec-dependent manner and assemble into fully functioning secretion machinery within the inner membrane. In case of the flagellar core export apparatus, assembly of the FliP protein-conducting pore is facilitated by an integral membrane chaperone, FliO (Fabiani et al. 2017; Fukumura et al. 2017).

The assembly and function of the T3SS is a highly regulated process, both on the genetic and protein level. Flagellar assembly begins with the transcription of the flagellar master regulator *flhDC* and is under complex genetic regulation (Mouslim and Hughes 2014). The heteromultimeric $FlhD_4C_2$ complex then activates transcription from class 2 promoters, which gene products are the building blocks of the rod and hook structures. The most striking example of the tight regulation of flagellar assembly is the so-called substrate specificity switch, which occurs after hook completion. The hook grows to a relatively controlled length and extends to 55 ± 6 nm from the cell surface in *Salmonella* (Hirano et al. 1994). The switch in substrate specificity divides the many flagellar buildings blocks into two distinct substrate classes: early and late substrates. Hook polymerization stops when the export apparatus protein FlhB switches secretion specificity from early to late substrates (Fraser et al. 2003). In the flagellum, early substrates make up the rod and hook (expressed from class 2 promoters), whereas the late substrates assemble the hook–filament junction and the filament (expressed primarily from class 3 promoters) (Kutsukake et al. 1990). An infrequent molecular ruler mechanism controls this secretion switch: the ruler protein FliK is secreted intermittently and triggers a conformational change in FlhB only when hook length reaches that of the unfolded, N-terminal domain of FliK, which gives the C-terminal domain of FliK enough time to interact with FlhB and to flip the switch in substrate specificity (Erhardt et al. 2011).

The ordered secretion of the multiple structural substrates through the T3SS is truly essential for proper assembly of these complex nanomachines. This regulation is highly complex and relies on multiple steps including (1) substrate-specific molecular properties for recognition by the export machinery, which favours secretion of the required substrates at a given stage of assembly, (2) enrichment of the substrates by the C-ring near the export gate for efficient export, (3) targeted delivery of substrate cargo to the export gate by the ATPase complex, and (4) docking of substrates to the export gate components $FlhA_C$ and $FlhB_C$ before injection into the secretion pore made of multiple subunits of FliPQR.

3 Signals for Type-III Secretion

Unlike proteins that are recognized and secreted by generic secretion machineries, such as the Sec-system, T3SS substrates do not share a well-defined N-terminal peptide secretion signal. It remains unclear whether the targeting and recognition of substrates by the export gate is a single or a multi-step process. Singer et al. investigated possible secretion signals of fT3SS substrates in detail by determining the individual contributions of the N-terminal peptide sequence and of the 5′-UTR of secretion substrate mRNAs (Singer et al. 2014). It appears not only that multiple signals are needed for efficient substrate secretion via the fT3SS but also that not all substrates require the same types of targeting signals and levels of regulation, even when they belonged to the same substrate class. For instance, secretion of the early substrate FlgB decreased by several orders of magnitude after replacing its 5′-UTR with that of the arabinose operon, while replacing the 5′-UTR had little effect on the secretion efficiency of other substrates. It thus appears possible that the optimal subset of secretion signals evolved individually for each substrate depending on its respective secretory requirements.

Late substrates such as the hook–filament junction proteins FlgKL, the filament cap FliD and the filament subunits FljB/FliC additionally harbour a C-terminal binding site for their cognate secretion chaperones—FlgN, FliT and FliS, respectively—which function is to prevent aggregation and degradation within the cytoplasm (Fraser et al. 1999; Auvray et al. 2001). The reason why only late, but not early substrates require specific secretion chaperones remains unclear, but it might confer a specificity for late substrates as only chaperone-bound substrates can associate with the FlhA docking platform (Bange et al. 2010; Kinoshita et al. 2013; Portaliou et al. 2017; Xing et al. 2018). The N-terminal peptide sequences are most likely recognized by the fT3SS, though it is unclear which component is specifically responsible for this recognition step. Whether the substrates are initially recognized by the ATPase complex and then delivered to the export gate or independently recognized by the docking platform or the secretion pore made up by FliPQR remains a major question to be investigated. The different secretion signals and cognate chaperones might constitute a regulatory mechanism, which controls the spatio-temporal export of flagellar substrates and thus ensures a stepwise targeting and ordered assembly process.

In addition to the multi-component secretion signal, the distinct substrate classes feature different binding affinities to the export gate, which might contribute to the correct order of export during assembly. Prior to the secretion and polymerization of the filament, two components need to be assembled. The hook–filament junction (FlgKL), which polymerizes on top of the hook structure and constitutes a platform to anchor the filament and the filament cap, made up of five subunits of FliD, which acts as a foldase to drive polymerization of the flagellin (FljB or FliC) (Macnab 2003). As those components are only required in low copy numbers (5–11), but yet need to be assembled prior to the export of the up to 20,000 filament subunits, additional regulatory mechanisms have evolved to ensure their proper secretion.

FlgKL, FliD and their cognate chaperones, FlgN and FliT, are expressed from both class 2 and class 3 promoters, while the flagellins are exclusively expressed from class 3 promoters (Kutsukake et al. 1990). The synthesis of the hook–filament junction and filament cap proteins during hook basal body formation might confer an advantage for more rapid and efficient export of these rather low copy substrates upon substrate specificity switching from early to late substrate secretion. This temporal advantage is accompanied by a higher binding affinity to the export gate docking platform and the ATPase complex. FlgN–FlgKL and FliT–FliD chaperone–substrate complexes exhibit a 14-fold higher binding affinity to $FlhA_C$ compared to the FliS–FliC complex (Kinoshita et al. 2013). Also, only FlgN–FlgKL and FliT–FliD complexes bind to the ATPase complex component FliJ for delivery to the export gate (Evans et al. 2006; Bange et al. 2010). Both, earlier transcription and binding advantages provide a more efficient recognition and secretion of the low copy late substrates prior to the high copy filament subunits and thus ensure proper assembly of the flagellar structure.

4 Cargo Transfer and Accumulation

The fT3SS is accompanied by two non-essential components, which accumulate substrates near the export gate in the cytoplasm and enhance the efficiency of the export machinery. One of these components is the associated ATPase complex (FliHIJ), composed of an ATPase (FliI), a regulator of ATPase activity (FliH) and an adaptor protein (FliJ) necessary for FliI ring formation and interactions with substrate–chaperone complexes (Minamino and Macnab 1999, 2000a; Ibuki et al. 2011). This ATPase complex acts as a pilot for protein delivery to the export gate. FliI is present in two different conformations, either a FliH dimer binds to FliI ($FliH_2FliI$) in the cytoplasm to inhibit ring formation, and ATPase activity or FliI forms a homohexameric ring structure in cooperation with FliJ, which binds to the central pore of the hexameric FliI ring. $FliH_2FliI$ is proposed to act as a dynamic carrier for substrates and chaperone–substrate complexes for delivery to the $FlhA_C$–$FlhB_C$ docking platform (Ibuki et al. 2011; Bai et al. 2014). Thereby, the extreme N-terminal region of FliH interacts with FlhA and FliN. FliN homologues in the vT3SS (SctQ) interact alike with a number of injectisome components including homologues of FliI and FliH (Minamino et al. 2009; Hara et al. 2012). It has been shown that FliI exhibits rapid turnovers between the basal body and the cytoplasmic pool, which might deliver bound substrates from the cytoplasm to the export gate. In contrast, the $FliH_{12}FliI_6FliJ$ ring is located at the base of the flagellum and might act as a static loader of secretion substrates to the secretion channel (Bai et al. 2014). The second accessory component, which is required for rapid export by the fT3SS, is the cytoplasmic ring (C-ring), which is located underneath the MS-ring. The C-ring provides binding sites for $FliH_2FliI$ substrate–chaperone complexes and is thereby involved in the accumulation of substrates near the export gate (González-Pedrajo et al. 2006). Over-expression of the flagellar master regulator, *flhDC*, and thus over-production of secretion substrates overcomes a C-ring deficient mutant, highlighting, that the C-ring is not strictly required for recognition and secretion of flagellar building blocks by the fT3SS (Erhardt and Hughes 2010). Even though the ATPase and the C-ring are not essential for protein secretion through the fT3SS and assembly of the flagellar structure, their presence enhances the efficiency of this secretion machinery remarkably.

5 Substrate Docking at the Export Gate

Once substrates are delivered and accumulated near the export gate, the last step before export is binding to the docking platform formed by the fT3SS components $FlhA_C$ and $FlhB_C$, which sets off injection into the secretion channel formed by FliP, FliQ and FliR (Ward et al. 2018; Kuhlen et al. 2018). FlhA, FlhB and FliPQR are essential for the secretion of flagellar building blocks by the fT3SS. Over the years, a fundamental understanding of substrate recognition and secretion through the fT3SS has been achieved, but the molecular mechanisms and functions are still poorly understood. While $FlhA_C$ and $FlhB_C$ have been proposed to form the export gate docking platform, the exact mechanism of substrate binding and injection into the secretion channel remains debated. Furthermore, it should be investigated, whether FlhA and FlhB bind different substrates, and which properties of the fT3SS proteins and substrates confer this difference in binding affinities.

FlhA is composed of three distinct regions. The N-terminus consists of eight transmembrane α-helices and is connected to a large C-terminal cytoplasmic domain by a flexible linker. The C-terminal domain of FlhA ($FlhA_C$) forms a nonameric ring structure beneath the secretion channel and provides several functions. $FlhA_C$ consists of four domains, which are involved in different steps of substrate recognition. The two subdomains on the N-terminal side of $FlhA_C$ provide binding sites for chaperone–substrate complexes and are directly involved in the translocation of flagellar proteins. Upon binding of a late substrate to its cognate chaperone, the chaperone exhibits a conformational change, which releases it from its autoinhibitory conformation. This allows for the presentation of a recognition helix on the chaperone structure, which is recognized by the FlhA binding cleft between the interface of the D1 and D2 domains (Xing et al. 2018). It should be noted that only FliT and FliS adopt an autoinhibitory conformation in the absence of their cognate substrates, while FlgN alone is capable of binding to FlhA (Kinoshita et al. 2013). The substrate–chaperone binding cleft on FlhA is not conserved in the FlhA homologue SctV, though SctV also possesses highly conserved regions that could serve as chaperone–substrate binding sites (Abrusci et al. 2013; Xing et al. 2018). Truncations of the most C-terminal domain of $FlhA_C$ (D4) affect the strict order of secretion as they allow for the premature secretion of late substrates prior to hook completion (McMurry et al. 2004; Bange et al. 2010; Minamino et al. 2010; Kinoshita et al. 2013). Therefore, D4 might act as an inhibitor to suppress certain interactions within $FlhA_C$ or with other T3SS components such as FlhB to ensure the correct secretion order.

FlhB is responsible for substrate specificity switching and thus plays a regulatory role both on a genetic and protein level. Newly synthesized FlhB undergoes autocleavage in a flexible loop connecting two subdomains of the large cytoplasmic C-terminal domain, $FlhB_{CN}$ and $FlhB_{CC}$, which remain closely associated after cleavage. This cleavage is crucial for the recognition of late substrates, as it has been shown that a cleavage-deficient mutant is unable to control hook length and constitutively secretes early substrates (Minamino and Macnab 2000b; Fraser et al. 2003; Ferris et al. 2005; Ferris and Minamino 2006). This raises the question of why this autocleavage event has such a large impact on the substrate recognition abilities of the fT3SS. Major structural rearrangements occur within the loop connecting $FlhB_{CN}$ and $FlhB_{CC}$, as well as in the flexible linker region connecting $FlhB_C$ with the transmembrane domain of FlhB ($FlhB_{TM}$) after autocleavage (Lountos et al. 2009; Meshcheryakov et al. 2013; Minamino 2014). Interestingly, Monjarás Feria et al. showed that the precise position of the autocleavage site is not important in the FlhB homologue SctU (SpaS in the *Salmonella* SPI-1 vT3SS) by introducing an artificial cleavage site at a different position and obtaining a fully functional protein (Monjarás Feria et al. 2015).

In which way the interplay between FlhA and FlhB modulates substrate recognition remains to be investigated. The fact that $FlhA_C$ provides binding sites exclusively for chaperone–substrate complexes suggests a dual recognition mechanism at the export gate, where FlhA would only recognize late substrates, while FlhB would be specific to early substrates. One hypothesis to explain this mechanism is that the C-terminal domain of FlhBc ($FlhB_{CC}$) might act as an inhibitor for late substrate secretion by concealing potential binding sites on FlhA. The conformational change of $FlhB_{CC}$ upon the substrate specificity switch might then reveal late substrate binding sites on FlhA (Fig. 2a). It would be important to understand if a tight association of $FlhB_{CN}$ and $FlhB_{CC}$ is crucial for the ordered recognition of flagellar substrates.

The docking mechanism of the fT3SS does not seem to be conserved in the vT3SS. For instance, SctU exhibits some important differences in *Salmonella*. The autocleavage of SctU does not appear to have the same regulatory role in substrate specificity switching as autocleavage of FlhB. Autocleavage of SctU occurs after folding and before incorporation into the needle complex, and thus, an autocleavage-deficient $SpaS^{N258A}$ mutant has no effect on needle length in contrast to an autocleavage-deficient $FlhB^{N269A}$ mutant, which displays a polyhook phenotype (Monjarás Feria et al. 2015; Fraser et al. 2003). It will be interesting to investigate how the mechanisms of substrate specificity switching differ between FlhB and SctU.

While in flagellar assembly, the fT3SS machinery exhibits only one substrate specificity switching event, the vT3SS possesses two substrate switching mechanisms. The first substrate switching mechanism from early to middle or translocator substrates is similar to that observed in the fT3SS, while the second switching mechanism from the middle to late or effector substrates relies on a protein absent in

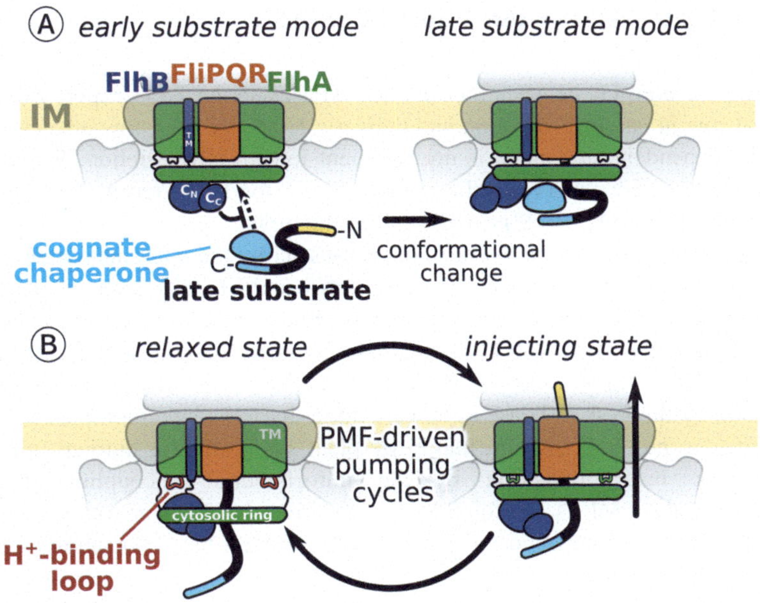

Fig. 2 Models of the role of FlhB and FlhA in switching between early and late mode of secretion, and coupling of the PMF, respectively. **a** A possible mechanism for differential substrate recognition by the export gate involves a conformational change in the C-terminus of FlhB. In early substrate mode, prior to hook completion, the conformation of $FlhB_C$ would inhibit recognition of chaperone–substrate complexes by FlhA and only allow secretion of early substrates (left panel). After hook completion, substrate specificity switches from early to late substrates due to a conformational change of $FlhB_C$, exposing the D4 domain of FlhA for recognition and secretion of chaperone–substrate complexes (right panel). **b** Model of a proton/protein antiporter mechanism for FlhA. Proton-actuated conformational changes of the H^+-binding loop of FlhA is thought to induce cyclic movements of the cytoplasmic domain, which would pump substrates across the IM. These cyclic movements could be synchronized with binding (left panel) and release of substrates for injection into the secretion channel (right panel)

the flagellar system. This protein, known as gatekeeper, is associated with the membrane and C-terminal domains of SctV (Portaliou et al. 2017). The tight binding allows for the recognition at the docking platform of SctV and subsequent secretion of translocators, but blocks the premature secretion of effectors prior to host cell contact. Deletion of the gatekeeper protein results in the abolished secretion of translocators and increased secretion of effectors (Portaliou et al. 2016, 2017). Depending on the species, dissociation and subsequent secretion or degradation of the gatekeeper is essential for the second substrate switching mechanism. Release of the gatekeeper is controlled by environmental cues, such as calcium depletion (e.g. in EPEC) or pH changes (e.g. SPI-2 injectisome in *Salmonella*) (Gaytán et al. 2018; Yu et al. 2018).

6 Transport Across the Inner Membrane

The complex interplay between the soluble ATPase and the T3SS export gate plays a critical role in cargo recognition and delivery. However, it is now also clear that this ATP-dependent step alone is not sufficient for type-III secretion. While most early studies considered ATP as the primary source for T3SS, recent developments have identified that the PMF primarily plays this role.

The PMF results from a movement of protons across the inner membrane driven by a combination of two forces: (1) diffusion caused by the gradient of proton concentration (ΔpH) and (2) the electrostatic force provoked by the gradient of electrical potential ($\Delta\psi$). The PMF is crucial to many biological processes in all domains of life and is used for instance to perform chemical work (e.g. ATP production by the F_OF_1 ATPase in mitochondria, bacteria and chloroplasts), active transport (e.g. proton/potassium import by plant symporters or nutrient uptake in bacteria) or heat production by the mitochondrial uncoupling proteins. Additionally, the PMF is the force known to drive the flagellar rotary motor by coupling proton transport and mechanical work through the stator proteins MotA and MotB (Blair and Berg 1990).

Galperin and co-workers suggested already in 1982 that the PMF plays an important role in the assembly of the flagellum (Galperin et al. 1982). This was later confirmed for the vT3SS (Wilharm et al. 2004). While both studies proposed the hypothesis that the PMF could drive type-III secretion independently of ATP and set the basis for the existence of a proton-driven protein antiporter, they did not provide the demonstration that transport across the membrane would depend solely on the PMF.

In 2008, it was reported independently by Paul et al. and Minamino and Namba that the T3SS ATPase complex, although important for the efficiency of trans-membrane protein secretion, is not strictly required for the protein export process and that the fT3SS is intrinsically a PMF-driven protein export device (Paul et al. 2008; Minamino and Namba 2008). Further, it was shown that both over-expression of flagellar substrates and an increase in the cell's available PMF were sufficient to drive protein secretion in a PMF-dependent and ATP-independent way (Erhardt et al. 2014). This was also confirmed in the vT3SS, where additional levels of regulation control the conversion of the PMF to mechanical work (Erhardt et al. 2014; Lee et al. 2014).

Further investigation of both chemical (ΔpH) and electrical ($\Delta\psi$) components of the PMF revealed that $\Delta\psi$ constitutes the main, indispensable, force for flagellar protein export, while ΔpH becomes essential only when the ATPase complex is absent (Minamino et al. 2011). This observation suggests a possible dual role of the ATPase. In addition to shuttling cargo to the export gate, it is possible that the ATPase would also participate in further conversion of the PMF to enhance the efficiency of the export apparatus.

Similarly to the flagellar motors, which can both utilize protons and other ions, the versatility of the T3SS energization was revealed with the observation that, in the absence of the ATPase, the fT3SS can shift from using PMF to using the sodium-motive force (SMF) to fuel protein export (Minamino et al. 2016). The intricate relationship between PMF-driven export and the soluble ATPase is further highlighted by the report that sodium utilization is prevented upon the association of the ATPase complex with the export gate, suggesting that this interaction could control a switch between the two ion-specific modes (Minamino et al. 2016).

How exactly the export apparatus itself couples PMF and protein translocation remains unknown; however, several studies have implicated FlhA, a large transmembrane (TM) component of the fT3SS, as being directly involved in the coupling (Saijo-Hamano et al. 2010; Hara et al. 2011; Minamino et al. 2016; Erhardt et al. 2017). FlhA consists of three major domains: a membrane-anchored part comprising several transmembrane segments, a flexible cytoplasmic loop between the TM4 and TM5, and a C-terminal, cytosolic rigid domain, which forms a nonameric ring (Minamino et al. 2010; Abrusci et al. 2012; Morimoto et al. 2014; Zilkenat et al. 2016). Identification and characterization of a mutant in the flexible loop (G368C in *S. Typhimurium*) revealed the importance of conformational rearrangements in FlhA and suggested for the first time that the protein and its C-terminal domain are directly involved in substrate translocation (Minamino et al. 2010). A further indication that FlhA could act as a proton/protein antiporter is the conservation of several ionizable residues as it would be expected for proton-conducting proteins (Lancaster 2003), for instance like in the flagellar motor (Zhou et al. 1998). Initial mutagenesis of FlhA charged residues *in Salmonella* suggested that D208, present at the end of a cytoplasmic flexible loop internal to the transmembrane part of the protein, would be a good candidate for proton binding (Hara et al. 2011; Minamino et al. 2010). However, this residue, although important, was demonstrated not to be strictly essential by a more extensive, mutagenesis study (Erhardt et al. 2017). In this second work, three residues: R147, D154, D158—present at the beginning of the loop between TM4 and TM5—were found to be essential for FlhA function and thus hypothesized to constitute the site of coupling to the PMF. Taken together, the probable organization of FlhA in two, membrane-anchored and cytosolic, ring-structured domains associated by a flexible linker, and the identification of critical charged residues within an internal loop of the membrane domain led to the hypothesis that the energy from the PMF would be converted to mechanical work by conformational changes within FlhA. Back-and-forth movements of the cytosolic ring upon proton binding would turn FlhA into a protein pump (Fig. 2b). In the vT3SS, the FlhA homologue SctV would play a similar role and also couple the proton flux and protein export (Lee et al. 2014).

Interestingly, the role of FlhA as an ion/protein antiporter is consistent with the reported ability of the T3SS to use both protons and sodium as coupling ions. Over-expression of FlhA induces an accumulation of intracellular Na^+ in the presence of a NaCl gradient, establishing that the protein indeed constitutes an ion channel (Minamino et al. 2016).

7 Transport and Assembly Beyond the Inner Membrane

Once translocated across the inner membrane, the T3SS substrate proteins travel through a narrow channel within the respective nanomachines until they reach their site of assembly or, in the case of the vT3SS-secreted effectors, the host cell. This journey can be reasonably short for early substrates like the rod and ring proteins (tens of nanometres), or rather long for the late substrates, which are translocated along several hundreds of nanometres or even several micrometres.

The variability in the length and environments of the periplasmic and extracellular components of the T3SS-assembled nanomachines raises several questions. Does transport beyond the inner membrane rely on further energy sources? Can protein refolding participate to energize their transport? Are there active mechanisms controlling the length of the successive elements of the nanomachines?

The kinetics and energization of the assembly of the several micrometres long flagellar filament have been debated over the years, with conflicting mechanisms proposed that would permit flagellin secretion over such a long distance (Iino 1974; Aizawa and Kubori 1998; Turner et al. 2012; Evans et al. 2013; Renault et al. 2017). Recently, it was hypothesized that flagellin subunits could form a continuous chain in the channel of the growing filament and that folding of the emerging subunit could provide sufficient force to pull the rest of the chain, energizing transport over the long filament (Evans et al. 2013). Contribution of inter-subunits chains was later ruled out, and it was demonstrated that injection from the T3SS, followed by diffusion along the hollow structure is sufficient to grow filaments of more than ten micrometres length (Renault et al. 2017).

While converting the potential energy of protein folding into a dragging force for subunit transport does not seem to be important for the overall mechanism of protein export, it is possible that folding plays a role in the export dynamics of specific substrates, e.g. in case of the length-measuring mechanism of the hook ruler protein FliK. The dynamics of FliK secretion appear to directly control the substrate specificity switch mediated by an interaction with FlhB: the ruler is exported at a fast speed when the hook is still too short and at a slower speed when the hook reaches its optimal length (Erhardt et al. 2011). FliK has a peculiar structure consisting of two globular domains linked by a loop (Kodera et al. 2015). Fast folding of the N-terminal domain and quick export when the hook is short would result in fast unfolding of the C-terminal domain and a low probability of contact with its target protein FlhB, while slower export upon completion of the hook would give the C-terminal domain enough time to trigger the secretion switch (Erhardt et al. 2011). Thus, protein folding is mechanistically important to modulate the FliK export dynamics, which permits a timely controlled regulation of hook length; however, it does not energize the export in itself.

As we just described, control over the length of the flexible hook is direct and relies on the expression of a ruler protein. On the other side, the other two flagellar structural components of variable length did not evolve a specific ruler mechanism to regulate their assembly and prevent unconstrained elongation. The flagellar rod length is controlled by an indirect, although striking, mechanism: growing until it hits the outer membrane (Cohen et al. 2017). This mechanism is indirect in the sense that it takes advantage of the control of the width of the periplasmic space by Braun's lipoprotein and does not require a specialized ruler molecule. Finally, the flagellar filament is perhaps the most intriguing given its extreme length. While potential active length control mechanisms have been debated (Hughes 2017), it occurs that filament length would not rely on any control mechanism per se, but rather derives from its quadratically decreasing elongation kinetics, giving an intrinsic statistical limit to how long it can grow (Renault et al. 2017).

8 Conclusion

The T3SS of both the flagellum and injectisome have evolved from common ancestral secretion machinery to permit the assembly of highly specialized nanomachines comprised of dozens of different proteins. This requires the precise orchestration of substrate recognition and high-speed export in a time-dependent manner. A series of secretion signals, docking complexes and switch mechanisms enable the chronological delivery of substrates to the T3SS export gate. Substrate translocation across the inner membrane is achieved by a PMF-coupled fast protein pump, which permits secretion rates of more than a thousand amino acids per second. The T3SS have proven to be not only versatile and efficient, but also robust secretion machineries, capable of overcoming environmental changes and energy limitation.

While a fundamental understanding of the core mechanisms of type-III secretion has emerged over the last decades, several questions are still left unanswered as to how the secretion machinery works on the molecular level. In particular, the atomic organization of the T3SS, the molecular mechanisms of the substrate specificity switch and the coupling of the PMF to protein transport are challenges for the years to come.

References

Abrusci P, Vergara-Irigaray M, Johnson S et al (2012) Architecture of the major component of the type III secretion system export apparatus. Nat Struct Mol Biol 20:99–104. https://doi.org/10.1038/nsmb.2452

Abrusci P, Vergara-Irigaray M, Johnson S et al (2013) Architecture of the major component of the type III secretion system export apparatus. Nat Struct Mol Biol 20:99–104. https://doi.org/10.1038/nsmb.2452

Aizawa SI, Kubori T (1998) Bacterial flagellation and cell division. Genes Cells 3:625–634. https://doi.org/10.1046/j.1365-2443.1998.00219.x

Auvray F, Thomas J, Fraser GM, Hughes C (2001) Flagellin polymerisation control by a cytosolic export chaperone. J Mol Biol 308:221–229. https://doi.org/10.1006/jmbi.2001.4597

Bai F, Morimoto YV, Yoshimura SDJ et al (2014) Assembly dynamics and the roles of FliI ATPase of the bacterial flagellar export apparatus. Sci Rep 4:1–7. https://doi.org/10.1038/srep06528

Bange G, Kümmerer N, Engel C et al (2010) FlhA provides the adaptor for coordinated delivery of late flagella building blocks to the type III secretion system. Proc Natl Acad Sci U S A 107:11295–11300. https://doi.org/10.1073/pnas.1001383107

Berg HC, Anderson RA (1973) Bacteria swim by rotating their flagellar filaments. Nature 245:380–382. https://doi.org/10.1038/245380a0

Blair DF, Berg HC (1990) The MotA protein of *E. coli* is a proton-conducting component of the flagellar motor. Cell 60:439–449

Blocker A, Komoriya K, Aizawa S-I (2003) Type III secretion systems and bacterial flagella: insights into their function from structural similarities. Proc Natl Acad Sci U S A 100:3027–3030. https://doi.org/10.1073/pnas.0535335100

Chevance FFV, Hughes KT (2008) Coordinating assembly of a bacterial macromolecular machine. Nat Rev Microbiol 6:455–465. https://doi.org/10.1038/nrmicro1887

Cohen EJ, Ferreira JL, Ladinsky MS et al (2017) Nanoscale-length control of the flagellar driveshaft requires hitting the tethered outer membrane. Science (80-) 356:197–200. https://doi.org/10.1126/science.aam6512

Cornelis GR (2006) The type III secretion injectisome. Nat Rev Microbiol 4:811–825. https://doi.org/10.1038/nrmicro1526

Diepold A, Wagner S (2014) Assembly of the bacterial type III secretion machinery. FEMS Microbiol Rev 38:802–822. https://doi.org/10.1111/1574-6976.12061

Erhardt M, Hughes KT (2010) C-ring requirement in flagellar type III secretion is bypassed by FlhDC upregulation. Mol Microbiol 75:376–393. https://doi.org/10.1111/j.1365-2958.2009.06973.x

Erhardt M, Namba K, Hughes KT (2010) Bacterial nanomachines: the flagellum and type III injectisome. Cold Spring Harb Perspect Biol 2. https://doi.org/10.1101/cshperspect.a000299

Erhardt M, Singer HM, Wee DH et al (2011) An infrequent molecular ruler controls flagellar hook length in Salmonella enterica. EMBO J 30:2948–2961. https://doi.org/10.1038/emboj.2011.185

Erhardt M, Mertens ME, Fabiani FD, Hughes KT (2014) ATPase-independent type-III protein secretion in Salmonella enterica. PLoS Genet 10. https://doi.org/10.1371/journal.pgen.1004800

Erhardt M, Wheatley P, Kim EA et al (2017) Mechanism of type-III protein secretion: regulation of FlhA conformation by a functionally critical charged-residue cluster. Mol Microbiol 104:234–249. https://doi.org/10.1111/mmi.13623

Evans LDB, Stafford GP, Ahmed S et al (2006) An escort mechanism for cycling of export chaperones during flagellum assembly. Proc Natl Acad Sci U S A 103:17474–17479. https://doi.org/10.1073/pnas.0605197103

Evans LDB, Poulter S, Terentjev EM et al (2013) A chain mechanism for flagellum growth. Nature 504. https://doi.org/10.1038/nature12682

Fabiani FD, Renault TT, Peters B et al (2017) A flagellum-specific chaperone facilitates assembly of the core type III export apparatus of the bacterial flagellum. PLoS Biol 15:e2002267. https://doi.org/10.1371/journal.pbio.2002267

Ferris HU, Minamino T (2006) Flipping the switch: bringing order to flagellar assembly. Trends Microbiol 14:519–526. https://doi.org/10.1016/j.tim.2006.10.006

Ferris HU, Furukawa Y, Minamino T et al (2005) FlhB regulates ordered export of flagellar components via autocleavage mechanism. J Biol Chem. https://doi.org/10.1074/jbc.m509438200

Fraser GM, Bennett JCQ, Hughes C (1999) Substrate-specific binding of hook-associated proteins by FlgN and FliT, putative chaperones for flagellum assembly. Mol Microbiol 32:569–580. https://doi.org/10.1046/j.1365-2958.1999.01372.x

Fraser GM, Hirano T, Ferris HU et al (2003) Substrate specificity of type III flagellar protein export in Salmonella is controlled by subdomain interactions in FlhB. Mol Microbiol 48:1043–1057. https://doi.org/10.1046/j.1365-2958.2003.03487.x

Fukumura T, Makino F, Dietsche T et al (2017) Assembly and stoichiometry of the core structure of the bacterial flagellar type III export gate complex. PLoS Biol 15:e2002281. https://doi.org/10.1371/journal.pbio.2002281

Galan J, Wolf-Watz H (2006) Protein delivery into eukaryotic cells by type III secretion machines. Nature 444:567–573. https://doi.org/10.1038/nature05272

Galperin MY, Dibrov PA, Glagolev AN (1982) ΔμH+ is required for flagellar growth in *Escherichia coli*. FEBS Lett 143:319–322. https://doi.org/10.1016/0014-5793(82)80125-7

Gaytán MO, Monjarás Feria J, Soto E et al (2018) Novel insights into the mechanism of SepL-mediated control of effector secretion in enteropathogenic *Escherichia coli*. Microbiologyopen 7:1–17. https://doi.org/10.1002/mbo3.571

González-Pedrajo B, Minamino T, Kihara M, Namba K (2006) Interactions between C ring proteins and export apparatus components: a possible mechanism for facilitating type III protein export. Mol Microbiol 60:984–998. https://doi.org/10.1111/j.1365-2958.2006.05149.x

Hara N, Namba K, Minamino T (2011) Genetic characterization of conserved charged residues in the bacterial flagellar type III export protein FlhA. PLoS One 6. https://doi.org/10.1371/journal.pone.0022417

Hara N, Morimoto YV, Kawamoto A et al (2012) Interaction of the extreme N-terminal region of flih with FlhA is required for efficient bacterial flagellar protein export. J Bacteriol 194:5353–5360. https://doi.org/10.1128/JB.01028-12

Hirano T, Yamaguchi S, Oosawa K, Aizawal S-I (1994) Roles of FliK and FlhB in determination of flagellar hook length in Salmonella typhimunum. J Bacteriol 176:5439–5449

Hueck CJ (1998) Type III protein secretion systems in bacterial pathogens of animals and plants. Microbiol Mol Biol Rev 62:379–433. https://doi.org/10.1111/j.1365-2958.2006.05301.x

Hughes KT (2017) Flagellum length control: how long is long enough? Curr Biol 27:R413–R415. https://doi.org/10.1016/j.cub.2017.04.008

Ibuki T, Imada K, Minamino T et al (2011) Common architecture of the flagellar type III protein export apparatus and F- and V-type ATPases. Nat Struct Mol Biol 18:277–282. https://doi.org/10.1038/nsmb.1977

Iino T (1974) Assembly of Salmonella flagellin in vitro and in vivo. J Supramol Struct 384:372–384

Kinoshita M, Hara N, Imada K et al (2013) Interactions of bacterial flagellar chaperone-substrate complexes with FlhA contribute to co-ordinating assembly of the flagellar filament. Mol Microbiol 90:1249–1261. https://doi.org/10.1111/mmi.12430

Kodera N, Uchida K, Ando T, Aizawa SI (2015) Two-ball structure of the flagellar hook-length control protein flik as revealed by high-speed atomic force microscopy. J Mol Biol 427:406–414. https://doi.org/10.1016/j.jmb.2014.11.007

Kubori T, Matsushima Y, Nakamura D et al (1998) Supramolecular structure of the *Salmonella typhimurium* type III protein secretion system. Science (80-) 280:602–605. https://doi.org/10.1126/science.280.5363.602

Kuhlen L, Abrusci P, Johnson S et al (2018) Structure of the core of the type III secretion system export apparatus. Nat Struct Mol Biol 1–8

Kutsukake K, Ohya Y, Tetsuo A (1990) Transcriptional analysis of the flagellar regulon of *Salmonella typhimurium*. J Bacteriol 172:741–747

Lancaster CRD (2003) The role of electrostatics in proton-conducting membrane protein complexes. FEBS Lett 545:52–60. https://doi.org/10.1016/S0014-5793(03)00393-4

Lee PC, Zmina SE, Stopford CM et al (2014) Control of type III secretion activity and substrate specificity by the cytoplasmic regulator PcrG. Proc Natl Acad Sci U S A 111:E2027–E2036. https://doi.org/10.1073/pnas.1402658111

Lountos GT, Austin BP, Nallamsetty S, Waugh DS (2009) Atomic resolution structure of the cytoplasmic domain of Yersinia pestis YscU, a regulatory switch involved in type III secretion. Protein Sci 18:467–474. https://doi.org/10.1002/pro.56

Macnab RM (2003) How bacteria assemble flagella. Annu Rev Microbiol 57:77–100. https://doi.org/10.1146/annurev.micro.57.030502.090832

Macnab RM (2004) Type III flagellar protein export and flagellar assembly. Biochim Biophys Acta Mol Cell Res 1694:207–217

McMurry JL, Van Arnam JS, Kihara M, Macnab RM (2004) Analysis of the cytoplasmic domains of Salmonella FlhA and interactions with components of the flagellar export machinery. J Bacteriol 186:7586–7592. https://doi.org/10.1128/JB.186.22.7586-7592.2004
Meshcheryakov VA, Kitao A, Matsunami H, Samatey FA (2013) Inhibition of a type III secretion system by the deletion of a short loop in one of its membrane proteins. Acta Cryst 69:812–820. https://doi.org/10.1107/S0907444913002102
Minamino T (2014) Protein export through the bacterial flagellar type III export pathway. Biochim Biophys Acta Mol Cell Res 1843:1642–1648. https://doi.org/10.1016/j.bbamcr.2013.09.005
Minamino T, Macnab RM (1999) Components of the Salmonella flagellar export apparatus and classification of export substrates. J Bacteriol 181:1388–1394
Minamino T, Macnab RM (2000a) FliH, a soluble component of the type III flagellar export apparatus of Salmonella, forms a complex with FliI and inhibits its ATPase activity. Mol Microbiol 37:1494–1503. https://doi.org/10.1046/j.1365-2958.2000.02106.x
Minamino T, Macnab RM (2000b) Domain structure of Salmonella FlhB, a flagellar export component responsible for substrate specificity switching. J Bacteriol 182:4906–4914
Minamino T, Namba K (2008) Distinct roles of the FliI ATPase and proton motive force in bacterial flagellar protein export. Nature 451:485–488. https://doi.org/10.1038/nature06449
Minamino T, Yoshimura SDJ, Morimoto YV et al (2009) Roles of the extreme N-terminal region of FliH for efficient localization of the FliH-FliI complex to the bacterial flagellar type III export apparatus. Mol Microbiol 74:1471–1483. https://doi.org/10.1111/j.1365-2958.2009.06946.x
Minamino T, Shimada M, Okabe M et al (2010) Role of the C-terminal cytoplasmic domain of FlhA in bacterial flagellar type III protein export. J Bacteriol 192:1929–1936. https://doi.org/10.1128/JB.01328-09
Minamino T, Morimoto YV, Hara N, Namba K (2011) An energy transduction mechanism used in bacterial flagellar type III protein export. Nat Commun 2:475. https://doi.org/10.1038/ncomms1488
Minamino T, Morimoto YV, Hara N et al (2016) The bacterial flagellar type III export gate complex is a dual fuel engine that can use both H+ and Na+ for flagellar protein export. PLoS Pathog 12:1–18. https://doi.org/10.1371/journal.ppat.1005495
Monjarás Feria JV, Lefebre MD, Stierhof YD et al (2015) Role of autocleavage in the function of a type III secretion specificity switch protein in Salmonella enterica serovar Typhimurium. MBio 6:e01459-15. https://doi.org/10.1128/mbio.01459-15
Morimoto YV, Ito M, Hiraoka KD et al (2014) Assembly and stoichiometry of FliF and FlhA in Salmonella flagellar basal body. Mol Microbiol 91:1214–1226. https://doi.org/10.1111/mmi.12529
Mouslim C, Hughes KT (2014) The effect of cell growth phase on the regulatory cross-talk between flagellar and Spi1 virulence gene expression of FlhD 4 C 2 transcriptional activation into an early cell growth phase role for flagellar production from a late cell growth phase role in virulence gene expression. PLoS Pathog 10. https://doi.org/10.1371/journal.ppat.1003987
Paul K, Erhardt M, Hirano T et al (2008) Energy source of flagellar type III secretion. Nature 451:489–492. https://doi.org/10.1038/nature06497
Portaliou AG, Tsolis KC, Loos MS et al (2016) Type III secretion: building and operating a remarkable nanomachine from essential to specialized bacterial protein secretion systems. Trends Biochem Sci 41:175–189. https://doi.org/10.1016/j.tibs.2015.09.005
Portaliou AG, Tsolis KC, Loos MS et al (2017) Hierarchical protein targeting and secretion is controlled by an affinity switch in the type III secretion system of enteropathogenic *Escherichia coli*. EMBO J 36:3517–3531. https://doi.org/10.15252/embj
Renault TT, Abraham AO, Bergmiller T et al (2017) Bacterial flagella grow through an injection-diffusion mechanism. Elife 6:1–22. https://doi.org/10.7554/elife.23136
Saijo-Hamano Y, Imada K, Minamino T et al (2010) Structure of the cytoplasmic domain of FlhA and implication for flagellar type III protein export. Mol Microbiol 76:260–268. https://doi.org/10.1111/j.1365-2958.2010.07097.x

Singer HM, Erhardt M, Hughes KT (2014) Comparative analysis of the secretion capability of early and late flagellar type III secretion substrates. Mol Microbiol 93:505–520. https://doi.org/10.1111/mmi.12675

Terashima H, Kawamoto A, Tatsumi C et al (2018) In vitro reconstitution of functional type III protein export and insights into flagellar assembly. MBio 9. https://doi.org/10.1128/mbio.00988-18

Turner L, Stern AS, Berg HC (2012) Growth of flagellar filaments of *Escherichia coli* is independent of filament length. J Bacteriol 194:2437–2442. https://doi.org/10.1128/JB.06735-11

Ward E, Renault TT, Kim EA et al (2018) Type-III secretion pore formed by flagellar protein FliP. Mol Microbiol 107:94–103. https://doi.org/10.1111/mmi.13870

Wilharm G, Lehmann V, Krauss K et al (2004) Yersinia enterocolitica type III secretion depends on the proton motive force but not on the flagellar motor components MotA and MotB. Infect Immun 72:4004–4009. https://doi.org/10.1128/IAI.72.7.4004-4009.2004

Xing Q, Shi K, Portaliou A et al (2018) Structures of chaperone-substrate complexes docked onto the export gate in a type III secretion system. Nat Commun 9. https://doi.org/10.1038/s41467-018-04137-4

Yu XJ, Grabe GJ, Liu M et al (2018) Ssav interacts with SsaL to control the translocon-to-effector switch in the salmonella SPI-2 type three secretion system. MBio 9. https://doi.org/10.1128/mbio.01149-18

Zhou J, Sharp LL, Tang HL et al (1998) Function of protonatable residues in the flagellar motor of *Escherichia coli*: a critical role for Asp 32 of MotB. J Bacteriol 180:2729–2735

Zilkenat S, Franz-Wachtel M, Stierhof Y-D et al (2016) Determination of the stoichiometry of the complete bacterial type III secretion needle complex using a combined quantitative proteomic approach. Mol Cell Proteomics 15:1598–1609. https://doi.org/10.1074/mcp.M115.056598

Flagellar Hook/Needle Length Control and Secretion Control in Type III Secretion Systems

Shin-Ichi Aizawa

Contents

Abstract The flagellum is a motile organ, and the needle complex is a type III secretion apparatus for pathogenesis. There are more similarities than differences between the two structures at the molecular level. Here I focus on the hook and the needle and discuss their length control mechanism. The hook is a substructure of the flagellum and the needle is a part of the needle complex. Both structures are tubular structures that have a central channel for protein secretion. Their lengths are controlled by an intriguing mechanism involving a ruler protein and a switchable gate of the protein secretion system. A model for length control is proposed.

S.-I. Aizawa (✉)
Prefectural University of Hiroshima, Hiroshima, Japan
e-mail: aizawa@pu-hiroshima.ac.jp

Current Topics in Microbiology and Immunology (2020) 427: 161–172
https://doi.org/10.1007/82_2019_169
© Springer Nature Switzerland AG 2019
Published Online: 11 June 2019

1 Introduction

Flagellum is a motile apparatus (Macnab 2003; Aizawa 2013, 2014), and the needle is used in pathogenesis (Kubori et al. 1998, 2000). Both supramolecular structures secrete proteins in a unique manner; they secrete proteins as they are without modification and are called the type III secretion system. The secretion apparatus is used for the formation of extracellular flagellar filaments, or for injection of effectors to the host cells. The construction mechanism of the secretion apparatus shares much more common features than differences between the two apparatuses (Kawamoto et al. 2013; Zhong et al. 2012).

Here I am going to discuss length control of the flagellar hook and the needle, but before that I want to describe the construction steps of the flagellum and the needle complex, because this length control mechanism is a complicated and thus intriguing phenomenon including secretion process. Since research on the flagella has a long history and precedes that on the needle complex, I will describe the flagellum first and then the needle complex in comparison with the flagellum in each section.

Flagellum is a gigantic protein complex composed of ca. 20 proteins. Ten of them are used for construction of the extracellular part of the flagellum (the rod–hook–filament) and thus must be secreted from the cytoplasm. The other ten proteins are localized in the cytoplasm and/or in the inner membrane and directly or indirectly serve in the secretion event. It is often convenient to use names of substructures rather than protein names derived from the gene names. They are C ring, MS ring, rod, hook, and filament, from those in the cytoplasm to outside the cell (Fig. 1). Proteins required for motor function are not included in the flagellum but localized in the periphery. Without motor function, flagellum is still functional as a secretion apparatus. Here, I will not discuss about the motor proteins of the flagellum.

The needle complex is an injection apparatus of pathogenic effectors into the eukaryotic host cells (Galan and Collmer 1999). The nomenclature of protein components of the needle complex has been different from species to species for a long time, but now the unified names (Sct) are proposed and being used (Diepold and Wagner 2014; Wagner et al. 2018). In *Salmonella*, there are even two kinds of needle complexes: SPI1 and SPI2. Since SPI1 needle complex was first discovered prior to the others in *Yersinia*, *Shigella*, *EPEC,* and so on, I will mainly employ the conventional nomenclature for SPI1 needle complex to avoid confusion. The SPI1 needle complex is a gigantic protein complex composed of ca. 16 proteins. Although the needle complex resembles the flagellum in appearance, terminology for substructures of the needle complex is a little different from that of the flagellum.

2 Morphology of the Flagella/Needle Structure

2.1 Components Outside the Cell

Extracellular substructures of the flagellum are mostly filamentous—the rod, the hook, and the filament. Lengths of the rod and the hook are regulated to be constant, but filament length varies depending on the physiological conditions of growth. The filament distally grows; new subunits (flagellins) are added at the distal tip of a pre-existing filament, and thus it is called distal growth. A protein complex called the cap helps polymerization of flagellins at the growing tip; without the cap, flagellins leak as monomer into the media. The hook and the rod grow in the same manner.

The needle part of the needle complex is a straight filament with a constant length. It is believed to be a counterpart of the hook from its position in the structure, but it could correspond to the rod (see next section). It was recently shown that a cap exists in the needle growth (Kato et al. 2018). At the tip of the mature needle, a protein complex called translocon or translocase is attached and is deployed to make a hole in the plasma membrane of the host cell.

2.2 Periplasmic Components

The periplasmic space, a space between the inner membrane and outer membrane, is a unique place where a group of sugar-carrier proteins works and is a gap for a flagellum to inevitably cross. The outer membrane provides a rigid ground for the flagellum to rotate and exert propelling force by the helical filament. The rod consists of two parts, the proximal and distal rod, and passes through the gap to keep the distance constant. The proximal rod is composed of four proteins (FliE, FlgB, FlgC, and FlgF), and the distal rod is composed of a single protein, FlgG. FlgG is the major component of the rod, and the rest are minor components made of only five or six subunits (Jones et al. 1990). Rod length is limited by the width of the periplasmic space, which is determined by the length of Braun's lipoprotein to tether the outer membrane to the peptidoglycan layer (Cohen et al. 2017).

Single-point mutations in FlgG give rise to an abnormally long rod. Length of the mutant rod is controlled by FliK in a similar manner as in the hook (Chevance et al. 2007). FliK is soluble and one of the proteins secreted during flagellum formation (Minamino et al. 1999). In the absence of FliK, the rod elongates to uncontrolled length (Williams et al. 1996). FlgG resembles FlgE in amino acid sequence. A sequence of 18 amino acid residues that exists in FlgG but not in FlgE seems important to give different physical properties to these similar structures. Insertion of the sequence in FlgE makes the hook straight like a rod (Hiraoka et al. 2017). These facts suggest that the hook might be developed from the outer rod in evolution.

In the needle complex, the rod is made of a few (not more than six) copies of only one protein, PrgJ (SctI), which is called the inner rod (Kubori et al. 2000, Zilkenat et al. 2016). The inner rod is a short structure corresponds to the flagellar proximal rod. Then, the needle (PrgI, SctF) could correspond to the distal rod but not the hook, as guessed from the homology between the rod and the hook just mentioned. There is another substructure composed of InvG (SctC), one of the secretin families, that covers the base of the needle and inner rod. It is not known if there are any specific roles of InvG during needle construction (Hu et al. 2018).

The flagellum exists in both Gram-negative and Gram-positive bacteria species. Flagella in a Gram-positive *Bacillus subtilis* smoothly rotate without so-called bushing in the cell wall; that is, the outer membrane is not essential for construction of intact flagella (Kubori et al. 1997). However, the needle complex exists only in Gram-negative species. Gram-positive species that retain the needle have not been found yet so far. There are no good reasons why Gram-positive species cannot have the needle and the type III secretion system.

2.3 Components in the Cytoplasm and the Inner Membrane

In the construction steps of the flagellum, which proteins come first? There are two candidates: (1) a ring structure called the MS ring made of a single kind of membrane protein FliF, and (2) the export gate made of FliP, FliQ, FliR, FlhA, and FlhB. The hypothesis (1) is derived from an experiment, in which overexpression of the *Salmonella fliF* gene in *Escherichia coli* results in overproduction of the MS rings in the inner membrane in the absence of all other flagellar proteins (Ueno et al. 1994). The hypothesis (2) claims that three membrane proteins (FliP/FliQ/FliR) form the secretion channel and then FliF assembles around the structure to form the MS ring. This question is still under debate.

Two more membrane components, FlhA and FlhB, are attached to the secretion channel. In contrast to the passive role of the channel proteins (FliP/FliQ/FliR), the roles of these two proteins are active and serve as the secretion gate to open and close the channel. Mutations in FlhA and FlhB affect switch mechanism, and thus, hook length will be changed. Beneath the MS ring, a cup-shaped structure is formed with three proteins (FliG/FliM/FliN) and called the C (from cytoplasmic) ring. This structure is multi-functional; it is necessary for secretion; it works as the switch of rotational direction of the motor; and it is the rotor (Yamaguchi et al. 1986; Macnab 2003). The C ring, a mysterious structure, is also involved in hook length control as will be shown later.

In the needle complex, PrgK (SctJ) together with PrgH (SctD) forms a ring structure (the inner ring) similar to the MS ring. The cytoplasmic domain of PrgH connects to the sorting platform (composed of SpaO (SctQ)/OrgA (SctK)/OrgB (SctL)), which is corresponding to the C ring. The sorting platform is much smaller than the C ring, since the latter retains a large extra domain for motor and switching function. The central channel components (SpaP (SctR)/SpaQ (SctS)/SpaR (SctT)) are highly homologous with FliP/FliQ/FliR in the amino acid sequence (Hueck 1998). Two other proteins, InvA (SctV) and SpaS (SctU), form the secretion gate and also highly homologous with FlhA and FlhB, respectively. This high homology between these membrane proteins is widespread among other species having the type III secretion systems, suggesting that the flagellum and the needle complex might be come from the same origin in evolution.

3 Hook/Needle Length

3.1 Length Measurements

Accurate measurements of length are essential for discussing length control. Measurement of hook length is technically not easy, because the hook is much shorter than the filament, and because the helical curvature is much large than that of the filament. We usually purify the whole flagellar structure and prepare the hook–basal body (HBB). We also have to make hook shape straight by low temperature and by acidic stain solution (Hirano et al. 1994).

For the needle, there is no such trouble because it is straight. But it is still necessary to purify the basal body attached with needles for accurate measurements. In *Yersinia*, the needles are spontaneously released from the basal body, and thus even the purification step can be skipped.

3.2 The Average and the Standard Deviation

Giving only the average value of hook length sometimes misleads readers. In the previously published literatures, we often come across a sentence like "hook length is controlled at 55 nm," which gives a strong impression that hook length is tightly controlled by a ruler. This is not true. Hook length is average 55 nm with the standard deviation 6 nm. The standard deviation (10% of the average) tells us how tightly or loosely hook length is controlled and thus very important to elucidate the mechanism of length control (Aizawa 2013, 2014; Hughes 2012a, b). The standard deviation of needle length is also about 10% of the average length. As we will see later, flagellum/needle length is not tightly controlled by a ruler but a consequence of a sporadic event.

The first example of length determination mechanism using molecular ruler was shown with tobacco mosaic virus (TMV). The component proteins (capsid proteins) polymerize into a tubular rod by tightly binding with its viral RNA. The defined length of RNA determines the defined length of the rod at 300 ± 5 nm. The standard deviation of the length distribution of the rod was 1.7% of the average length, which is presumably derived from measurement error. Thus, we can say that the rod length is rather accurately determined by a ruler molecule RNA. In contrast, hook length is loosely controlled. We can find very long or short hooks, albeit few, in a population of the hooks in wild-type strain. It should be noted that TMV particles self-assemble within the cell, whereas the hook is formed outside the cell. Thus, the hook component FlgE molecules are transported one by one from the cytoplasm to the assembly site through the central channel of pre-existing structures (the rod and the hook itself). Therefore, it is natural to assume that hook length could be controlled at the base.

3.3 Mutants with Deviated Hook Length

Flagellar length varies from species to species. In a simple case where a flagellum consists of a single kind of flagellin, length is determined by the amount of flagellin expressed. Overexpression of the flagellin gene results in a longer flagellum or flagella. However, hook length is determined by a different mechanism. Overexpression of the hook protein (FlgE) did not change hook length (Muramoto et al. 1998). Overproduction of FliK did not affect hook length (Muramoto et al. 1999). Deletion mutants of FlgK or FlgL that interact with the hook distal tip did not change hook length. In contrast, the component proteins of the C ring affect hook length. Mutations in switch proteins (FliG/FliM/FliN) gave rise to short hooks with a defined length (Makishima et al. 2001). Deletion mutants of these genes can still produce hooks with filaments, if FliI, an ATPase, is overproduced (Konishi et al. 2009). But lengths of the hooks formed under the artificial conditions are widely distributed, indicating that the C ring is required for the tight regulation of hook length. Components of the secretion gate, FlhA and FlhB, also affect hook length. Point mutants of FlhA or FlhB give rise to polyhooks (Barker et al. 2016; Williams et al. 1996). In short, proteins neighboring to the hook structure do not affect hook length, while proteins locating at the base and involved in secretion event do affect hook length.

The needle consists of a single kind of protein PrgI(SctF) (80 aa), and needle length is controlled by InvJ (SctP) (336 aa), which is homologous with FliK (402 aa). The molecular length of InvJ is proportional to needle length (Wee and Hughes 2015). Polyneedles were also found in the *spa32* (*sctP*) deletion mutant in *Shigella* species and in the *yscP* (*sctP*) deletion mutant in *Yersinia* species, suggesting that needle length is controlled by a general mechanism using FliK homologues. The length control mechanism of the hook and the needle seems to be, if not identical, very similar.

4 A Ruler Protein SctP/FliK

4.1 Ruler

A direct evidence of a molecular ruler was first shown in the needle complex of *Yersinia* species. A *fliK* homologue in *Yersinia* species is *yscP* (*sctP*). Analysis of mutants having a set of truncated *yscP* genes showed that needle length was linearly proportional to the molecular sizes of the truncated YscP (Journet et al. 2003).

We repeated the same experiment using flagellar system and obtained the same results; hook length was linearly proportional to the molecular length of engineered FliK (Shibata et al. 2007). FliK is not included in the flagellar structure but secreted

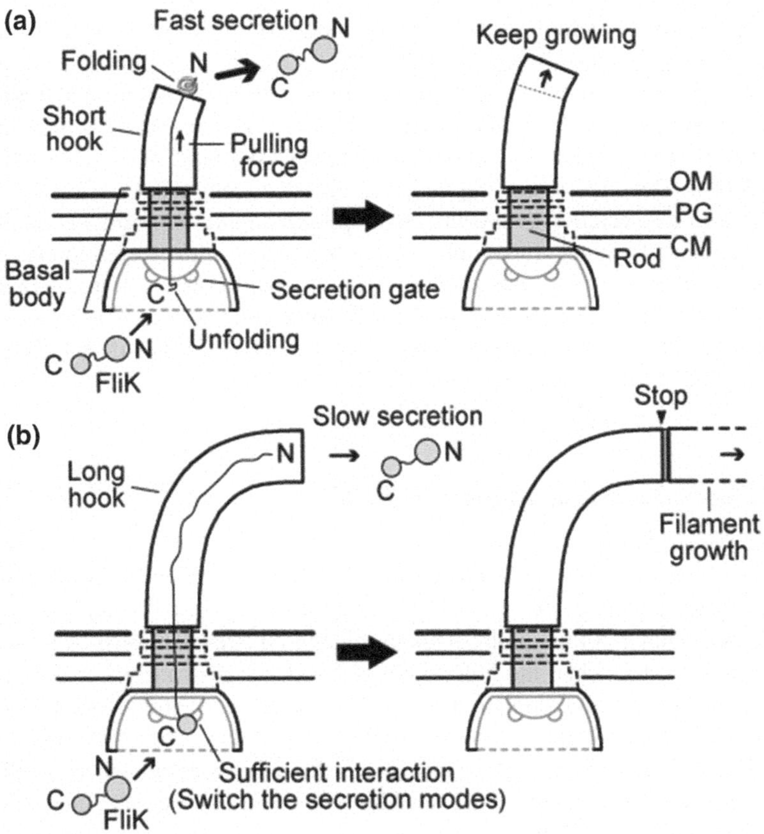

Fig. 1 A model of length control by FliK and switchable secretion gate. **a** When hook length is short, the FliK N-terminal region goes out of the hook channel and folds into a ball before the C-terminal domain reaches at the secretion gate. The ball quickly moves away from the hook tip by Brownian motion and pulls the rest of FliK molecule, which leaves little time for the C-terminal domain to interact with the secretion gate. The hook keeps growing till **b** occurs. **b** When hook length is long, FliK molecule moves slowly in the hook channel, which allows enough time for the C-terminal domain to interact with the secretion gate and to switch the secretion modes. After the secretion mode changed, hook elongation stops and filament growth starts

into the medium until the hook elongates to a certain length (Minamino et al. 1999). Therefore, FliK is not a ruler in a conventional sense. We look into the molecular structure of FliK in detail.

4.2 Two Separate Functions of FliK

Historically speaking, genetic analysis of *fliK* mutants precedes structural analysis of FliK. Many spontaneous and genetically engineered mutants were analyzed, and we learned that the N-terminal region is necessary for determining hook length, and that the C-terminal region is essential for switching the secretion modes of substrates (Williams et al. 1996; Minamino et al. 2004, 2006; Hirano et al. 2005). As mentioned above, switching of the secretion modes of the secretion gate is the key event to form a complete flagellum. When the C-terminal region is deleted, hooks keep growing to be polyhooks. However, even when the N-terminal region is deleted, as long as the C-terminal domain is intact, a filament is formed onto a polyhook, which is called the polyhook–filament (Williams et al. 1996; Hirano et al. 2005). The N-terminal region goes out first, and the C-terminal region interacts with the secretion gate (FlhB) to stop FlgE supply (Kinoshita et al. 2017). Therefore, the C-terminal domain plays the primary role in the secretion event, and the N-terminal region plays the secondary role in control of secretion. In *Yersinia*, the YscP-YscU (corresponding to FliK-FlhB) interaction was shown by means of nuclear magnetic resonance spectroscopy (Ho et al. 2017). It is still not clear how the C ring affects the interaction between FliK and FlhB.

4.3 The N-Terminal Region of FliK

How does FliK molecule sense hook length? FliK is infrequently secreted during hook elongation (Minamino et al. 1999; Erhardt et al. 2011). When transported through the channel, a FliK molecule is unfolded in a stretched form. A prevailing hypothesis says that when the N-terminus reached the growing tip, it transmits a signal to the C-terminal domain through the stretched peptide, and the signal enhances the interaction between the C-terminal domain and FlhB (Minamino et al. 2009). But this hypothesis had a problem; if the hook was elongated a little bit longer than the average length, FliK would miss a chance to interact with FlhB. An experiment solved this problem; when FliK on an arabinose-induced plasmid was expressed in a polyhook (deletion of *fliK*) mutant, all polyhooks on a cell instantly produced filaments by arabinose addition (Uchida et al. 2014). This indicates that FliK C-terminal domain almost always interacts with FlhB and switching occurs when hook length is longer than the predetermined length. Then we have to elucidate how FliK passes through the gate without switching to occur when hook length is still shorter than the predetermined length.

4.4 *The Molecular Structure of FliK*

FliK is a highly soluble protein consisting of 405 amino acids. Secondary structure prediction indicates that the N-terminal half of FliK is non-structural and the C-terminal half has a compact domain. NMR study on FliK in solution showed that this prediction was largely correct; the N-terminal half was non-structured, and the C-terminal has a compact domain composed of two alpha-helices and four beta-sheets (Mizuno et al. 2011). However, a high-speed AFM, which was developed for the direct observation of a single particle in solution, showed that N-terminal region actually takes on a ball shape, which is a little larger than the ball shape of the C-terminal domain (Kodera et al. 2015). What is the role of the ball shape in the N-terminal region?

4.5 *A Model for FliK Function in Length Control*

In the hook elongation process, FliK N-terminus goes first into the channel, and the C-terminal domain remaining in the cytoplasm still retains a compact structure for some period of time. The interaction between the C-terminal domain and the secretion gate occurs in a brief time, because the C-terminal domain will, sooner or later, be unfolded as FliK molecule goes out the channel into the medium.

As mentioned above when the hook elongates longer than the predetermined length, the N-terminal region will stay longer in time in the channel of the hook as a stretched peptide chain, allowing the FliK C-terminal domain to keep its intact globular shape, thus increasing the chance of its interaction with the secretion gate. When the hook is still short, as soon as FliK N-terminus reaches the tip of a growing hook, it will start folding into a ball shape. Once the N-terminal region folds into a ball, the ball will quickly move away from the tip by the Brownian motion. The quick exit of the N-terminal ball will rapidly pull the rest of the molecule in the channel of the hook to unfold the C-terminal domain, leaving a small chance for the C-terminal domain to interact with the secretion gate (Fig. 1).

Statistical analysis of length distribution of polyhooks (hooks grown in the absence of FliK) indicates that the hook grows very rapidly until it reaches a peak near the predetermined length. After the peak, the hook will grow at a slower but steady speed (Koroyasu et al. 1998). This kinetic change of growth speed will give more chances for FliK to interact with the secretion gate at the proper length of the hook, even though FliK is infrequently secreted (Erhardt et al. 2011; Ho et al. 2017).

5 Conclusions

After all, controlling hook/needle length is controlling amounts of the hook/needle protein to be secreted by the secretion gate (FlhA and FlhB) in cooperation with FliK. FlhB/SctU autocleavage is essential for this event to occur (Ferris et al. 2005; Monjaras et al. 2015). There are some other proteins such as the C ring that play secondary but important roles in regulating punctual switching of the secretion gate.

Deletion of the ruler results in polyhooks and polyneedles in corresponding species. Cells with polyhooks or polyneedles are deteriorated in function; polyhooks do not convey torque properly to the filament; cells with polyneedles are less infectious. In short, length of the hook and the needle is controlled to a certain value for exerting the optimal function.

References

Aizawa SI (2012a) Mystery of FliK in length control of the flagellar hook. J Bacteriol 194:4798–4800

Aizawa SI (2012b) Rebuttal: flagellar hook length is controlled by a secreted molecular ruler. J Bacteriol 194:4797

Aizawa SI (2013) Flagella. In: Schaechter M (ed) Brennan's online encyclopedia of genetics, 2nd ed. Elsevier, Academic Press, Oxford

Aizawa SI (2014) The flagellar World. Elsevier, Oxford

Barker CS, Inoue T, Meshcheryakova IV, Kitanobo S, Samatey FA (2016) Function of the conserved FHIPEP domain of the flagellar type III export apparatus, protein FlhA. Mol Microbiol 100:278–288

Chevance FF, Takahashi N, Karlinsey JE, Gnerer J, Hirano T, Samudrala R, Aizawa SI, Hughes KT (2007) The mechanism of outer membrane penetration by the eubacterial flagellum and implications for spirochete evolution. Genes Dev 21:2326–2335

Cohen EJ, Ferreira JL, Ladinsky MS, Beeby M, Hughes KT (2017) Nanoscale-length control of the flagellar driveshaft requires hitting the tethered outer membrane. Science (New York, NY) 356(6334):197–200

Diepold A, Wagner S (2014) Assembly of the bacterial type III secretion machinery. FEMS Microbiol Rev 38(4):802–822

Erhardt M, Singer HM, Wee DH, Keener JP, Hughes KT (2011) An infrequent molecular ruler controls flagellar hook length in *Salmonella enterica*. EMBO J 30:2948–2961

Ferris HU, Furukawa Y, Minamino T, Kroetz MB, Kihara M, Namba K, Macnab RM (2005) FlhB regulates ordered export of flagellar components via autocleavage mechanism. J Biol Chem 280(50):41236–41242

Galan JE, Collmer A (1999) Type III secretion machines: bacterial devices for protein delivery into host cells. Science 284:1322–1328

Hirano T, Yamaguchi S, Oosawa K, Aizawa SI (1994) Roles of FliK and FlhB in determination of flagellar hook length in *Salmonella typhimurium*. J Bacteriol 176:5439–5449

Hirano T, Shibata S, Ohnishi K, Tani T, Aizawa SI (2005) N-terminal signal region of FliK is dispensable for length control of the flagellar hook. Mol Microbiol 56:346–360

Hiraoka KD, Morimoto YV, Inoue Y, Fujii T, Miyata T, Makino F, Minamino T, Namba K (2017) Straight and rigid flagellar hook made by insertion of the FlgG specific sequence into FlgE. Sci Rep 7:46723

Ho O, Rogne P, Edgren T, Wolf-Watz H, Login FH, Wolf-Watz M (2017) Characterization of the ruler protein interaction interface on the substrate specificity switch protein in the *Yersinia* type III secretion system. J Biol Chem 292(8):3299–3311

Hu J, Worrall LJ, Hong C, Vuckovic M, Atkinson CE, Caveney N et al (2018) Cryo-EM analysis of the T3S injectisome reveals the structure of the needle and open secretin. Nat Commun 9 (1):3840

Hueck CJ (1998) Type III protein secretion systems in bacterial pathogens of animals and plants. Microbiol Mol Biol Rev 62:379–433

Hughes KT (2012a) Flagellar hook length is controlled by a secreted molecular ruler. J Bacteriol 194:4793–4796

Hughes KT (2012b) Rebuttal: Mystery of FliK in length control of the flagellar hook. J Bacteriol 194:4801

Jones CJ, Macnab RM, Okino H, Aizawa SI (1990) Stoichiometric analysis of the flagellar hook (basal body) complex of *Salmonella typhimurium*. J Mol Biol 212:377–387

Journet L, Agrain C, Broz P, Cornelis GR (2003) The needle length of bacterial injectisomes is determined by a molecular ruler. Science 302:1757–1760

Kato J, Dey S, Soto JE, Butan C, Wilkinson MC, De Guzman RN, Galán JE (2018) A protein secreted by the Salmonella type III secretion system controls needle filament assembly. eLife 7:6889

Kawamoto A, Morimoto YV, Miyata T, Minamino T, Hughes KT, Kato T, Namba K (2013) Common and distinct structural features of *Salmonella* injectisome and flagellar basal body. Sci Rep 3:3369

Kinoshita M, Aizawa SI, Inoue Y, Namba K, Minamino T (2017) The role of intrinsically disordered C-terminal region of FliK in substrate specificity switching of the bacterial flagellar type III export apparatus. Mol Microbiol 105(4):572–588

Kodera N, Uchida K, Ando T, Aizawa SI (2015) Two-ball structure of the flagellar hook-length control protein FliK as revealed by high-speed atomic force microscopy. J Mol Biol 427:406–414

Konishi M, Kanbe M, McMurry JL, Aizawa SI (2009) Flagellar formation in C-ring defective mutants by overproduction of FliI, the ATPase specific for the flagellar type III secretion. J Bacteriol 191:6186–6191

Koroyasu S, Yamazato M, Hirano T, Aizawa SI (1998) Kinetic analysis of the growth rate of the flagellar hook in *Salmonella typhimurium* by the population balance method. Biophysical J 74:436–443

Kubori T, Okumura M, Kobayashi N, Nakamura D, Iwakura M, Aizawa SI (1997) Purification and characterization of the flagellar hook-basal body complex of *Bacillus subtilis*. Mol Microbiol 24:399–410

Kubori T, Matsushima Y, Nakamura D, Urali J, Lara-Tejero M, Sukhan A, Galan JE, Aizawa SI (1998) Supremolecular structure of the *Salmonella Typhimurium* type III protein secretion system. Science 280:602–605

Kubori T, Aukhan A, Aizawa SI, Galan JE (2000) Molecular characterization and assembly of the needle complex of the *Salmonella typhimurium* type III protein secretion system. Proc Natl Acad Sci USA 97:10225–10230

Macnab RM (2003) How bacteria assemble flagella. Annu Rev Microbiol 57:77–100

Makishima S, Komoriya K, Yamaguchi S, Aizawa SI (2001) Length of the flagellar hook and the capacity of the type III export apparatus. Science 291:2411–2413

Minamino T, González-Pedrajo B, Yamaguchi K, Aizawa S, Macnab RM (1999) FliK, the protein responsible for flagellar hook length control in Salmonella, is exported during hook assembly. Mol Microbiol 34:295–304

Minamino T, Saijo-Hamano Y, Furukawa Y, González-Pedrajo B, Macnab RM, Namba K (2004) Domain organization and function of Salmonella FliK, a flagellar hook-length control protein. J Mol Biol 341:491–502

Minamino T, Ferris HU, Moriya N, Kihara M, Namba K (2006) Two parts of the T3S4 domain of the hook-length control protein FliK are essential for the substrate specificity switching of the flagellar type III export apparatus. J Mol Biol 362:1148–1158

Minamino T, Moriya N, Hirano T, Hughes KT, Namba K (2009) Interaction of FliK with the bacterial flagellar hook is required for efficient export specificity switching. Mol Microbiol 74:239–251

Mizuno S, Amida H, Kobayashi N, Aizawa SI, Tate S (2011) The NMR structure of FliK, the trigger for the switch of substrate specificity in the flagellar type III secretion apparatus. J Mol Biol 409:558–573

Monjarás Feria JV, Lefebre MD, Stierhof YD, Galán JE, Wagner S (2015) Role of autocleavage in the function of a type III secretion specificity switch protein in *Salmonella enterica* serovar Typhimurium. mBio 6(5):e01459–15

Muramoto K, Makishima S, Aizawa SI, Macnab RM (1998) Effect of cellular level of FliK on flagellar hook and filament assembly in *Salmonella typhimurium*. J Mol Biol 277:871–882

Muramoto K, Makishima S, Aizawa SI, Macnab RM (1999) Effect of hook subunit concentration on assembly and control of length of the flagellar hook of *Salmonella*. J Bacteriol 181:5808–5813

Shibata S, Takahashi N, Chevance FF, Karlinsey JE, Hughes KT, Aizawa SI (2007) FliK regulates flagellar hook length as an internal ruler. Mol Microbiol 64:1404–1415

Uchida K, Aizawa SI (2014) The flagellar soluble protein FliK determines the minimal length of the hook in *Salmonella enterica* serovar Typhimurium. J Bacteriol 196:1753–1758

Ueno T, Oosawa K, Aizawa SI (1994) Domain structures of the MS ring component protein (FliF) of the flagellar basal body of *Salmonella typhimurium*. J Mol Biol 236:546–555

Wagner S, Grin I, Malmsheimer S, Singh N, Torres-Vargas CE, Westerhausen S (2018) Bacterial type III secretion systems: a complex device for the delivery of bacterial effector proteins into eukaryotic host cells. FEMS Microbiol Lett 365. https://doi.org/10.1093/femsle/fny201

Wee DH, Hughes KT (2015) Molecular ruler determines needle length for the *Salmonella* Spi-1 injectisome. Proc Natl Acad Sci USA 112(13):4098–4103

Williams AW, Yamaguchi S, Togashi F, Aizawa SI, Kawagishi I, Macnab RM (1996) Mutations in *fliK* and *flhB* affecting flagellar hook and filament assembly in *Salmonella typhimurium*. J Bacteriol 178:2960–2970

Yamaguchi S, Fujita H, Ishihara H, Aizawa SI, Macnab RM (1986) Subdivision of flagellar genes of *Salmonella typhimurium* into regions responsible for assembly, rotation, and swithcing. J Bacteriol 166:187–193

Zhong D, Lefebre M, Kaur K, McDowell MA, Gdowski C, Jo S, Wang Y, Benedict SH, Lea SM, Galan JE, De Guzman RN (2012) The *Salmonella* type III secretion system inner rod protein PrgJ is partially folded. J BC 287:25303–25311

Zilkenat S, Franz-Wachtel M, Stierhof YD, Galán JE, Macek B, Wagner S (2016) Determination of the stoichiometry of the complete bacterial type III secretion needle complex using a combined quantitative proteomic approach. Mol Cell Proteomics 15(5):1598–1609

The Tip Complex: From Host Cell Sensing to Translocon Formation

William D. Picking and Michael L. Barta

Contents

Abstract Type III secretion systems are used by some Gram-negative bacteria to inject effector proteins into targeted eukaryotic cells for the benefit of the bacterium. The type III secretion injectisome is a complex nanomachine comprised of four main substructures including a cytoplasmic sorting platform, an envelope-spanning basal body, an extracellular needle and an exposed needle tip complex. Upon contact with a host cell, secretion is induced, resulting in the formation of a translocon pore in the host membrane. Translocon formation completes the conduit needed for effector secretion into the host cell. Control of type III secretion occurs in response to environmental signals, with the final signal being host cell contact. Secretion control

W. D. Picking (✉)
Department of Pharmaceutical Chemistry, University of Kansas, 2030 Becker Drive, Lawrence 66047, KS, USA
e-mail: picking@ku.edu

M. L. Barta
Higuchi Biosciences, 2099 Constant Ave., Lawrence 66047, KS, USA

Present Address:
M. L. Barta
Catalent Pharma Solutions, 10245 Hickman Mills Drive, Kansas City 64137, MO, USA

Current Topics in Microbiology and Immunology (2020) 427: 173–199
https://doi.org/10.1007/82_2019_171
© Springer Nature Switzerland AG 2019
Published Online: 06 June 2019

occurs primarily at two sites—the cytoplasmic sorting platform, which determines secretion hierarchy, and the needle tip complex, which is critical for sensing and responding to environmental signals. The best-characterized injectisomes are those from *Yersinia*, *Shigella* and *Salmonella* species where there is a wealth of information on the tip complex and the two translocator proteins. Of these systems, the best characterized from a secretion regulation standpoint is *Shigella*. In the *Shigella* system, the tip complex and the first secreted translocon both contribute to secretion control and, thus, both are considered components of the tip complex. In this review, all three of these type III secretion systems are described with discussion focused on the structure and formation of the injectisome tip complex and what is known of the transition from nascent tip complex to assembled translocon pore.

1 Introduction

The importance of type III secretion systems (T3SS) for the pathogenesis of Gram-negative human pathogens was acknowledged before they were recognized as actual secretion systems and long before visualization of the type III secretion nanomachine (the injectisome). The effects of calcium as it is related to the V antigen in *Yersinia* (*Pasteurella*) *pestis* were described more than a half-century ago (Bacon and Burrows 1956; Lawton et al. 1963) with similar observations in *Yersinia pseudotuberculosis* (Burrows and Bacon 1960). The observed calcium effects came to be known as the low calcium response (LCR) and it was found to have profound effects on the expression of the *Yersinia* V antigen. For the LCR, it was determined that in the absence of calcium *Y. pestis* growth was arrested at 37 °C, with no such growth arrest at 26 °C (Kupferberg and Higuchi 1958). This effect was reversed by including calcium at millimolar concentrations in the growth medium, restoring production of the *Yersinia* W and V antigens (Brubaker and Surgalla 1964; Lawton et al. 1963). Excitingly, the V antigen had been demonstrated to provide protection against bacterial challenge (Lawton et al. 1963), however, it was not immediately understood why or how this secreted antigen was protective. The V antigen later came to be known as LcrV and it was found to be an essential virulence factor for *Y. pestis* as well as other *Yersinia* species known to infect humans (i.e., *Y. enterocolitica* and *Y. pseudotuberculosis*) (Bhaduri et al. 1990; Goguen et al. 1986; Sample et al. 1987; Skrzypek and Straley 1995). Later studies described bacteria that had lost a large plasmid and become avirulent and no longer responded to the LCR (Ferber and Brubaker 1981). This long line of research would eventually lead investigators to identify the role of the plasmid-encoded *Yersinia* T3SS in evasion of host innate immunity (Michiels et al. 1990; Rosqvist et al. 1991) and the eventual study of the injectisome from both a functional and structural perspective. Injectisome needles were observed about a decade later (Hoiczyk and Blobel 2001) with higher resolution models of the *Yersinia* injectisome described shortly thereafter (Cornelis 2002). *Yersinia* continues to be a prime model system for studying type III secretion and it was in this system that identification of the first injectisome needle tip complex was made (described in more detail below).

Like *Yersinia*, *Shigella* species possess a plasmid that is essential for virulence (Sansonetti et al. 1981) and their ability to invade host cells within the gastrointestinal tract (Sansonetti et al. 1982). As with *Yersinia*, key *Shigella* antigens were found to be secreted into culture supernatants under conditions in which the virulence plasmid was active and the organism fully invasive (e.g., growth at 37 °C). In this case, the identified secreted proteins were called the invasion plasmid antigen (Ipa) proteins and they were considered to be potential protective antigens based on anecdotal information provided at that time (Buysse et al. 1987; Oaks et al. 1986). The first reports regarding the potentially complex structures of the *Shigella* injectisome structure were in 1999 (Blocker et al. 1999) with much more detailed structural information soon following (Blocker et al. 2001). Along with the *Yersinia* T3SS, this secretion system in *Shigella* has become a second model for improving our understanding of the structural and functional features of this important virulence determinant. Identification of the *Shigella* injectisome needle tip complex was reported shortly after identification of the analogous structure in *Yersinia* (Espina et al. 2006a). This is also discussed in further detail below.

A third pathogen that has become a paradigm for the study of type III secretion and injectisome structure–function relationships is *Salmonella enterica*. Unlike *Yersinia* and *Shigella*, *Salmonella* encodes many of its virulence determinants on pathogenicity islands (SPI) that are located chromosomally rather than on a large virulence plasmid. Two of these, SPI-1 and SPI-2, were found to encode T3SS that contribute to this pathogen's complex lifestyle as an extracellular and intracellular pathogen (Mills et al. 1995; Ochman et al. 1996; Shea et al. 1996). Kaniga and colleagues identified homologs of the secreted Ipa proteins from *Shigella* that were expressed from SPI-1, which were important for *Salmonella* entry into host cells (Kaniga et al. 1995a, b). This observation was followed up by the Galan and Miller groups who reported the first high-quality images of the *Salmonella* injectisome (Kimbrough and Miller 2000; Kubori et al. 1998). Taken together, the *Shigella*, *Yersinia* and *Salmonella* T3SS arguably represent the best-characterized T3SS with regard to our understanding of architecture, substructure and function. From these organisms, there is an evolving understanding of the assembly and dynamics of the individual components, including the most exposed portions of the apparatus, the needle and its associated tip complex (TC). The TC is the major focus of this review.

2 The Type III Secretion Apparatus or Injectisome (The Injectisome)

The injectisome is a complex nanomachine that functions through the combined actions of distinct substructures to promote the secretion of proteins (termed effectors) into target eukaryotic cells for the benefit of the infecting pathogen. These systems appear to be used exclusively by bacteria to communicate with eukaryotic cells for the benefit of the bacteria and have been identified in many plant- and animal-associated Gram-negative bacteria (Bergman et al. 1994; Galan et al. 1992; Gough et al. 1993; Wei and Beer 1993). It was initially unclear what types of structures were responsible for type III secretion, however, filamentous surface appendages (diameter of 6–8 nm) associated with pathogenesis were observed in the plant pathogen *Pseudomonas syringae* and called Hrp pili because of their roles in the plant hypersensitivity response and pathogenicity (Roine et al. 1997). Likewise, extracellular appendages that might be associated with type III secretion were identified in the human pathogen *Salmonella* following bacterial contact with host cells (Ginocchio et al. 1994). From various types of transmission electron micrograph analyses and high-resolution structure studies, we now have a reasonable understanding of the overall architecture of the injectisome (see Fig. 1) (Hu et al. 2015; Deng et al. 2017). While T3SS fall into broad groups that include multiple pathogen types (e.g., the Mxi/Spa/Ipa family includes *Salmonella* and *Shigella*), there are some general features common among all the known injectisomes. Each system contains a fixed basal body spanning the entire Gram-negative cell envelope (IM-cell wall-OM) (Marlovits et al. 2004). The basal body is sandwiched between an extracellular needle composed of a polymer of a small, helix-turn-helix needle protein (Cordes et al. 2005; Deane et al. 2006a; Demers et al. 2014; Fujii et al. 2012; Zhang et al. 2006; Wang et al. 2008) and a cytoplasmic sorting platform. The sorting platform is comprised of the export gate, an energy source (ATPase) and an associated hexameric unit that is the structural equivalent of the flagellar C ring (Hu et al. 2017; Lara-Tejero et al. 2011; Hu et al. 2015). Recognition of secretion substrates and determination of secretion hierarchy most likely occurs within this cytoplasmic portion of the injectisome. Induction of secretion occurs as a result of external signals such as changes in the bacterium's environment (e.g., host cell membrane contact, change in pH or changes extracellular calcium levels) which potentially implicates the needle and/or the complex of proteins located at the exposed end of the injectisome needle, the tip complex (TC), in sensing such signals. Once the environmental signals are encountered, however, the switch to effector secretion involves complexes that contain gatekeeper proteins such as SsaL for the SPI-2 injectisome of *Salmonella* (Yu et al. 2010, 2018), SepL in enteropathogenic *Escherichia coli* (Shaulov et al. 2017) or YopN in *Yersinia* (Bamyaci et al. 2018). The regulators of needle length have also been implicated in these switches to effector protein secretion. The focus here will be a description of the TC and a summary of how it may be involved in secretion control and sensing the environment.

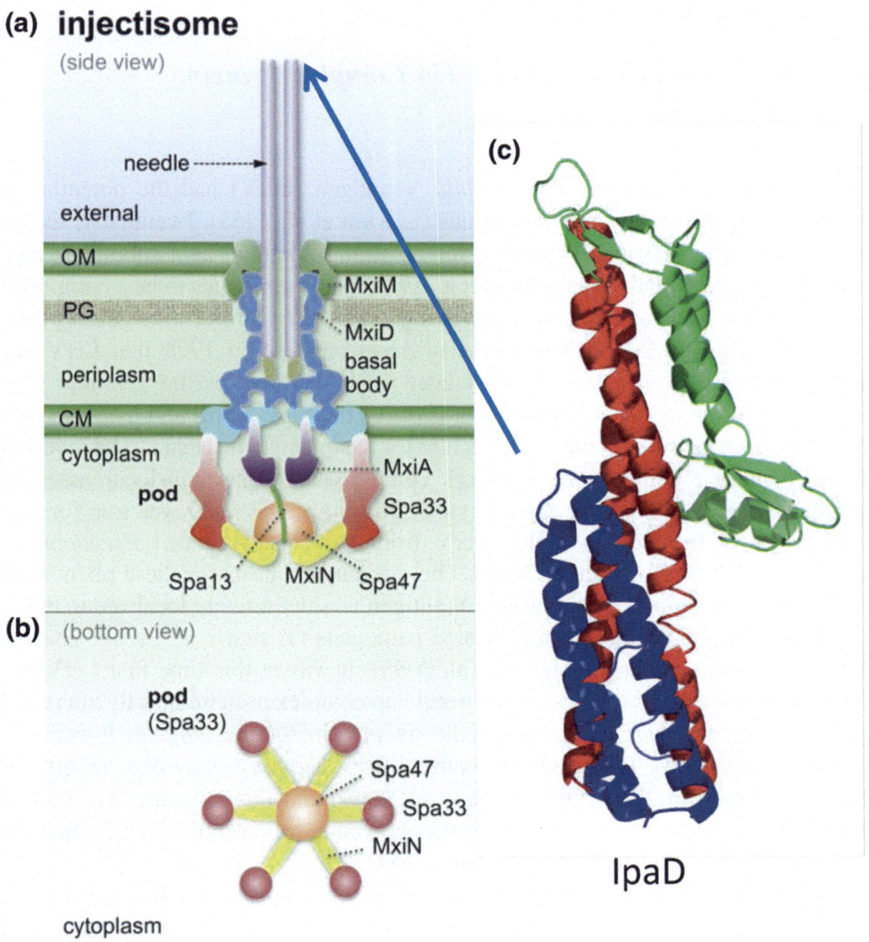

Fig. 1 The proposed architecture of the *Shigella* injectisome. In panel A, a side view of the injectisome is shown with the major basal body components indicated (in blue hues). Beneath the bacterial inner membrane is the cytoplasmic sorting platform with the main components indicated. At the top is the extracellular needle comprised of a polymer of MxiH. Panel B shows a bottom view of the sorting platform depicting the sixfold symmetry of this complex. The tip complex is not indicated here, but panel C shows the crystal structure of IpaD, which is the main component of the tip complex. IpaD resides as the most distal structure from the surface of the bacterium—presumably as a pentamer in the nascent injectisome. Panels A and B are adapted from (Hu et al. 2015). Copyright National Academy of Sciences. Panel C is PDB ID: 2J0O, (Johnson et al. 2007)

3 The Injectisome Needle Tip Complex

3.1 Discovery of the Nascent Tip Complex Protein and Overall Structure

It was noted long ago that the *Yersinia* V antigen (LcrV) had the potential for inducing protective immunity in animals (Lawton et al. 1963). Eventually, recombinant LcrV was specifically studied for its strong potential as a subunit vaccine component (Leary et al. 1995; Motin et al. 1994) and it continues to be a component of some of the lead candidate vaccines being explored for protection against plague (Verma and Tuteja 2016). The Cornelis group reported in 1998 that LcrV was required for proper secretion of what later were identified as the *Yersinia* T3SS translocator proteins (the Yersinia Outer Proteins or Yops) YopB and YopD (Sarker et al. 1998). Once the structural architecture for the injectisome began to be revealed, it was found that antibodies against LcrV could disrupt proper translocon assembly in targeted cells (in this case erythrocytes). Strikingly, the same was true for antibodies against the LcrV-homolog PcrV from the closely related *Pseudomonas aeruginosa* T3SS (Goure et al. 2005). The mechanistic basis for these phenomena did not become entirely clear until the V antigen was definitively localized to the tip of the injectisome needle in three related pathogens (*Y. pestis*, *P. aeruginosa* and *Aeromonas salmonicida*) (Mueller et al. 2005). It was at this time that LcrV was designated as the *Yersinia* injectisome needle tip complex protein. Shortly afterward, IpaD was identified as being the needle tip protein for the *Shigella* injectisome (Espina et al. 2006b). The crystal structure of LcrV had previously been reported to 2.2 Å resolution in 2004 (Derewenda et al. 2004) and this structure was used to model the LcrV tip protein as a pentameric needle "tip complex" (TC) atop the *Yersinia* YscF needle filament (Broz et al. 2007).

While advances were being made regarding LcrV as the *Yersinia* needle TC, parallel achievements were occurring for a needle TC from a distinct injectisome family. IpaD was identified as the *Shigella* needle tip protein in 2006 (Espina et al. 2006b) and its crystal structure was determined shortly thereafter (see Figs. 1 and 2), along with a TC homolog (BipD) from the same injectisome family and produced by *Burkholderia pseudomallei* (Erskine et al. 2006; Johnson et al. 2007). In addition to LcrV and IpaD, a number of homologous TC proteins have now been described and compared using biophysical analyses to identify their shared structural features (Deng et al. 2017; Espina et al. 2007; Sato and Frank 2011). One major feature of all the TC proteins characterized to date is the presence of an extended coiled-coil (green region within Fig. 2) that provides the scaffold upon which the rest of the protein is built. While the coiled-coil domain appears important for protein–protein interactions across all T3SS-possessing pathogens, the remainder of the TC protein structure appears to vary somewhat and may be adapted to more pathogen-specific functions.

The structural basis for the interaction between cognate needle tip protein and the actual needle proteins has been described for some of these systems, most

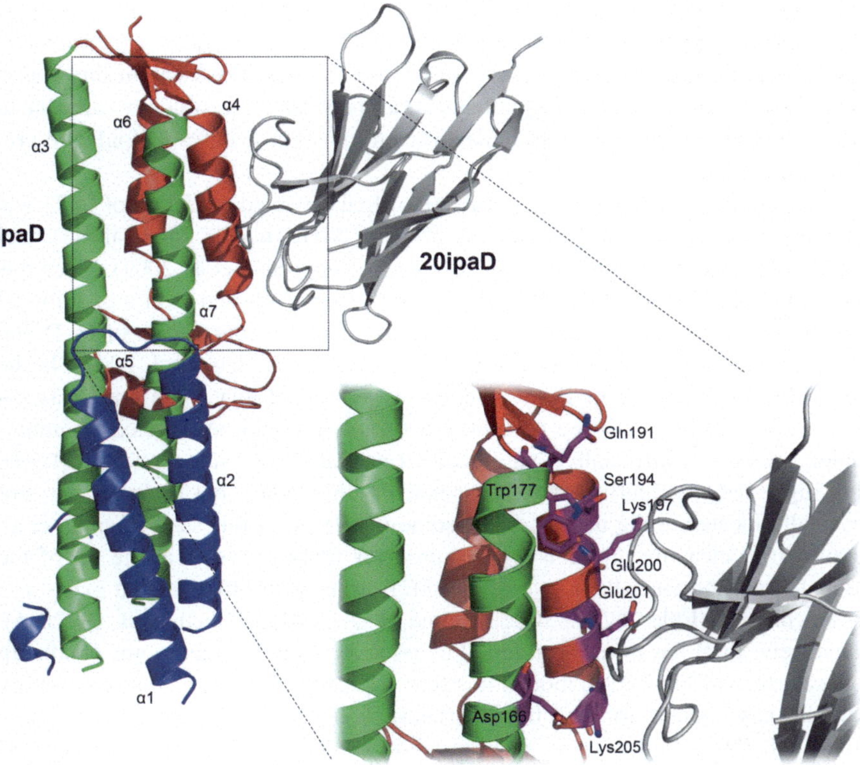

Fig. 2 Crystal structure of IpaD and the camelid single chain (VHH) antibody designated 20ipaD. The crystal structure of IpaD (colored based on individual domains: N-domain (blue), central coiled-coil (green), distal domain (red) (Johnson et al. 2007) is depicted in cartoon ribbon format in complex with the neutralizing VHH 20ipaD (colored gray). Complex [PDB ID: 5VXM, (Barta et al. 2017b)]. *Inset*, IpaD residues within hydrogen bonding distance (2.5–3.5 Å) are depicted as balls-and-sticks (magenta)

notably for the IpaD-MxiH combination, respectively, from *Shigella* and the homologous combination (SipD-PrgI) from *Salmonella* (Rathinavelan et al. 2011, 2014; Zhang et al. 2007). It is likely that the interactions between the TC and needle itself play an instrumental role in regulation of type III secretion, which is clearly the case for *Shigella* (Deane et al. 2006b; Picking et al. 2005; Kenjale et al. 2005). It has been shown that mutations within the MxiH needle protein of the *Shigella* T3SS can have profound effects on secretion kinetics and, in some cases, secretion substrate hierarchy (Deane et al. 2006b; Kenjale et al. 2005). Additional studies have shown that the C-terminus of IpaD is required for anchoring it to the injectisome needle tip (Picking et al. 2005) with deletion of only a few C-terminal amino acids resulting in secretion of uncontrolled amounts of IpaD into the culture supernatant. The stabilizing coiled-coil of IpaD has been implicated in generating secretion signals (Barta et al. 2012b; Roehrich et al. 2013; Stensrud et al. 2008) along with anchoring it to the tip of the *Shigella* injectisome needle (the IpaD C-terminus is part of the coiled-coil). If the IpaD coiled-coil does indeed interact with the needle protein, then it is not surprising that mutations within this region that influence contact (and thus communication) between these structures are able to influence secretion control. Additional structure/function features of IpaD are presented in Sect. 4.

Elucidation of the atomic-level structure of IpaD enabled modeling of IpaD as a pentameric TC at the distal end of the *Shigella* MxiH needle where it was later directly observed by transmission electron microscopy (Sani et al. 2007). Since this work, multiple models of the *Shigella* TC have been proposed with an additional hypothesis that the TC is a heteropentamer consisting of four copies of IpaD and one of the first secreted translocator protein IpaB (Veenendaal et al. 2007). In the initial description of the IpaD tip complex, it was noted that while IpaB was not detected on the surface of the nascent injectisome, it could be detected by immunoblot analysis of artificially long needles (generated by overexpressing MxiH). This suggested that IpaB could be associated with needles, but probably was not physically located at the nascent injectisome needle tip at this stage (Espina et al. 2006b). It is clear that IpaB is not required for IpaD to reside at the tip of the nascent injectisome needle (Olive et al. 2007) and in *ipaB* null strains it exists as a homopentamer (Epler et al. 2012; Espina et al. 2006b; Cheung et al. 2015). Conversely, IpaB has never been found to associate with the injectisome needle tip in the absence of IpaD, even though it is secreted at elevated levels, thus confirming that IpaD is the anchor for IpaB binding (Dickenson et al. 2013a; Espina et al. 2007; Olive et al. 2007). Furthermore, even for wild-type *Shigella*, the majority of the TC appear to exist as a pentamer of IpaD (Cheung et al. 2015), perhaps indicating that detection of IpaB within some TC may be due to an intermediate stage of the secretion process.

3.2 Maturation of the Tip Complex in Shigella

As described above, the IpaD TC protein and IpaB, a *Shigella* T3SS translocator protein, have long been known to control *Shigella* secretion. This role is seen as substantially elevated levels of Ipa and effector proteins in culture supernatants for null mutants of either (Menard et al. 1994). It is worth noting, however, that *Shigella* does not control its T3SS as stringently as many other systems do in the absence of extracellular signals for secretion induction. Such signals for *Shigella* could include incubation with the dye Congo red or contact with a host cell. This low level of background or steady-state secretion has not been specifically described for other T3SS, which implies that there is phylogenetic variation in the control mechanisms used by these pathogens. Consistent with system-to-system variation, similar roles in secretion control have been described for IpaD and its homolog (SipD) from *Salmonella* (Kaniga et al. 1995b), however, no such regulatory role has been reported for SipB, the IpaB-homolog in *Salmonella* (Kaniga et al. 1995b). This suggests that differences in the way that type III secretion is controlled can be seen even for closely related pathogens within the same T3SS family (e.g., *Shigella* and *Salmonella* SPI-1). As more distantly related injectisomes are considered, even greater differences in the mechanisms by which secretion is induced are observed. For example, *Yersinia* expresses their T3SS at 37 °C but only secretes when depletion of calcium is detected and this process may be under the control of multiple genes including the injectisome needle protein YscF (Torruellas et al. 2005). This is distinct from *Shigella* where cation concentrations in the media have not been reported to influence type III secretion, while the addition of small amphipathic dyes such as Congo red or Evans blue elicits a strong secretion phenotype (Bahrani et al. 1997).

Thus far, the discrete steps of secretion induction have only been described for the T3SS of *Shigella* species. For the *Shigella* injectisome, the nascent apparatus is primarily comprised of a pentameric IpaD TC population (Cheung et al. 2015; Espina et al. 2006b). As mentioned above, there may also be a minor TC population that is composed of only four copies of IpaD with the fifth position being filled with IpaB (Cheung et al. 2015; Veenendaal et al. 2007). Neither complex composition can be ruled out as being physiologically relevant, but within the context of the low background secretion phenotype of *Shigella* alluded to above, it is possible that this heteropentameric TC represents an intermediate state where IpaB is being recruited into the complex. Active recruitment of IpaB into the TC appears to require additional external signals, and these will be described in more detail below.

Enteric bacteria have evolved numerous mechanisms to allow survival and colonization of the human gastrointestinal tract (Merritt and Donaldson 2009). Among the many chemical defenses in the GI tract are bile salts that act to reduce the bacterial burden in the small intestine in addition to their contributions to nutrient uptake and metabolism (Schubert et al. 2017). It has been shown that bile salts influence a variety of *Shigella* behaviors, including increased: adherence to mammalian cells, invasiveness and protein secretion during laboratory propagation (Olive et al. 2007; Pope et al. 1995; Faherty et al. 2012). We were able to demonstrate that after a short (≤ 30 min) incubation in the presence of the bile salt deoxycholate (DOC), *Shigella* invasiveness increased significantly (Olive et al. 2007). Unexpectedly, this DOC-mediated virulence increase did not correlate with upregulated protein effector secretion, but in fact, resulted in the recruitment of IpaB to the tip of the *Shigella* injectisome needle (Olive et al. 2007) where it was stably maintained. IpaB was unable to localize to the needle tip for an *ipaD* null mutant, strongly suggesting an interaction within the TC for these two proteins. Meanwhile, in an *ipaB* null mutant, IpaD could still associate with the needle tip, however, incubation of these bacteria with DOC resulted in decreased levels of IpaD at this site (Olive et al. 2007). Considered in toto, these observations suggested that DOC acts directly on IpaD. Furthermore, the fact that IpaD and IpaB were both genetically identified as major players controlling the *Shigella* T3SS suggests they are able to physically communicate with each other and this correlates with identification of a stable interaction at the exposed injectisome tip.

3.3 The Translocon

The ultimate goal of the T3SS is the formation of a pore (the translocon) in the target cell membrane, thereby allowing the subsequent passage of effector proteins into the cell where they commandeer normal cellular functions for the benefit of the pathogen (Pizarro-Cerda et al. 2016). In generating the translocon, a difficulty that must be overcome by the pathogen is maintaining proteins having a significant hydrophobic character in the cytoplasm until they can be recognized for secretion, at which point they become imbedded within the targeted cell's membrane. Preventing interactions with the bacterial cytoplasmic membrane is critical and this is ensured by maintaining stability and solubility as a complex with a chaperone. For example, in *Shigella* the chaperone for the IpaB and IpaC is IpgC (Birket et al. 2007) and in *Salmonella* the chaperone for SipB and is SicA (Tucker and Galan 2000). In addition to this feature, it has been shown that the proteins secreted by the T3SS that are destined to become transmembrane proteins in the host cell possess an inherent balance of their transmembrane segments that prevent them from targeting to the bacterial cytoplasmic membrane (Krampen et al. 2018).

In *Shigella*, IpaB is the first of the two translocators to be secreted. As such it seems to have two roles—as a TC protein following exposure to bile salts and as a component of the translocon pore following host cell contact. A hallmark of all T3SS is the presence of a TC protein and two translocator proteins that function together to form the translocon pore (Deng et al. 2017). In the *Shigella* system, the translocon is formed by IpaB and IpaC (Blocker et al. 1999; Terry et al. 2008) and one assay for monitoring the formation of the complete translocon is measuring contact-mediated hemolysis (Picking et al. 2005). Based on its hydrophobicity, a potential role of IpaB in translocon formation from its position at the injectisome needle tip is sensing contact with the host cell membrane. Indeed, IpaB has been shown to interact with lipid membranes in vitro (De Geyter et al. 2000), as has the second hydrophobic translocator IpaC (Kueltzo et al. 2003). Furthermore, contact with host cells was found to trigger the release of Ipa translocators as a prelude to invasion (Watarai et al. 1995) and lipid-based signaling has been shown to trigger the secretion of effector proteins. These host-pathogen interactions were proposed to occur at cholesterol-rich cellular lipid rafts (van der Goot et al. 2004) and are consistent with the final step of TC maturation/activation being triggered by host cell lipids.

Red blood cell ghosts and liposomes having a defined composition were used to demonstrate that the final step of T3SS TC maturation is triggered by tip-localized IpaB sensing contact with host cell membranes (Epler et al. 2009). Treating *Shigella* with DOC followed by liposomes resulted in the recruitment of IpaC to the bacterial surface. Not surprisingly, recruitment of IpaC to the surface coincided with full induction of type III secretion. Unlike IpaB, which appears largely to be limited to the TC once it is recruited to the *Shigella* surface, the secreted IpaC was found to stick to many surfaces once it was in the extracellular milieu. IpaC was found as part of the needle TC, however, it was also found in complexes not associated with the bacteria, on the bacterial membrane surface and bound to almost anything with which it could interact. In these studies, it was demonstrated that maximal liposome-induced secretion occurred when cholesterol and sphingomyelin were present, which is in agreement with previous observations that IpaB is a cholesterol-binding protein (Hume et al. 2003) and *Shigella* invasion most likely occurs at lipid rafts (van der Goot et al. 2004). Similar to IpaB, purified IpaC is able to insert spontaneously into phospholipid membranes (Kueltzo et al. 2003; De Geyter et al. 1997) and has been shown to possess important effector functions that contribute to host cell entry (Jaumouille et al. 2008; Marquart et al. 1996; Terry et al. 2008; Tran Van Nhieu et al. 1999). IpaC's effector function is essential for *Shigella* entry into host cells (Terry et al. 2008), while its ability to insert into and disrupt membranes may also be responsible for *Shigella* escape into the cytoplasmic niche in which it replicates (Osiecki et al. 2001; Du et al. 2016). Unfortunately, most of the information obtained on the translocon is from indirect biochemical analyses of purified or partially purified proteins because visualization of the translocon has been difficult. This may be changing since Park and colleagues have now used cryo-electron tomography of mini-cells interacting with host cells to generate images of the in situ *Salmonella* translocon at the pathogen-host interface (Park et al. 2018).

Little information has been obtained on the step-wise TC maturation or secretion induction for other T3SS, however, interaction between the *Yersinia* TC protein LcrV and one of the *Yersinia* translocon proteins (YopD) has been reported as being important for type III secretion by this pathogen (Costa et al. 2010). Additionally, LcrV has been implicated in control of type III secretion, but this control has not been described as occurring from the injectisome needle tip (Hamad and Nilles 2007; Matson and Nilles 2001), but possibly via interactions with LcrG in the bacterial cytoplasm which serve to influence the mobilization of the YopB translocator (Nilles et al. 1998). Most importantly, LcrV is implicated in directing the formation of translocon pores in *Yersinia* (Mota 2006), suggesting that intermediates between LcrV TC formation and translocon assembly most likely do exist. It should be noted that alternative pathways for effector entry into target cells by *Y. pseudotuberculosis* have been proposed that might not necessarily even require TC contact with the traditional model of a translocon (Akopyan et al. 2011). However, this review will be limited to events expected to occur only at the injectisome needle tip and that are involved in the regulation of secretion induction. Because of this and the fact that the best-understood sequence of events related to T3SS induction are those that have been observed in the *Shigella* system, the remainder of this review will mostly focus on the *Shigella* TC.

3.4 Where Does the Tip Complex End and the Translocon Begin

In many cases, the proteins that make up the needle TC and the hydrophobic proteins that give rise to the translocon pore are collectively referred to as the "translocator proteins" (Deng et al. 2017). This is understandable since all three proteins are needed for proper translocation of effector proteins into target cells. In the same respect, the conduit that assembles to become the injectisome needle is also needed for proper translocation. Thus, for the purpose of this review, a distinction is made between the needle, the TC and the translocon pore. IpaD clearly falls into the category of TC protein for the *Shigella* injectisome since it is the protein initially placed at the tip of the MxiH needle once it is formed and, perhaps more importantly, because it is essential for controlling or limiting secretion from this position. In this respect, IpaB in the *Shigella* system could also be considered a needle TC protein because it is also essential for controlling secretion. Moreover, IpaB can also be found stably associated with the injectisome needle tip prior to secretion induction. It is in this role that IpaB most likely triggers type III secretion induction by sensing contact with the host cell membrane. The presence of translocator proteins as part of the TC has only been described in detail for the *Shigella* system (Cheung et al. 2015; Murillo et al. 2016; Olive et al. 2007). Simultaneous with or in transition from its function as a regulatory TC component, IpaB becomes essential for the formation of a fully functional translocon in conjunction with IpaC, which has no role in the control of type III secretion in *Shigella*. In fact, once IpaC secretion is triggered, the full cascade of secretion induction is completed. For these reasons, IpaD and IpaB are grouped here as the two components of the *Shigella* regulatory TC and IpaB and IpaC are separately grouped as the two translocator proteins in this organism. These are not arbitrary groupings from the *Shigella* perspective, but they may be difficult to apply precisely to other T3SS. While IpaD homologs from other systems have been implicated in controlling type III secretion and are found at the needle tip, this cannot be said of IpaB homologs. Nevertheless, the sections that follow will focus on IpaD as the initial needle TC protein and IpaB, which joins IpaD once the TC is primed to sense contact with host cells.

4 Sensing the Signals Responsible for Type III Secretion Induction in Shigella

4.1 Invasion Plasmid Antigen D

By now it is clear that IpaD is an essential virulence determinant for *Shigella* with a role in controlling type III secretion (Menard et al. 1993). Once nonpolar knockouts of the Ipa proteins became available it was learned that IpaB, along with IpaD, was responsible for controlling type III secretion in this pathogen (Menard et al. 1994). It was not until more than 10 years later, however, that IpaD was recognized as being a controlling unit that resides atop the *Shigella* injectisome needle (Espina et al. 2006b; Sani et al. 2007). Structurally, IpaD was initially proposed to form a homopentamer at the needle tip much like LcrV is proposed to do at the tip of the

Yersinia injectisome needle (Deane et al. 2006b; Epler et al. 2012), however, alternative TC compositions have been suggested that include four copies of IpaD and one copy of IpaB (Blocker et al. 2008). While most of the TC found on the *Shigella* surface are clearly composed of five copies of IpaD (Cheung et al. 2015), it cannot be ruled out that alternative states exist that represent intermediates related to TC maturation and the onset of secretion induction. Unlike the T3SS described for many other bacteria, *Shigella* displays a low level of background secretion that is readily measured by monitoring Ipa protein secretion in overnight cultures. Under such conditions, a small percentage of injectisomes could be in a state that lies between quiescence (having five copies of IpaD making up the TC) and primed (having a TC complex that contains a reduced number of IpaD moieties in combination with one or more copies of IpaB). The physiological importance for such intermediates cannot be ruled out because of the *Shigella* background secretion phenotype, however, it does appear that the first static state or checkpoint in the assembly or maturation of the newly made *Shigella* injectisome TC gives rise to a TC comprised of a pentamer of IpaD.

Maturation or priming of the *Shigella* needle TC, as defined by the recruitment of IpaB to become a major component of the TC, can be elicited by exposure to bile salts such as DOC (Olive et al. 2007). A mechanistic equivalent step for *Salmonella* has not been described, mostly because bile salts appear to actually inhibit SPI-1 expression (Eade et al. 2016). Nevertheless, bile salts have been shown to bind to the *Salmonella* TC protein SipD (Chatterjee et al. 2011). In *Shigella*, the mechanistic basis for bile salt-induced recruitment of IpaB into the TC appears to be a direct interaction between DOC and IpaD (Stensrud et al. 2008) with this binding causing a change in the structural features of the central coiled-coil of IpaD (Barta et al. 2012b). DOC binding occurs at the hydrophobic interface between helix 3 and helix 7 (the stabilizing coiled-coil that has also been implicated in anchoring IpaD at the needle tip) involving residues L134, K137, I138 and L315. Furthermore, mutation of some of these residues (L134 and L315) was shown to eliminate the organism's enhanced invasiveness that was seen following incubation with DOC (Barta et al. 2012b). Interestingly, the binding of DOC by IpaD was found to occur concomitant with an exacerbation of a kink found in helix 3, which results in a movement of ~ 10 Å for the end of helix 7 near the C-terminus of the protein. Such a change in the conformational dynamics at this region is expected to significantly affect the interaction between IpaD and the underlying needle assembly, as well as between the IpaD subunits within the TC. Mutagenesis studies have also implicated the C-terminal helix (helix 7) of IpaD in its ability to control type III secretion (Roehrich et al. 2013), however, the residues implicated here were further away from IpaD's proposed C-terminal anchor than were the residues involved in DOC binding. In this case, the mutations were selected for their ability to resist rapid induced secretion caused by incubation with Congo red.

As repeatedly mentioned in the above sections, IpaD has been proposed to exist as a pentamer as part of the *Shigella* needle TC. This is based on the structures seen for *Yersinia*, *P. aeruginosa* and *A. salmonicida* (Mueller et al. 2005) and what was initially proposed for Shigella (Deane et al. 2006b; Sani et al. 2007). Variations on the initially proposed IpaD pentamer include a four (IpaD) plus one (IpaB) model, along with a model in which an IpaD pentamer is located at the needle tip with the so-called distal domain (red domain in Fig. 2) extended outward (Epler et al. 2012). An interesting finding related to this last model is that when cleavage sites for tobacco etch virus (TEV) protease were placed on both sides of the distal domain and the distal domain was removed by site-specific cleavage, a loss of secretion control was observed, thus indicating that the distal domain has some role in

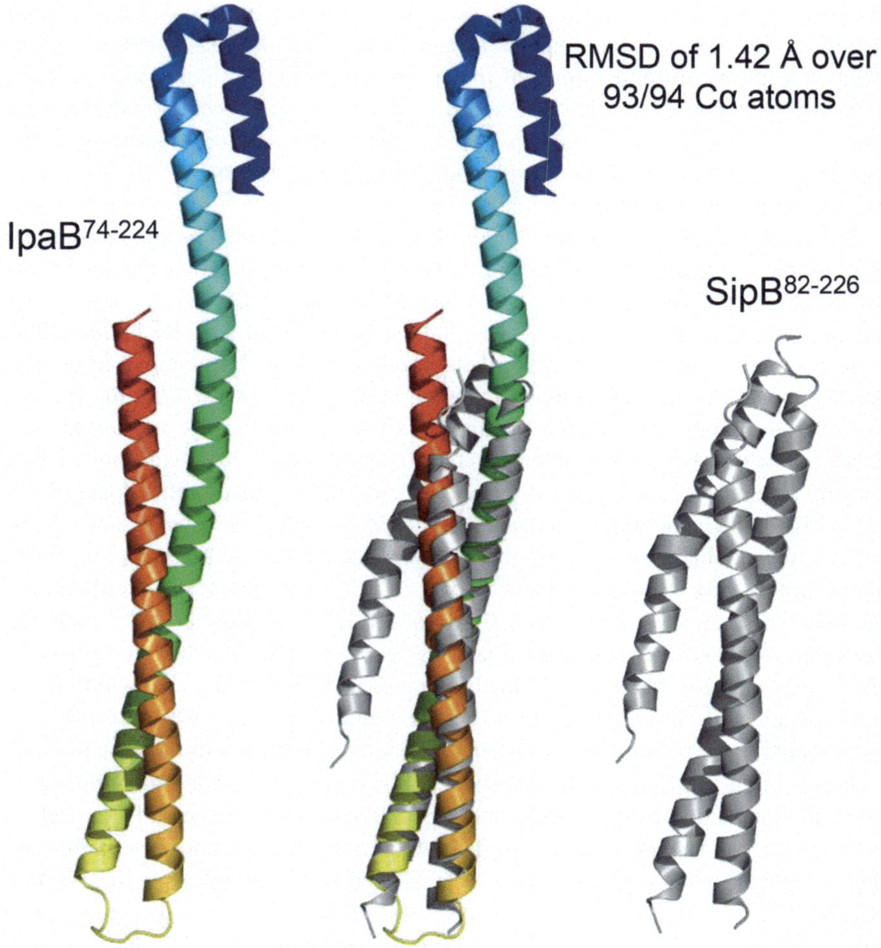

Fig. 3 Crystal Structure of T3SS Translocator Coiled-coil Fragments. *Left*, Cartoon ribbon diagram of the IpaB$^{74-224}$ crystal structure [PDB ID: 3U0C, (Barta et al. 2012a)], colored blue (N-terminus) to red (C-terminus). *Middle*, Structural alignment of the IpaB (residues 120-224) and SipB (residues 126-226) coiled-coil motifs align with an RMSD of 1.42 Å over 93/94 Cα atoms. *Right*, Cartoon ribbon diagram of the SipB$^{82-226}$ crystal structure [PDB ID: 3TUL, (Barta et al. 2012a)], colored gray

controlling secretion. While alone this result is not confirmatory with regard to the position of the distal domain within the TC, it does provide something to consider when considered alongside two other findings. First, IpaD was shown to be able to associate with a stable coiled-coil derived from the N-terminal part of IpaB whose crystal structure has been solved (Barta et al. 2012a) (see Fig. 3), but only if bile salts are present (Dickenson et al. 2013a). More importantly, association of IpaD with the IpaB fragment resulted in movement of a fluorescent probe located within the distal domain away from a second probe located at IpaD position 322, near the region that anchors IpaD to the tip of the needle (Dickenson et al. 2013a). These findings not only implicate the distal domain in interactions with IpaB, but also suggest that IpaD has a dynamic structure that can accommodate multiple conformations, including one in which the distal domain moves relative to the rest of the protein. IpaD had already been shown to consist of multiple folding units (Espina et al. 2006a), but this was the first biochemical evidence that there is flexibility in the distal domain. A second piece of evidence for the IpaD distal domain being important for type III signaling processes was derived from its interaction with specific camelid single-domain (VHH) antibodies generated by vaccination against recombinant IpaD, which behaves as a highly soluble monomer in solution. From a panel of IpaD-specific VHH antibodies, two populations were identified—one population exhibited significant neutralizing activity with regard to cellular invasion and contact-hemolysis while the other population was non-neutralizing. The former were found to uniformly recognize the IpaD distal domain (helix 4) as shown for one of the VHH (called 20ipaD) in Fig. 2 (Barta et al. 2017b). Recently, NMR studies have further indicated that interactions occur between the distal domain of IpaD and the purified N-terminal coiled-coil of IpaB (McShan et al. 2016). This study also found that the same phenomenon occurred for the Ipa homologs from *Salmonella*, SipD and SipB, and that mutations within the SipD distal domain equivalent (e.g., helix 4 in IpaD) led to a reduced ability to invade cultured Henle 407 cells (McShan et al. 2016).

4.2 *Invasion Plasmid Antigen B*

Because of its hydrophobic nature, IpaB has been more difficult to work with as a recombinant protein than IpaD (Barta et al. 2017a) and this has slowed efforts to fully appreciate its biochemistry and prevented determination of its atomic-level structure. It has been possible, however, to purify N-terminal IpaB fragments that have contributed substantially to our current level of understanding of IpaB. Crystal structures of stable (and soluble) N-terminal domains of both IpaB (residues 74-224) and SipB (residues 82-226) were determined and are shown in Fig. 3 (Barta et al. 2012a). Both fragments have been shown to be capable of interacting with their cognate needle tip protein as mentioned above (McShan et al. 2016), although no crystal structure is available to describe this interaction. A slightly longer IpaB fragment (residues 28-226) has been shown to strongly associate with its chaperone IpgC (Adam et al. 2012). It does appear that chaperone binding alters the IpaB fragment's structure, however, the precise influence of chaperone binding on the structure of the IpaB coiled-coil is not known. On the other hand, a co-crystal structure has been solved for an N-terminal fragment of the IpaB/SipB homolog AopB from *Aeromonas* in complex with its chaperone AcrH (Nguyen et al. 2015) and this structure indicates that the coiled-coil is specifically bent by association with the N-terminal groove of AcrH (colored purple in Fig. 4). This suggests that chaperone binding not only involves the immediate N-terminus of IpaB and its homologs, but it may also specifically perturb the stable coiled-coil structure found near the N-terminus. No structural information is yet available on the hydrophobic portions of this translocator protein family and it is this portion that is expected to be involved in membrane recognition and penetration. Intriguingly, current data support a model in which IpaB association with its chaperone in the bacterial cytoplasm occurs through its N-terminal region, however, this would be expected to leave the hydrophobic portion of the protein exposed. Despite this, the IpaB-IpgC complex is unable to associate, even peripherally, with phospholipid membranes (Dickenson et al. 2013b). Thus, it is clear that there is still much to learn about IpaB structure and function, even before it leaves the *Shigella* cytoplasm.

Based on what is known about the stable IpaB N-terminal domain, it has been hypothesized that the coiled-coil allows IpaB (or SipB) interaction with IpaD (or SipD) via the distal domain of the later (Dickenson et al. 2013a; McShan et al. 2016). This would position the hydrophobic portion of IpaB (SipB) so that it is available for recognition of and interaction with the host cell membrane (Dickenson et al. 2013b). The previously described neutralizing anti-IpaD VHHs support this model because their interaction with the IpaD distal domain appears to be

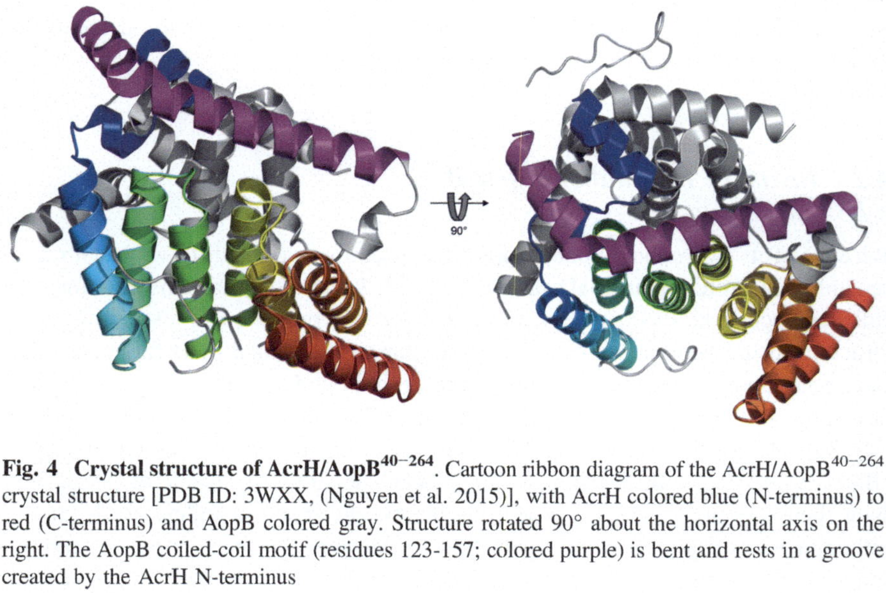

Fig. 4 Crystal structure of AcrH/AopB$^{40-264}$. Cartoon ribbon diagram of the AcrH/AopB$^{40-264}$ crystal structure [PDB ID: 3WXX, (Nguyen et al. 2015)], with AcrH colored blue (N-terminus) to red (C-terminus) and AopB colored gray. Structure rotated 90° about the horizontal axis on the right. The AopB coiled-coil motif (residues 123-157; colored purple) is bent and rests in a groove created by the AcrH N-terminus

interfering with steps involved in TC function such as IpaB recruitment or binding to the TC (Barta et al. 2017b). Other groups suggest a similar orientation for the IpaB hydrophobic domain, but favor a model where it is anchored to IpaD via its C-terminus based on decreased detection of IpaB within the TC as residues are removed from the end of the protein (Shen et al. 2010). To be clear, IpaB's initial interaction with the host cell membrane is a pivotal event in triggering the final step of type III secretion induction and for the recruitment of IpaC, which is needed for formation of the translocon pore. There will be additional discussion of IpaB's interaction with membranes below.

Based on homology with pore-forming toxins, it was proposed that the IpaB coiled-coil was involved in anchoring this protein to the needle TC following its recruitment to the bacterial surface. Such a scenario might equate to the role of these coiled-coil structures in a family of bactericidal toxins produced by some bacteria to gain an advantage against competing bacteria (Jakes 2012). For example, colicin E3 and colicin Ia share a similar coiled-coil structure with IpaB and SipB and these extended structures are used by the colicins to span the periplasm to bridge the domain needed for binding to an outer membrane receptor with the domain that interacts with the target cell cytoplasmic membrane following an outer membrane translocation step (Wiener et al. 1997). Because of the parallel need for presentation of a membrane active moiety at a distance, it would seem logical that the translocator coiled-coil could be used in a similar manner. Intriguingly, this same coiled-coil structure has been heavily implicated in the control of *Shigella* type III secretion (Murillo et al. 2016). In an extensive mutagenesis study, seven mutations were identified within the IpaB N-terminal portion that affected *Shigella*'s ability to respond to Congo red induction of secretion and all but one of these resided within the coiled-coil (Murillo et al. 2016). These secretion phenotypes fell into two groups, those having mild and those having strong defects in sensing Congo red. As with other aspects of IpaB function, however, there is still much to learn about these phenotypes since defects in sensing Congo red did not strictly correlate with defects in contact-mediated hemolysis or invasion of HeLa cells (Murillo et al. 2016).

While a combination of Congo red phenotypes found for IpaD and IpaB mutants can be used to develop a model for secretion control in *Shigella*, it should be kept in mind that the way Congo red works is still not entirely clear. Congo red has been known to induce protein unfolding in some cases, possibly after penetrating regions possessing a somewhat intrinsically unfolded state (Zhang et al. 2009; Kim et al. 2003). It is this propensity for binding to unfolded regions of proteins which has made it useful in staining of amyloid fibrils for diagnostic purposes (Serpell et al. 1997). Furthermore, in preferentially associating with partially unfolded regions of proteins, Congo red can shift a protein's folding toward an unfolded state. If this is occurring within the *Shigella* injectisome needle TC, then it is difficult to confidently assign a physiological role to Congo red-induced secretion. Any disruptions within the TC, especially where there are interfacial interactions, such as between IpaD and IpaB might well be expected to give rise to changes in secretion status. This is only said as a cautionary statement in considering TC functions using artificial inducers and to point out that there continues to be much to learn about the dynamics of IpaD and IpaB within the TC. A similar statement might be said with regard to studying translocon pore formation using contact-mediated hemolysis, however, such artificial systems have thus far been invaluable in reaching our current level of understanding of type III secretion in *Shigella*.

In addition to being a TC component, IpaB is also a translocator protein. While this review does not consider the translocon pore as a part of the injectisome needle TC, it probably is important to consider the potential state of IpaB as it encounters the host cell membrane. *Shigella* mutants harboring an *ipaC* null mutation do not suffer defects in type III secretion control, however, they are completely noninvasive. The inability to invade cells is due to the inability to form a functional translocon for effector delivery and to the absence of IpaC's early effector functions (Terry et al. 2008). Nevertheless, *ipaC* null mutants are still able to induce a low level of contact-mediated hemolysis ($\sim$10% relative to wild-type), indicating that something has been inserted into the target cell membrane that is able to compromise membrane integrity (Blocker et al. 1999). Because a functional translocon does not form for these mutants, IpaB can be considered to continue to be a part of the needle TC following host cell contact, but it should be noted that it is still potentially membrane active. It may be here that the ability to work with purified IpaB has proven most useful. IpaB was first purified efficiently as a complex with its chaperone IpgC (Birket et al. 2007), but it is readily separated from its chaperone using mild detergents (Barta et al. 2017a). It is important, however, that these detergents be continually present in the preparation to maintain IpaB solubility. While this has made detailed structural analysis of IpaB difficult, it was recently found that the detergent used to prepare IpaB can have a profound effect on its biochemical properties.

IpaB clearly has an intrinsic ability to interact with membranes and this property is shared with its homolog from *Salmonella* (De Geyter et al. 2000; Hume et al. 2003). When prepared in the detergent lauryl-dimethylamine-*N*-oxide (LDAO), IpaB exists as a monomer in solution, however, when prepared in the alternative detergent *N*-octyl-poly-oxyethylene (OPOE) it forms a tetramer (Dickenson et al. 2013b). Both of these detergents are so mild that they do not disrupt phospholipid vesicles at or slightly above their critical micelle concentrations, which allows them to be present when looking at IpaB-membrane interactions. In either detergent, IpaB can associate with phospholipid vesicles, however, only the oligomeric form of IpaB is able to cause the release of small molecules from these liposomes and this release shows the hallmarks of being the result of pore formation (Adam et al. 2014; Dickenson et al. 2013b). Thus, even before orchestrating the formation of an active translocon pore through interactions with IpaC, IpaB itself is proposed to insert into target cell membranes. This provides the trigger for IpaC recruitment, translocon formation and secretion induction (Epler et al. 2009; van der Goot et al. 2004). However, based on background hemolysis levels and biochemical analysis of the IpaB-membrane interaction, it is possible that this insertion event results in the formation of what might be termed a pre-translocon pore composed of an oligomeric complex of IpaB (Dickenson et al. 2013b). While such a pore would not be a fully functional translocon, it could provide a platform into which IpaC can be incorporated, leading to formation of the active translocon pore. It is at this point in wild-type *Shigella* that IpaB becomes an integral part of the translocon while maintaining a bridge to IpaD at the tip of the injectisome needle.

5 Concluding Remarks

Much of what is known about type III secretion control from the TC has been determined based upon the tractability of the *Shigella* system. It is clear that, despite a high degree of conservation within the injectisomes from a wide array of pathogens, there are still differences in the cargo they inject into host cells, subtle variations in their assembly and function of their component pieces, and in how they respond to external stimuli. Perhaps their greatest unifying feature is the formation of a translocon pore upon host cell contact, enabling translocation of host altering effector proteins that are maintained within the bacterial cytoplasm in association with cognate chaperones. The chaperone then is instrumental in targeting the effectors to the sorting platform for ultimate delivery through the translocon and into the cytoplasm of the host cell. The resulting host-pathogen intercommunication can lead to a variety of outcomes: colonization of a host cell surface through effacing lesions; invasion of macrophages, epithelial cells or lymphocytes; or killing of macrophages. Yet, despite their many differences, it makes sense that there are significant mechanistic similarities that are guided by the general architecture of these amazing nanomachines. The four major macromolecular assemblies within the injectisome (sorting platform with ATPase, envelope-spanning basal body, extracellular needle and needle TC) may display subtle differences but they largely appear to be consistent from one system to the next.

It is generally accepted that all injectisome needle TC are essential for creating a continuous conduit from the bacterium through the host cell membrane, suggesting they share important mechanistic characteristics. This would seem to be borne out of the conservation of key structural features for all the extracellular portions of the injectisome that are involved in controlling secretion (Deng et al. 2017). The structure of the needle protein monomer has consistently been shown to be a relatively small helix-turn-helix protein (Deane et al. 2006b; Zhang et al. 2007; Wang et al. 2007). The initial needle TC protein can vary in overall structure, but is consistently built upon a stable anti-parallel helical coiled-coil scaffold (Derewenda et al. 2004; Espina et al. 2007; Johnson et al. 2007). And finally, it is the first hydrophobic translocator protein that, at least for *Shigella* and *Salmonella*, possesses an elongated helical coiled-coil that may provide a contact interface with the nascent TC protein (Barta et al. 2012a). Altogether, it appears that these protein–protein interfaces, likely involving extensive contributions from coiled-coil motifs, are critical to the assembly, communication and overall function of the injectisome.

Thus, while a significant portion of this review has focused on the *Shigella* injectisome needle tip complex and its role in regulating type III secretion, it is likely that there are lessons to be learned here that will hold true in many, or perhaps most, other systems. Nevertheless, it is certainly a mistake to view the injectisome needle TC as a static object that progresses from one rigid form to the next. For *Shigella*, the TC is a fluid structure that has yielded transitions that can be teased apart to reveal what can be described as discrete steps in type III secretion induction. Part of what has made this so, may be the one confounding feature of the *Shigella* T3SS—the fact that it is in a low, but continual, steady-state secretion state. The same progression of steps may very well occur for other T3SS, but due to stringent control prior to host cell contact, these steps may be difficult, if not impossible, to dissect. Over time, other systems will be discovered or methods devised to determine how universal the process of step-wise type III secretion induction actually is.

References

Adam PR, Dickenson NE, Greenwood JC 2nd, Picking WL, Picking WD (2014) Influence of oligomerization state on the structural properties of invasion plasmid antigen B from *Shigella flexneri* in the presence and absence of phospholipid membranes. Proteins 82(11):3013–3022. https://doi.org/10.1002/prot.24662

Adam PR, Patil MK, Dickenson NE, Choudhari S, Barta M, Geisbrecht BV, Picking WL, Picking WD (2012) Binding affects the tertiary and quaternary structures of the Shigella translocator protein IpaB and its chaperone IpgC. Biochemistry 51(19):4062–4071. https://doi.org/10.1021/bi300243z

Akopyan K, Edgren T, Wang-Edgren H, Rosqvist R, Fahlgren A, Wolf-Watz H, Fallman M (2011) Translocation of surface-localized effectors in type III secretion. Proc Natl Acad Sci USA 108(4):1639–1644. https://doi.org/10.1073/pnas.1013888108

Bacon GA, Burrows TW (1956) The basis of virulence in *Pasteurella pestis*: an antigen determining virulence. Br J Exp Pathol 37(5):481–493

Bahrani FK, Sansonetti PJ, Parsot C (1997) Secretion of Ipa proteins by *Shigella flexneri*: inducer molecules and kinetics of activation. Infect Immun 65(10):4005–4010

Bamyaci S, Ekestubbe S, Nordfelth R, Erttmann SF, Edgren T, Forsberg A (2018) YopN is required for efficient effector translocation and virulence in *Yersinia pseudotuberculosis*. Infect Immun 86(8). https://doi.org/10.1128/iai.00957-17

Barta ML, Adam PR, Dickenson NE (2017a) Recombinant expression and purification of the Shigella Translocator IpaB. Methods Mol Biol 1531:173–181. https://doi.org/10.1007/978-1-4939-6649-3_15

Barta ML, Dickenson NE, Patil M, Keightley A, Wyckoff GJ, Picking WD, Picking WL, Geisbrecht BV (2012a) The structures of coiled-coil domains from type III secretion system translocators reveal homology to pore-forming toxins. J Mol Biol 417(5):395–405. https://doi.org/10.1016/j.jmb.2012.01.026

Barta ML, Guragain M, Adam P, Dickenson NE, Patil M, Geisbrecht BV, Picking WL, Picking WD (2012b) Identification of the bile salt binding site on IpaD from *Shigella flexneri* and the influence of ligand binding on IpaD structure. Proteins 80(3):935–945

Barta ML, Shearer JP, Arizmendi O, Tremblay JM, Mehzabeen N, Zheng Q, Battaile KP, Lovell S, Tzipori S, Picking WD, Shoemaker CB, Picking WL (2017b) Single-domain antibodies pinpoint potential targets within Shigella invasion plasmid antigen D of the needle tip complex for inhibition of type III secretion. J Biol Chem 292(40):16677–16687. https://doi.org/10.1074/jbc.M117.802231

Bergman T, Erickson K, Galyov E, Persson C, Wolf-Watz H (1994) The lcrB (yscN/U) gene cluster of *Yersinia pseudotuberculosis* is involved in Yop secretion and shows high homology to the spa gene clusters of *Shigella flexneri* and *Salmonella typhimurium*. J Bacteriol 176 (9):2619–2626

Bhaduri S, Turner-Jones C, Taylor MM, Lachica RV (1990) Simple assay of calcium dependency for virulent plasmid-bearing clones of *Yersinia enterocolitica*. J Clin Microbiol 28(4):798–800

Birket SE, Harrington AT, Espina M, Smith ND, Terry CM, Darboe N, Markham AP, Middaugh CR, Picking WL, Picking WD (2007) Preparation and characterization of translocator/chaperone complexes and their component proteins from *Shigella flexneri*. Biochemistry 46(27):8128–8137. https://doi.org/10.1021/bi700099c

Blocker A, Gounon P, Larquet E, Niebuhr K, Cabiaux V, Parsot C, Sansonetti P (1999) The tripartite type III secreton of *Shigella flexneri* inserts IpaB and IpaC into host membranes. J Cell Biol 147(3):683–693

Blocker A, Jouihri N, Larquet E, Gounon P, Ebel F, Parsot C, Sansonetti P, Allaoui A (2001) Structure and composition of the *Shigella flexneri* "needle complex", a part of its type III secreton. Mol Microbiol 39(3):652–663

Blocker AJ, Deane JE, Veenendaal AK, Roversi P, Hodgkinson JL, Johnson S, Lea SM (2008) What's the point of the type III secretion system needle? Proc Natl Acad Sci USA 105 (18):6507–6513. https://doi.org/10.1073/pnas.0708344105

Broz P, Mueller CA, Muller SA, Philippsen A, Sorg I, Engel A, Cornelis GR (2007) Function and molecular architecture of the Yersinia injectisome tip complex. Mol Microbiol 65 (5):1311–1320. https://doi.org/10.1111/j.1365-2958.2007.05871.x

Brubaker RR, Surgalla MJ (1964) The effect of Ca++ and Mg++ on lysis, growth, and production of virulence antigens by *Pasteurella pestis*. J Infect Dis 114:13–25

Burrows TW, Bacon GA (1960) V and W antigens in strains of *Pasteurella pseudotuberculosis*. Br J Exp Pathol 41:38–44

Buysse JM, Stover CK, Oaks EV, Venkatesan M, Kopecko DJ (1987) Molecular cloning of invasion plasmid antigen (ipa) genes from *Shigella flexneri*: analysis of ipa gene products and genetic mapping. J Bacteriol 169(6):2561–2569

Chatterjee S, Zhong D, Nordhues BA, Battaile KP, Lovell S, De Guzman RN (2011) The crystal structures of the Salmonella type III secretion system tip protein SipD in complex with deoxycholate and chenodeoxycholate. Protein Sci 20(1):75–86. https://doi.org/10.1002/pro.537

Cheung M, Shen DK, Makino F, Kato T, Roehrich AD, Martinez-Argudo I, Walker ML, Murillo I, Liu X, Pain M, Brown J, Frazer G, Mantell J, Mina P, Todd T, Sessions RB, Namba K, Blocker AJ (2015) Three-dimensional electron microscopy reconstruction and cysteine-mediated crosslinking provide a model of the type III secretion system needle tip complex. Mol Microbiol 95(1):31–50. https://doi.org/10.1111/mmi.12843

Cordes FS, Daniell S, Kenjale R, Saurya S, Picking WL, Picking WD, Booy F, Lea SM, Blocker A (2005) Helical packing of needles from functionally altered Shigella type III secretion systems. J Mol Biol 354(2):206–211. https://doi.org/10.1016/j.jmb.2005.09.062

Cornelis GR (2002) The Yersinia Ysc-Yop 'type III' weaponry. Nat Rev Mol Cell Biol 3(10):742–752. https://doi.org/10.1038/nrm932

Costa TR, Edqvist PJ, Broms JE, Ahlund MK, Forsberg A, Francis MS (2010) YopD self-assembly and binding to LcrV facilitate type III secretion activity by *Yersinia pseudotuberculosis*. J Biol Chem 285(33):25269–25284. https://doi.org/10.1074/jbc.M110.144311

De Geyter C, Vogt B, Benjelloun-Touimi Z, Sansonetti PJ, Ruysschaert JM, Parsot C, Cabiaux V (1997) Purification of IpaC, a protein involved in entry of *Shigella flexneri* into epithelial cells and characterization of its interaction with lipid membranes. FEBS Lett 400(2):149–154

De Geyter C, Wattiez R, Sansonetti P, Falmagne P, Ruysschaert JM, Parsot C, Cabiaux V (2000) Characterization of the interaction of IpaB and IpaD, proteins required for entry of *Shigella flexneri* into epithelial cells, with a lipid membrane. Eur J Biochem 267(18):5769–5776

Deane JE, Cordes FS, Roversi P, Johnson S, Kenjale R, Picking WD, Picking WL, Lea SM, Blocker A (2006a) Expression, purification, crystallization and preliminary crystallographic analysis of MxiH, a subunit of the *Shigella flexneri* type III secretion system needle. Acta Crystallogr Sect F Struct Biol Cryst Commun 62(Pt 3):302–305. https://doi.org/10.1107/S1744309106006555

Deane JE, Roversi P, Cordes FS, Johnson S, Kenjale R, Daniell S, Booy F, Picking WD, Picking WL, Blocker AJ, Lea SM (2006b) Molecular model of a type III secretion system needle: implications for host-cell sensing. Proc Natl Acad Sci USA 103(33):12529–12533. https://doi.org/10.1073/pnas.0602689103

Demers JP, Habenstein B, Loquet A, Kumar Vasa S, Giller K, Becker S, Baker D, Lange A, Sgourakis NG (2014) High-resolution structure of the Shigella type-III secretion needle by solid-state NMR and cryo-electron microscopy. Nat Commun 5:4976. https://doi.org/10.1038/ncomms5976

Deng W, Marshall NC, Rowland JL, McCoy JM, Worrall LJ, Santos AS, Strynadka NCJ, Finlay BB (2017) Assembly, structure, function and regulation of type III secretion systems. Nat Rev Microbiol 15(6):323–337. https://doi.org/10.1038/nrmicro.2017.20

Derewenda U, Mateja A, Devedjiev Y, Routzahn KM, Evdokimov AG, Derewenda ZS, Waugh DS (2004) The structure of *Yersinia pestis* V-antigen, an essential virulence factor and mediator of immunity against plague. Structure 12(2):301–306. https://doi.org/10.1016/j.str.2004.01.010

Dickenson NE, Arizmendi O, Patil MK, Toth RTt, Middaugh CR, Picking WD, Picking WL (2013a) N-terminus of IpaB provides a potential anchor to the Shigella type III secretion system tip complex protein IpaD. Biochemistry 52(49):8790–8799. https://doi.org/10.1021/bi400755f

Dickenson NE, Choudhari SP, Adam PR, Kramer RM, Joshi SB, Middaugh CR, Picking WL, Picking WD (2013b) Oligomeric states of the Shigella translocator protein IpaB provide structural insights into formation of the type III secretion translocon. Protein Sci 22(5):614–627. https://doi.org/10.1002/pro.2245

Du J, Reeves AZ, Klein JA, Twedt DJ, Knodler LA, Lesser CF (2016) The type III secretion system apparatus determines the intracellular niche of bacterial pathogens. Proc Natl Acad Sci USA 113(17):4794–4799. https://doi.org/10.1073/pnas.1520699113

Eade CR, GungCC, Bullard B, Gonzalez-Escobedo G, Gunn JS, Altier C (2016) Bile acids function synergistically to repress invasion gene expression in *Salmonella* by destabilizing the invasion regulator HilD. Infect Immun 84(8):2198–2208. https://doi.org/10.1128/IAI.00177-16

Epler CR, Dickenson NE, Bullitt E, Picking WL (2012) Ultrastructural analysis of IpaD at the tip of the nascent MxiH type III secretion apparatus of Shigella flexneri. J Mol Biol 420(1–2):29–39. https://doi.org/10.1016/j.jmb.2012.03.025

Epler CR, Dickenson NE, Olive AJ, Picking WL, Picking WD (2009) Liposomes recruit IpaC to the *Shigella flexneri* type III secretion apparatus needle as a final step in secretion induction. Infect Immun 77(7):2754–2761. https://doi.org/10.1128/IAI.00190-09

Erskine PT, Knight MJ, Ruaux A, Mikolajek H, Wong Fat Sang N, Withers J, Gill R, Wood SP, Wood M, Fox GC, Cooper JB (2006) High resolution structure of BipD: an invasion protein associated with the type III secretion system of *Burkholderia pseudomallei*. J Mol Biol 363(1):125–136. https://doi.org/10.1016/j.jmb.2006.07.069

Espina M, Ausar SF, Middaugh CR, Baxter MA, Picking WD, Picking WL (2007) Conformational stability and differential structural analysis of LcrV, PcrV, BipD, and SipD from type III secretion systems. Protein Sci 16(4):704–714. https://doi.org/10.1110/ps.062645007

Espina M, Ausar SF, Middaugh CR, Picking WD, Picking WL (2006a) Spectroscopic and calorimetric analyses of invasion plasmid antigen D (IpaD) from *Shigella flexneri* reveal the presence of two structural domains. Biochemistry 45(30):9219–9227. https://doi.org/10.1021/bi060625v

Espina M, Olive AJ, Kenjale R, Moore DS, Ausar SF, Kaminski RW, Oaks EV, Middaugh CR, Picking WD, Picking WL (2006b) IpaD localizes to the tip of the type III secretion system needle of *Shigella flexneri*. Infect Immun 74(8):4391–4400. https://doi.org/10.1128/IAI.00440-06

Faherty CS, Redman JC, Rasko DA, Barry EM, Nataro JP (2012) *Shigella flexneri* effectors OspE1 and OspE2 mediate induced adherence to the colonic epithelium following bile salts exposure. Mol Microbiol 85(1):107–121. https://doi.org/10.1111/j.1365-2958.2012.08092.x

Ferber DM, Brubaker RR (1981) Plasmids in *Yersinia pestis*. Infect Immun 31(2):839–841

Fujii T, Cheung M, Blanco A, Kato T, Blocker AJ, Namba K (2012) Structure of a type III secretion needle at 7-A resolution provides insights into its assembly and signaling mechanisms. Proc Natl Acad Sci USA 109(12):4461–4466. https://doi.org/10.1073/pnas.1116126109

Galan JE, Ginocchio C, Costeas P (1992) Molecular and functional characterization of the Salmonella invasion gene invA: homology of InvA to members of a new protein family. J Bacteriol 174(13):4338–4349

Ginocchio CC, Olmsted SB, Wells CL, Galan JE (1994) Contact with epithelial cells induces the formation of surface appendages on Salmonella typhimurium. Cell 76(4):717–724

Goguen JD, Walker WS, Hatch TP, Yother J (1986) Plasmid-determined cytotoxicity in *Yersinia pestis* and *Yersinia pseudotuberculosis*. Infect Immun 51(3):788–794

Gough CL, Genin S, Lopes V, Boucher CA (1993) Homology between the HrpO protein of *Pseudomonas solanacearum* and bacterial proteins implicated in a signal peptide-independent secretion mechanism. Mol Gen Genet 239(3):378–392

Goure J, Broz P, Attree O, Cornelis GR, Attree I (2005) Protective anti-V antibodies inhibit Pseudomonas and Yersinia translocon assembly within host membranes. J Infect Dis 192 (2):218–225. https://doi.org/10.1086/430932

Hamad MA, Nilles ML (2007) Roles of YopN, LcrG and LcrV in controlling Yops secretion by *Yersinia pestis*. Adv Exp Med Biol 603:225–234. https://doi.org/10.1007/978-0-387-72124-8_20

Hoiczyk E, Blobel G (2001) Polymerization of a single protein of the pathogen *Yersinia enterocolitica* into needles punctures eukaryotic cells. Proc Natl Acad Sci USA 98(8):4669–4674. https://doi.org/10.1073/pnas.071065798

Hu B, Lara-Tejero M, Kong Q, Galan JE, Liu J (2017) In situ molecular architecture of the Salmonella type III secretion machine. Cell 168(6):1065–1074, e1010. https://doi.org/10.1016/j.cell.2017.02.022

Hu B, Morado DR, Margolin W, Rohde JR, Arizmendi O, Picking WL, Picking WD, Liu J (2015) Visualization of the type III secretion sorting platform of *Shigella flexneri*. Proc Natl Acad Sci USA 112(4):1047–1052. https://doi.org/10.1073/pnas.1411610112

Hume PJ, McGhie EJ, Hayward RD, Koronakis V (2003) The purified Shigella IpaB and Salmonella SipB translocators share biochemical properties and membrane topology. Mol Microbiol 49(2):425–439

Jakes KS (2012) Translocation trumps receptor binding in colicin entry into *Escherichia coli*. Biochem Soc Trans 40(6):1443–1448. https://doi.org/10.1042/BST20120207

Jaumouille V, Francetic O, Sansonetti PJ, Tran Van Nhieu G (2008) Cytoplasmic targeting of IpaC to the bacterial pole directs polar type III secretion in Shigella. EMBO J 27(2):447–457. https://doi.org/10.1038/sj.emboj.7601976

Johnson S, Roversi P, Espina M, Olive A, Deane JE, Birket S, Field T, Picking WD, Blocker AJ, Galyov EE, Picking WL, Lea SM (2007) Self-chaperoning of the type III secretion system needle tip proteins IpaD and BipD. J Biol Chem 282(6):4035–4044. https://doi.org/10.1074/jbc.M607945200

Kaniga K, Trollinger D, Galan JE (1995a) Identification of two targets of the type III protein secretion system encoded by the inv and spa loci of *Salmonella typhimurium* that have homology to the Shigella IpaD and IpaA proteins. J Bacteriol 177(24):7078–7085

Kaniga K, Tucker S, Trollinger D, Galan JE (1995b) Homologs of the Shigella IpaB and IpaC invasins are required for *Salmonella typhimurium* entry into cultured epithelial cells. J Bacteriol 177(14):3965–3971

Kenjale R, Wilson J, Zenk SF, Saurya S, Picking WL, Picking WD, Blocker A (2005) The needle component of the type III secreton of Shigella regulates the activity of the secretion apparatus. J Biol Chem 280(52):42929–42937. https://doi.org/10.1074/jbc.M508377200

Kim YS, Randolph TW, Manning MC, Stevens FJ, Carpenter JF (2003) Congo red populates partially unfolded states of an amyloidogenic protein to enhance aggregation and amyloid fibril formation. J Biol Chem 278(12):10842–10850. https://doi.org/10.1074/jbc.M212540200

Kimbrough TG, Miller SI (2000) Contribution of *Salmonella typhimurium* type III secretion components to needle complex formation. Proc Natl Acad Sci USA 97(20):11008–11013. https://doi.org/10.1073/pnas.200209497

Krampen L, Malmsheimer S, Grin I, Trunk T, Luhrmann A, de Gier JW, Wagner S (2018) Revealing the mechanisms of membrane protein export by virulence-associated bacterial secretion systems. Nat Commun 9(1):3467. https://doi.org/10.1038/s41467-018-05969-w

Kubori T, Matsushima Y, Nakamura D, Uralil J, Lara-Tejero M, Sukhan A, Galan JE, Aizawa SI (1998) Supramolecular structure of the *Salmonella typhimurium* type III protein secretion system. Science 280(5363):602–605

Kueltzo LA, Osiecki J, Barker J, Picking WL, Ersoy B, Picking WD, Middaugh CR (2003) Structure-function analysis of invasion plasmid antigen C (IpaC) from *Shigella flexneri*. J Biol Chem 278(5):2792–2798. https://doi.org/10.1074/jbc.M208383200

Kupferberg LL, Higuchi K (1958) Role of calcium ions in the stimulation of growth of virulent strains of *Pasteurella pestis*. J Bacteriol 76(1):120–121

Lara-Tejero M, Kato J, Wagner S, Liu X, Galan JE (2011) A sorting platform determines the order of protein secretion in bacterial type III systems. Science 331(6021):1188–1191. https://doi.org/10.1126/science.1201476

Lawton WD, Erdman RL, Surgalla MJ (1963) Biosynthesis and purification of V and W antigen in *Pasteurella pestis*. J Immunol 91:179–184

Leary SE, Williamson ED, Griffin KF, Russell P, Eley SM, Titball RW (1995) Active immunization with recombinant V antigen from *Yersinia pestis* protects mice against plague. Infect Immun 63(8):2854–2858

Marlovits TC, Kubori T, Sukhan A, Thomas DR, Galan JE, Unger VM (2004) Structural insights into the assembly of the type III secretion needle complex. Science 306(5698):1040–1042. https://doi.org/10.1126/science.1102610

Marquart ME, Picking WL, Picking WD (1996) Soluble invasion plasmid antigen C (IpaC) from *Shigella flexneri* elicits epithelial cell responses related to pathogen invasion. Infect Immun 64(10):4182–4187

Matson JS, Nilles ML (2001) LcrG-LcrV interaction is required for control of Yops secretion in *Yersinia pestis*. J Bacteriol 183(17):5082–5091

McShan AC, Kaur K, Chatterjee S, Knight KM, De Guzman RN (2016) NMR identification of the binding surfaces involved in the Salmonella and Shigella Type III secretion tip-translocon protein-protein interactions. Proteins 84(8):1097–1107. https://doi.org/10.1002/prot.25055

Menard R, Sansonetti P, Parsot C, Vasselon T (1994) Extracellular association and cytoplasmic partitioning of the IpaB and IpaC invasins of *S. flexneri*. Cell 79(3):515–525

Menard R, Sansonetti PJ, Parsot C (1993) Nonpolar mutagenesis of the ipa genes defines IpaB, IpaC, and IpaD as effectors of *Shigella flexneri* entry into epithelial cells. J Bacteriol 175(18):5899–5906

Merritt ME, Donaldson JR (2009) Effect of bile salts on the DNA and membrane integrity of enteric bacteria. J Med Microbiol 58(Pt 12):1533–1541. https://doi.org/10.1099/jmm.0.014092-0

Michiels T, Wattiau P, Brasseur R, Ruysschaert JM, Cornelis G (1990) Secretion of Yop proteins by Yersiniae. Infect Immun 58(9):2840–2849

Mills DM, Bajaj V, Lee CA (1995) A 40 kb chromosomal fragment encoding *Salmonella typhimurium* invasion genes is absent from the corresponding region of the *Escherichia coli* K-12 chromosome. Mol Microbiol 15(4):749–759

Mota LJ (2006) Type III secretion gets an LcrV tip. Trends Microbiol 14(5):197–200. https://doi.org/10.1016/j.tim.2006.02.010

Motin VL, Nakajima R, Smirnov GB, Brubaker RR (1994) Passive immunity to Yersiniae mediated by anti-recombinant V antigen and protein A-V antigen fusion peptide. Infect Immun 62(10):4192–4201

Mueller CA, Broz P, Muller SA, Ringler P, Erne-Brand F, Sorg I, Kuhn M, Engel A, Cornelis GR (2005) The V-antigen of Yersinia forms a distinct structure at the tip of injectisome needles. Science 310(5748):674–676. https://doi.org/10.1126/science.1118476

Murillo I, Martinez-Argudo I, Blocker AJ (2016) Genetic dissection of the signaling cascade that controls activation of the Shigella Type III secretion system from the needle tip. Sci Rep 6:27649. https://doi.org/10.1038/srep27649

Nguyen VS, Jobichen C, Tan KW, Tan YW, Chan SL, Ramesh K, Yuan Y, Hong Y, Seetharaman J, Leung KY, Sivaraman J, Mok YK (2015) Structure of AcrH-AopB chaperone-translocator complex reveals a role for membrane hairpins in type III secretion system translocon assembly. Structure 23(11):2022–2031. https://doi.org/10.1016/j.str.2015.08.014

Nilles ML, Fields KA, Straley SC (1998) The V antigen of *Yersinia pestis* regulates Yop vectorial targeting as well as Yop secretion through effects on YopB and LcrG. J Bacteriol 180(13):3410–3420

Oaks EV, Hale TL, Formal SB (1986) Serum immune response to Shigella protein antigens in rhesus monkeys and humans infected with *Shigella* spp. Infect Immun 53(1):57–63

Ochman H, Soncini FC, Solomon F, Groisman EA (1996) Identification of a pathogenicity island required for Salmonella survival in host cells. Proc Natl Acad Sci USA 93(15):7800–7804

Olive AJ, Kenjale R, Espina M, Moore DS, Picking WL, Picking WD (2007) Bile salts stimulate recruitment of IpaB to the *Shigella flexneri* surface, where it colocalizes with IpaD at the tip of the type III secretion needle. Infect Immun 75(5):2626–2629. https://doi.org/10.1128/IAI.01599-06

Osiecki JC, Barker J, Picking WL, Serfis AB, Berring E, Shah S, Harrington A, Picking WD (2001) IpaC from Shigella and SipC from Salmonella possess similar biochemical properties but are functionally distinct. Mol Microbiol 42(2):469–481

Park D, Lara-Tejero M, Waxham MN, Li W, Hu B, Galan JE, Liu J (2018) Visualization of the type III secretion mediated Salmonella-host cell interface using cryo-electron tomography. Elife 7. https://doi.org/10.7554/elife.39514

Picking WL, Nishioka H, Hearn PD, Baxter MA, Harrington AT, Blocker A, Picking WD (2005) IpaD of *Shigella flexneri* is independently required for regulation of Ipa protein secretion and efficient insertion of IpaB and IpaC into host membranes. Infect Immun 73(3):1432–1440. https://doi.org/10.1128/IAI.73.3.1432-1440.2005

Pizarro-Cerda J, Charbit A, Enninga J, Lafont F, Cossart P (2016) Manipulation of host membranes by the bacterial pathogens Listeria, Francisella, Shigella and Yersinia. Semin Cell Dev Biol 60:155–167. https://doi.org/10.1016/j.semcdb.2016.07.019

Pope LM, Reed KE, Payne SM (1995) Increased protein secretion and adherence to HeLa cells by Shigella spp. following growth in the presence of bile salts. Infect Immun 63(9):3642–3648

Rathinavelan T, Lara-Tejero M, Lefebre M, Chatterjee S, McShan AC, Guo DC, Tang C, Galan JE, De Guzman RN (2014) NMR model of PrgI-SipD interaction and its implications in the needle-tip assembly of the Salmonella type III secretion system. J Mol Biol 426(16):2958–2969. https://doi.org/10.1016/j.jmb.2014.06.009

Rathinavelan T, Tang C, De Guzman RN (2011) Characterization of the interaction between the Salmonella type III secretion system tip protein SipD and the needle protein PrgI by paramagnetic relaxation enhancement. J Biol Chem 286(6):4922–4930. https://doi.org/10.1074/jbc.M110.159434

Roehrich AD, Guillossou E, Blocker AJ, Martinez-Argudo I (2013) Shigella IpaD has a dual role: signal transduction from the type III secretion system needle tip and intracellular secretion regulation. Mol Microbiol 87(3):690–706. https://doi.org/10.1111/mmi.12124

Roine E, Wei W, Yuan J, Nurmiaho-Lassila EL, Kalkkinen N, Romantschuk M, He SY (1997) Hrp pilus: an hrp-dependent bacterial surface appendage produced by *Pseudomonas syringae* pv. tomato DC3000. Proc Natl Acad Sci USA 94(7):3459–3464

Rosqvist R, Forsberg A, Wolf-Watz H (1991) Intracellular targeting of the Yersinia YopE cytotoxin in mammalian cells induces actin microfilament disruption. Infect Immun 59 (12):4562–4569

Sample AK, Fowler JM, Brubaker RR (1987) Modulation of the low-calcium response in *Yersinia pestis* via plasmid-plasmid interaction. Microb Pathog 2(6):443–453

Sani M, Botteaux A, Parsot C, Sansonetti P, Boekema EJ, Allaoui A (2007) IpaD is localized at the tip of the *Shigella flexneri* type III secretion apparatus. Biochim Biophys Acta 1770 (2):307–311. https://doi.org/10.1016/j.bbagen.2006.10.007

Sansonetti PJ, Kopecko DJ, Formal SB (1981) *Shigella sonnei* plasmids: evidence that a large plasmid is necessary for virulence. Infect Immun 34(1):75–83

Sansonetti PJ, Kopecko DJ, Formal SB (1982) Involvement of a plasmid in the invasive ability of *Shigella flexneri*. Infect Immun 35(3):852–860

Sarker MR, Neyt C, Stainier I, Cornelis GR (1998) The Yersinia Yop virulon: LcrV is required for extrusion of the translocators YopB and YopD. J Bacteriol 180(5):1207–1214

Sato H, Frank DW (2011) Multi-functional characteristics of the *Pseudomonas aeruginosa* type III needle-tip protein, PcrV; comparison to orthologs in other gram-negative bacteria. Front Microbiol 2:142. https://doi.org/10.3389/fmicb.2011.00142

Schubert K, Olde Damink SWM, von Bergen M, Schaap FG (2017) Interactions between bile salts, gut microbiota, and hepatic innate immunity. Immunol Rev 279(1):23–35. https://doi.org/10.1111/imr.12579

Serpell LC, Sunde M, Blake CC (1997) The molecular basis of amyloidosis. Cell Mol Life Sci 53 (11–12):871–887

Shaulov L, Gershberg J, Deng W, Finlay BB, Sal-Man N (2017) The ruler protein EscP of the enteropathogenic *Escherichia coli* type III secretion system is involved in calcium sensing and secretion hierarchy regulation by interacting with the gatekeeper protein SepL. MBio 8(1). https://doi.org/10.1128/mbio.01733-16

Shea JE, Hensel M, Gleeson C, Holden DW (1996) Identification of a virulence locus encoding a second type III secretion system in *Salmonella typhimurium*. Proc Natl Acad Sci USA 93 (6):2593–2597

Shen DK, Saurya S, Wagner C, Nishioka H, Blocker AJ (2010) Domains of the *Shigella flexneri* type III secretion system IpaB protein involved in secretion regulation. Infect Immun 78 (12):4999–5010. https://doi.org/10.1128/IAI.00470-10

Skrzypek E, Straley SC (1995) Differential effects of deletions in lcrV on secretion of V antigen, regulation of the low-Ca^{2+} response, and virulence of *Yersinia pestis*. J Bacteriol 177(9):2530–2542

Stensrud KF, Adam PR, La Mar CD, Olive AJ, Lushington GH, Sudharsan R, Shelton NL, Givens RS, Picking WL, Picking WD (2008) Deoxycholate interacts with IpaD of *Shigella flexneri* in inducing the recruitment of IpaB to the type III secretion apparatus needle tip. J Biol Chem 283(27):18646–18654. https://doi.org/10.1074/jbc.M802799200

Terry CM, Picking WL, Birket SE, Flentie K, Hoffman BM, Barker JR, Picking WD (2008) The C-terminus of IpaC is required for effector activities related to Shigella invasion of host cells. Microb Pathog 45(4):282–289. https://doi.org/10.1016/j.micpath.2008.06.003

Torruellas J, Jackson MW, Pennock JW, Plano GV (2005) The *Yersinia pestis* type III secretion needle plays a role in the regulation of Yop secretion. Mol Microbiol 57(6):1719–1733. https://doi.org/10.1111/j.1365-2958.2005.04790.x

Tran Van Nhieu G, Caron E, Hall A, Sansonetti PJ (1999) IpaC induces actin polymerization and filopodia formation during Shigella entry into epithelial cells. EMBO J 18(12):3249–3262. https://doi.org/10.1093/emboj/18.12.3249

Tucker SC, Galan JE (2000) Complex function for SicA, a *Salmonella enterica* serovar typhimurium type III secretion-associated chaperone. J Bacteriol 182(8):2262–2268

van der Goot FG, Tran van Nhieu G, Allaoui A, Sansonetti P, Lafont F (2004) Rafts can trigger contact-mediated secretion of bacterial effectors via a lipid-based mechanism. J Biol Chem 279 (46):47792–47798. https://doi.org/10.1074/jbc.M406824200

Veenendaal AK, Hodgkinson JL, Schwarzer L, Stabat D, Zenk SF, Blocker AJ (2007) The type III secretion system needle tip complex mediates host cell sensing and translocon insertion. Mol Microbiol 63(6):1719–1730. https://doi.org/10.1111/j.1365-2958.2007.05620.x

Verma SK, Tuteja U (2016) Plague vaccine development: current research and future trends. Front Immunol 7:602. https://doi.org/10.3389/fimmu.2016.00602

Wang Y, Ouellette AN, Egan CW, Rathinavelan T, Im W, De Guzman RN (2007) Differences in the electrostatic surfaces of the type III secretion needle proteins PrgI, BsaL, and MxiH. J Mol Biol 371(5):1304–1314. https://doi.org/10.1016/j.jmb.2007.06.034

Wang Y, Zhang L, Picking WL, Picking WD, De Guzman RN (2008) Structural dissection of the extracellular moieties of the type III secretion apparatus. Mol BioSyst 4(12):1176–1180. https://doi.org/10.1039/b808271p

Watarai M, Tobe T, Yoshikawa M, Sasakawa C (1995) Contact of Shigella with host cells triggers release of Ipa invasins and is an essential function of invasiveness. EMBO J 14(11):2461–2470

Wei ZM, Beer SV (1993) HrpI of Erwinia amylovora functions in secretion of harpin and is a member of a new protein family. J Bacteriol 175(24):7958–7967

Wiener M, Freymann D, Ghosh P, Stroud RM (1997) Crystal structure of colicin Ia. Nature 385 (6615):461–464. https://doi.org/10.1038/385461a0

Yu XJ, Grabe GJ, Liu M, Mota LJ, Holden DW (2018) SsaV interacts with SsaL to control the translocon-to-effector switch in the Salmonella SPI-2 type three secretion system. MBio 9(5). https://doi.org/10.1128/mbio.01149-18

Yu XJ, McGourty K, Liu M, Unsworth KE, Holden DW (2010) pH sensing by intracellular Salmonella induces effector translocation. Science 328(5981):1040–1043. https://doi.org/10.1126/science.1189000

Zhang L, Wang Y, Olive AJ, Smith ND, Picking WD, De Guzman RN, Picking WL (2007) Identification of the MxiH needle protein residues responsible for anchoring invasion plasmid antigen D to the type III secretion needle tip. J Biol Chem 282(44):32144–32151. https://doi.org/10.1074/jbc.M703403200

Zhang L, Wang Y, Picking WL, Picking WD, De Guzman RN (2006) Solution structure of monomeric BsaL, the type III secretion needle protein of *Burkholderia pseudomallei*. J Mol Biol 359(2):322–330. https://doi.org/10.1016/j.jmb.2006.03.028

Zhang YZ, Xiang X, Mei P, Dai J, Zhang LL, Liu Y (2009) Spectroscopic studies on the interaction of Congo Red with bovine serum albumin. Spectrochim Acta A Mol Biomol Spectrosc 72(4):907–914. https://doi.org/10.1016/j.saa.2008.12.007

Diversity and Evolution of Type III Secreted Effectors: A Case Study of Three Families

Donald Patrick Bastedo, Timothy Lo, Bradley Laflamme, Darrell Desveaux and David S. Guttman

Contents

Patrick Bastedo and Timothy Lo contributed equally to this work.

D. P. Bastedo · T. Lo · B. Laflamme · D. Desveaux (✉) · D. S. Guttman (✉)
Department of Cell & Systems Biology, University of Toronto, Toronto, ON, Canada
e-mail: darrell.desveaux@utoronto.ca

D. S. Guttman
e-mail: david.guttman@utoronto.ca

Current Topics in Microbiology and Immunology (2020) 427: 201–230
https://doi.org/10.1007/82_2019_165
© Springer Nature Switzerland AG 2019
Published Online: 26 June 2019

Abstract A broad range of Gram-negative bacteria employ a type III secretion system (T3SS) to deliver virulence proteins termed type III secreted effectors directly into the cytoplasm of eukaryotic host cells. While effectors can contribute to the colonization of eukaryotic hosts by bacterial symbionts and pathogens, they can also elicit host immune responses that restrict bacterial growth. These opposing selective pressures have shaped the evolution of effector families and may be responsible for their incredible diversity in biochemical function, mechanism of action, and taxonomic distribution. In this chapter, we focus on three distinct effector families whose members are distributed among both plant and animal pathogens. We first discuss the LRR-NEL and YopJ families of effectors. These two effector families possess ubiquitin ligase and acetyltransferase activity, respectively, which in both cases can be directed against host innate immune signal transduction pathways to promote infection. Finally, we discuss the TALE family of transcription activator-like effectors that serve to reprogram host immunity transcriptional responses. This chapter aims to highlight the diversity within these three effector families that results from the strong and dynamic evolutionary forces shaping the interface between host and bacterium.

1 Introduction

The type III secretion system (T3SS) is a needle-like nanostructure that allows Gram-negative bacteria to inject proteins, termed 'type III secreted effectors' (hereafter, simply 'effectors'), directly into the cytoplasm of host cells (Galán and Collmer 1999). Structural components of T3SSs are highly conserved across bacterial species and are utilized by both pathogens and symbionts to colonize plant, animal, and protist hosts (Notti and Stebbins 2016). T3SSs were first discovered in *Yersinia* and *Salmonella* (Michiels et al. 1990, 1991) and have since been reported in a wide range of Gram-negative bacteria including *Bordetella*, *Burkholderia*, *Chlamydia*, *Erwinia*, *Escherichia*, *Pseudomonas*, *Ralstonia*, *Rhizobium*, *Salmonella*, *Shigella*, *Vibrio*, and *Xanthomonas* (Galán and Collmer 1999). The T3SS plays a fundamental role in the lifestyle of many Gram-negative bacteria by allowing intercellular interactions between bacteria and their hosts.

A critical challenge faced by pathogenic bacteria is the need to avoid or subvert host immune responses (Lamkanfi and Dixit 2010; Rosadini and Kagan 2015). Human pathogens are recognized by the host immune system at the cell surface through pattern recognition receptors (PRRs) such as the Toll-like receptors, activating immune signal transduction cascades that involve mitogen-activated protein kinases (MAPKs), and the NF-κB family of transcription factors (Dong et al. 2002; O'Neill 2002). In addition, cytosolic complexes called 'inflammasomes' contain nucleotide-binding, LRR domain-containing NOD-like receptors (NLRs) that function as intracellular PRRs to sense a variety of different pathogen- and damage-associated molecular patterns (Lechtenberg et al. 2014; Hauenstein et al. 2015; Hu and Chai 2016).

Similar to the animal immune system, plants recognize phytopathogens through the use of membrane-spanning PRR complexes that transduce external stimuli through cytosolic kinase domains (Ronald and Beutler 2010). Propagation of these signals results in PRR-triggered immune responses dependent on both MAPK cascades and regulation of WRKY domain-containing transcription factors that modulate immunity-related gene expression. To recognize intracellular pathogen-associated molecules, plant immune systems also use NLR-like proteins (Maekawa et al. 2011; Jones et al. 2016). However, in contrast to the animal immune system where NLR proteins recognize broadly conserved bacterial molecules, plant NLRs detects pathogen effectors either directly, or indirectly, by monitoring host proteins for effector-induced perturbations (Jones et al. 2016; Khan et al. 2016). Plant NLR activation triggers immune signaling cascades that lead to a second layer of immunity termed effector-triggered immunity (Jones et al. 2016).

These highly effective and multi-layered immune systems present a daunting challenge for prospective bacterial pathogens. While effectors provide one of the most effective means for bacterial pathogens to subvert host immunity, they also frequently act as immune elicitors (Jones and Dangl 2006; Dodds and Rathjen 2010; Raymond et al. 2013; Stuart et al. 2013; Khan et al. 2018). These contrasting roles in host-pathogen interactions hint at the intensity and erratic nature of the selective pressures acting on effectors. It is likely that such strong evolutionary pressures are responsible for driving the emergence and diversification of the large number of existing effector families, which collectively encompass a wide range of biochemical functions (Scott and Hartland 2017). This diversity, which has evolved over the hundreds of millions of years since bacteria first began interacting with eukaryotes through translocated effectors, can act as a guide to the variety of mechanisms bacterial pathogens can use to manipulate their hosts.

This chapter aims to summarize recent advances in effector biology with a focus on the genetic and functional diversity of three distinct effector families. We first focus on two effector families (LRR-NEL ubiquitin ligases and YopJ-like acetyltransferases) that are broadly conserved not only among diverse bacterial pathogens of plants and animals, but in plant symbionts as well. Finally, we discuss a more restricted effector family, the transcription activator-like effectors (TALEs), originally identified in plant pathogens from the genus *Xanthomonas*, and provide a framework to illustrate their allelic diversity.

2 LRR-Associated Novel E3 Ubiquitin Ligases (LRR-NELs)

2.1 LRR-NELs: General Features

Many fundamental processes in eukaryotic host cells are regulated by ubiquitination (Yau and Rape 2016). A number of bacterial effectors have been described that co-opt the structure and function of eukaryotic E3 ubiquitin ligases despite lacking obvious sequence similarity, including AvrPtoB from *Pseudomonas syringae*, which adopts a fold that resembles eukaryotic RING-like E3 ligases (Dong et al. 2009), and SopA from *Salmonella typhimurium*, which structurally resembles HECT-like E3 s (Zhang et al. 2006; Diao et al. 2008). However, members of a third class of bacterial effectors with E3 ligase activity share a structurally unique alpha-helical fold referred to as 'novel E3 ubiquitin ligase' (NEL) domains, first described for IpaH1.4 of *Shigella flexneri* (Singer et al. 2008). NEL-containing proteins belong to a family whose members share a characteristic domain structure, an amino-terminal leucine-rich repeat (LRR) domain that is followed by a carboxy-terminal NEL domain (NCBI Conserved Domain Database entry cl26018; Marchler-Bauer et al. 2011, 2015). Such LRR-NELs are emerging as recurrent features of effector repertoires (Table 1).

The carboxy-terminal NEL domains of IpaH family proteins are highly similar, while their amino-terminal LRR domains are more variable and serve to define substrate specificity (Rohde et al. 2007). Full-length structures of the LRR-NELs IpaH3 and SspH2 (from *Shigella* and *Salmonella*, respectively; Zhu et al. 2008; Quezada et al. 2009), in vitro characterizations (Singer et al. 2008; Zhu et al. 2008; Quezada et al. 2009), and structure-guided mutagenesis (Chou et al. 2012) has shown that LRR domains also act to suppress auto-ubiquitination in the absence of substrate. This mechanism may serve to enhance the stability of these effectors in host cells while also limiting ubiquitination of 'off-target' substrates (Chou et al. 2012). Below, we briefly summarize functional characterizations of 'canonical' LRR-NELs like those of *Shigella* and *Salmonella* and then proceed to highlight family members that deviate from this canonical domain structure.

2.2 *Canonical LRR-NEL Effector Proteins of Shigella, Salmonella, and Rhizhobia*

Gastrointestinal pathogens in the genus *Shigella* (e.g., *S. dysenteriae*, *S. flexneri*, *S. boydii,* and *S. sonnei*) possess a 220 kb virulence plasmid that encodes a T3SS and a number of effectors, including LRR-NEL proteins of the IpaH family (IpaH1.4, IpaH2.5, IpaH4.5, IpaH7.8, and IpaH9.8) (Buchrieser et al. 2000). These effectors contribute to virulence by targeting various components of NF-κB signaling and inflammasome activation (Ashida and Sasakawa 2016). For example: IKKγ/NEMO, a regulatory component of the inhibitor of kappa-B kinase (IKK) complex (Israël 2010), is ubiquitinated and degraded by IpaH9.8 to dampen NF-κB signaling (Ashida et al. 2010); IpaH1.4 and IpaH2.5 ubiquitinate and cause degradation of HOIP, the catalytic component of the Linear Ubiquitin Chain Assembly Complex (LUBAC), which positively regulates NF-κB signaling (de Jong et al. 2016); IpaH4.5 targets the p65 subunit of NF-κB itself (Wang et al. 2013) (and also TANK-Binding Kinase 1, positive regulator of an IRF3-dependent cytokine expression pathway; Zheng et al. 2016); and IpaH7.8 targets glomulin (a cullin-RING ligase inhibitor which negatively

Table 1 Summary of effectors discussed in this chapter with known or proposed functions and/or targets

Effector family	Effector name	Organism	Effector function	Proposed target(s)	Host/niche	Reference(s)
LRR-NEL/YopM (CDD: cl26018)	IpaH1.4	*Shigella flexneri*	Ubiquitin ligase (LRR-NEL)	LUBAC (NF-κB activation)	Human	Singer et al. (2008), de Jong et al. (2016)
	IpaH2.5	*S. flexneri*	Ubiquitin ligase (LRR-NEL)	LUBAC (NF-κB activation)	Human	de Jong et al. (2016)
	IpaH3	*S. flexneri*	Ubiquitin ligase (LRR-NEL)	Unknown	Human	Zhu et al. (2008)
	IpaH4.5	*S. flexneri*	Ubiquitin ligase (LRR-NEL)	NF-κB p65, TBK1 (kinase)	Human	Zhu et al. (2008), Wang et al. (2013), Zheng et al. (2016)
	IpaH7.8	*S. flexneri*	Ubiquitin ligase (LRR-NEL)	glomulin (negative regulation of NLRP3/NLRC4)	Human	Zhu et al. (2008), Suzuki et al. (2014)
	IpaH9.8	*S. flexneri*	Ubiquitin ligase (LRR-NEL)	IKKγ/NEMO	Human	Zhu et al. (2008), Ashida et al. (2010)
	SspH1	*Salmonella enterica*	Ubiquitin ligase (LRR-NEL)	PKN1 (kinase)	Human	Keszei et al. (2014)
	SspH2	*S. enterica*	Ubiquitin ligase (LRR-NEL)	SGT1 (NLR co-chaperone)	Human	Bhavsar et al. (2013)
	SlrP	*S. enterica*	Ubiquitin ligase (LRR-NEL)	Thioredoxin	Human	Bernal-Bayard and Ramos-Morales (2009), Zouhir et al. (2014)
	SlrP	*S. enterica*	Ubiquitin ligase (LRR-NEL)	ERdj3 (chaperone)	Human	Bernal-Bayard et al. (2010)
	NopM	*Sinorhizobium fredii*	Ubiquitin ligase (LRR-NEL)	Unknown	Plants (root nodules)	Xin et al. (2012)
	YopM	*Yersinia pseudotuberculosis*	Protein-binding adaptor (LRR only)	Caspase-1	Human	Larock and Cookson (2012)
	YopM	*Y. enterocolitica*	Protein-binding adaptor (LRR only)	Pyrin kinases PRK2, RSK1	Human	Hentschke et al. (2010)
	YopM	*Y. pestis*	Ubiquitin ligase (LRR only)	NLRP3	Human	Wei et al. (2016)
	Pspto_1492	*Pseudomonas syringae*	Ubiquitin ligase (long LRR-NEL)	Unknown	Plants	Zhu et al. (2008)
	Pspto_4093	*P. syringae*	Ubiquitin ligase (long LRR-NEL)	Unknown	Plants	Zhu et al. (2008)
	PP_2566	*Pseudomonas putida*	Ubiquitin ligase (long LRR-NEL)	Unknown	Soil	Zhu et al. (2008)
	RipAR (Rsp1236)	*Ralstonia solanacearum*	Ubiquitin ligase (NEL only)	Unknown	Plants	Nakano et al. (2017)
	RipAW (Rsp1475)	*R. solanacearum*	Ubiquitin ligase (NEL only)	Unknown	Plants	Nakano et al. (2017)

(continued)

Table 1 (continued)

Effector family	Effector name	Organism	Effector function	Proposed target(s)	Host/niche	Reference(s)
YopJ/HopZ (CDD: cl07849)	YopJ	*Y. pseudotuberculosis*	S/T-acetyltransferase	MEK2, IKK	Human	Mittal et al. (2006)
	YopJ	*Y. pseudotuberculosis*	S/T-acetyltransferase	MAPKK6	Human	Mukherjee et al. (2006)
	YopJ	*Y. pestis*	S/T-acetyltransferase	RICK, TAK1 kinases	Human	Meinzer et al. (2012), Paquette et al. (2012)
	AopP	*Aeromonas salmonicida*	S/T-acetyltransferase	NF-kB pathway	Fish, Human	Fehr et al. (2006)
	VopA	*Vibrio parahaemolyticus*	S/T/K-acetyltransferase	JNK, p38 (MAP kinases)	Human	Trosky et al. (2007)
	AvrA	*S. enterica*	S/T-acetyltransferase	MAPK signaling, NF-kB activation	Human	Jones et al. (2008)
	HopZ1a	*P. syringae*	S/T-acetyltransferase	Pseudokinase ZED1	Plants	Ma et al. (2006), Lewis et al. (2013)
	HopZ3	*P. syringae*	S/T-acetyltransferase	RLCK VII kinases	Plants	Ma et al. (2006), Lee et al. (2015)
	PopP2	*R. solanacearum*	S/T/K-acetyltransferase	WRKY domain transcription factors	Plants	Tasset et al. 2010, Sarris et al. (2015), Le Roux et al. (2015)
	NopJ	*Rhizobium*	S/T-acetyltransferase	Unknown	Plants (root nodules)	Kambara et al. (2009)
TALE/TALE-like effectors (CDD: cl28750)	TALEs	*Xanthomonas spp.*	Transcription activator mimic	Host promoters	Plants	White et al. (2009), Erkes et al. (2017)
	truncTALES	*Xanthomonas spp.*	Block immunity-eliciting TALEs	Direct binding to TALEs	Plants	Read et al. (2016)
	iTALES	*Xanthomonas spp.*	Block immunity-eliciting TALEs	Direct binding to TALEs	Plants	Ji et al. (2016)
	RipTals	*R. solanacearum*	Transcription activator mimic	Host promoters	Plants	Cunnac et al. (2004), de Lange et al. (2013), Li et al. (2013a)
	Bats	*Burkholderia rhizoxinica*	Transcription activator mimic	Host promoters	Fungus	de Lange et al. (2014), Juillerat et al. (2014)
	MOrTLs	*Unknown marine bacteria*	Transcription activator mimic	Host promoters	Unknown	de Lange et al. (2015)

regulates NLRP3 and NLRC4-containing inflammasomes) to promote pyroptotic cell death in macrophages, enhancing subsequent invasion of epithelial cells (Suzuki et al. 2014). *Shigella* spp. also have chromosomally encoded LRR-NEL genes, but these likely provide overlapping/redundant functions; individual deletion mutants of the chromosomal loci showed no differences relative to a wild-type strain in a mouse infection model, while a deletion of all chromosomal genes (Δ*ipaH*-null) had attenuated virulence (Ashida et al. 2007).

Salmonella enterica delivers three LRR-NELs (SspH1, SspH2, and SlrP) into host cells. SspH1 modulates NF-κB pathway activation and binds to an inflammasome-regulatory kinase, PKN1 (Haraga and Miller 2006; Keszei et al. 2014). Like *Shigella* IpaH7.8, *Salmonella* SspH2 targets NLR activity, albeit indirectly through SGT1, an NLR co-chaperone (Bhavsar et al. 2013). SlrP targets both thioredoxin (Bernal-Bayard and Ramos-Morales 2009; Zouhir et al. 2014) and ERdj3, an endoplasmic reticulum luminal chaperone (Bernal-Bayard et al. 2010). Remarkably, a recent proteomic study found that the abundance of as many as 37 human proteins is affected by SlrP expression (Cordero-Alba et al. 2016). The fact that both *Shigella* and *Salmonella* possess multiple LRR-NELs that manipulate the NF-κB pathway and inflammasome activation illustrates the importance of these pathways for an effective anti-bacterial immune response.

LRR-NEL effectors have also been identified in rhizobial species, nitrogen-fixing symbionts required for formation of root nodules on leguminous plants. NopM is an LRR-NEL effector first identified in a mass spectrometry characterization of the secreted effectors of *Sinorhizobium fredii* HH103 (Rodrigues et al. 2007) and is closely related to effectors from *Bradyrhizobium* and *Rhizobium* spp. Functional characterization of the *Rhizobium* homologue showed that NopM can promote nodulation of *Lablab purpureus* (hyacinth bean), but can also induce defense responses in *Nicotiana benthamiana* (wild tobacco) (Xin et al. 2012). This result demonstrates context-dependent consequences of effector functions and illustrates the shifting (and often opposing) influences they have on the fitness of both bacteria and host.

2.3 Variations on a Theme: Non-canonical LRR-NELs

The LRR-NEL family also includes effectors that deviate from the canonical domain structure exemplified by the *Shigella* and *Salmonella* effectors described above. YopM effectors from *Yersinia* spp. lack the NEL domain, *Ralstonia* members lack LRR domains, and sequences from *P. syringae* and *Pseudomonas putida* have exceptionally long amino-terminal extensions that lack annotated functional domains (Rohde et al. 2007; Fig. 1a). We discuss each of these three domain structure variants below.

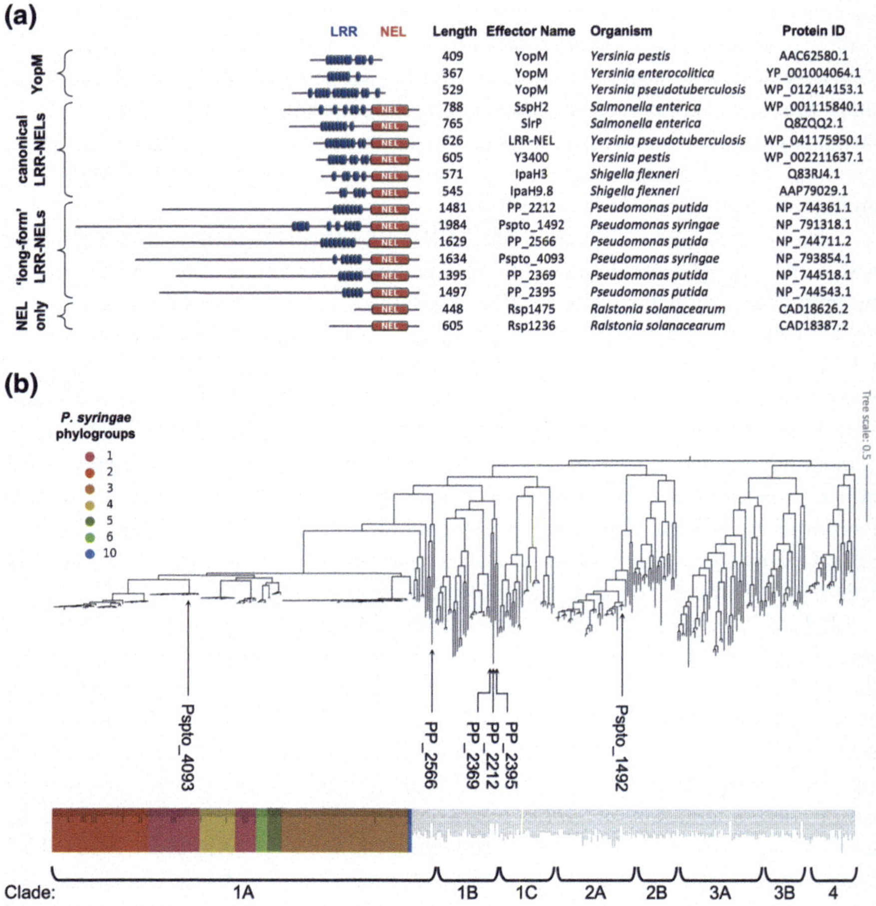

Fig. 1 Diversity of the LRR-associated Novel E3 ubiquitin Ligases (LRR-NELs). **a** Domain architecture of representative LRR-NEL proteins. Leucine-rich repeats are shown in blue, while NEL domains are shown in red. Note that this representation is not derived from alignment; rather, unaligned sequences were positioned by centering on the midpoint of NEL domains (where present) or on the LRR domain midpoint (where a NEL domain is absent). Domain boundaries were extracted from the GenBank accessions listed under 'Protein ID.' **b** Phylogenetic relationships of 'long-form' LRR-NEL proteins. Representative subsets of protein sequences related to Pspto_1492, Pspto_4093, PP_2212, PP_2369, PP_2395, and PP_2566 were identified by BLASTP search (Altschul et al. 1997). All of the hits shown have e-values $\leq 10^{e-120}$, span >70% of at least one of these six query sequences, and range in length from 1063 to 2004 amino acids. Multiple sequence alignment was performed using MUSCLE (Edgar 2004), and a maximum likelihood phylogenetic analysis was performed with PhyML (Guindon and Gascuel 2003; Guindon et al. 2005, 2010). Branch support was calculated by 'Approximate Likelihood-Ratio Test' (aLRT; Anisimova and Gascuel 2006), and only nodes with greater than 70% support are shown. Labels for *P. syringae* sequences from our in-house database are colored according to the phylogroup designation of the isolate encoding the protein

***LRR Only: YopM Proteins of Pathogenic Yersinia spp*.** Three species from the genus *Yersinia* have received considerable research attention due to their ability to cause disease in humans. Among these, *Yersinia pestis* is notorious for having caused past epidemics of bubonic and pneumonic plague, while *Yersinia enterocolitica* and *Yersinia pseudotuberculosis* can cause a range of gastrointestinal diseases (Hueck 1998). These three species differ from non-pathogenic *Yersinia* isolates by the presence of a 70 kb virulence plasmid, pCD (also known as pYV in *Y. pseudotuberculosis*). This virulence plasmid encodes structural components of the T3SS, regulatory proteins, and six effectors: YopE, YopH, YopJ, YopM, YopO (also known as YpkA), and YopT.

The plasmid-borne YopM effectors from pathogenic *Yersinia* spp. consist of variable numbers of leucine-rich repeat units, and all lack a carboxy-terminal NEL domain (Evdokimov et al. 2001; Fig. 1a). YopM effectors nevertheless play an important role in infection by pathogenic *Yersinia*. Deletion of *yopM* from *Yersinia* spp. results in attenuated virulence and reduced colonization of target organs in mice (Trülzsch et al. 2004). Experiments with cultured macrophages indicate that YopM contributes to modulation of the immune responses induced by other effectors. *Yersinia* strains engineered with a *yopM* deletion induce significantly more inflammasome activity (measured by outputs such as secretion of inflammatory cytokines) than otherwise isogenic double mutants that add mutations in *yopE* (Chung et al. 2016; Ratner et al. 2016a) and *yopT* (Chung et al. 2016). Conversely, a *yopM yopJ* double mutant results in higher levels of cytokine secretion than either of the single mutants (Ratner et al. 2016a, b). These studies provide interesting examples of effector cooperativity and underscore the complexity of effector-effector interactions.

Since YopM proteins lack additional domains or defined catalytic motifs, it is implied that they act through direct binding interactions with host immunity-related proteins. Indeed, several recent studies implicate components of the inflammasome as important targets of YopM. Proposed mechanisms of inflammasome inhibition vary and include: (1) direct binding of YopM to caspase-1 resulting in nuclear sequestration (Larock and Cookson 2012); (2) direct binding of YopM to host kinases PRK2 and RSK1 (Hentschke et al. 2010), which in turn phosphorylate the NLR inflammasome component Pyrin (Chung et al. 2016; Ratner et al. 2016a); and (3) induced ubiquitination of the NLR inflammasome component NLRP3 (Wei et al. 2016). Surprisingly, in this last report the authors present data that suggests an E3 ligase activity for YopM, despite lacking a NEL domain. They attribute catalytic function to a cysteine residue in a conserved amino-terminal 'CLD' motif that was identified based on similarity to the 'CxD' catalytic motif of canonical NEL domains. This unexpected result suggests a scenario whereby convergent evolution has restored E3 ligase function to YopM following its earlier loss from an ancestral LRR-NEL. Further study of other YopM sequences will clarify whether this enzymatic activity is unique to this particular allele or is instead a general feature of the *Yersinia* YopM family.

In addition to the plasmid-borne YopM effectors mentioned above, chromosomally encoded canonical LRR-NEL proteins are also present in both pathogenic

(*Y. pestis*, *Y. pseudotuberculosis*, but not *Y. enterocolitica*) and environmental (*Y. intermedia*, *Y. similis* and *Y. wautersii*) *Yersinia* species (Fig. 1a; Hu et al. 2016). These LRR-NELs are found at the same chromosomal site and share flanking genes, but there is evidence of plasticity at this locus—between one and three LRR-encoding genes are present in tandem (Hu et al. 2016). In all instances with three chromosomal LRR genes, the protein encoded by the second copy (and sometimes the third) lacks a NEL domain in the protein-coding sequence, although nucleotide sequences downstream of the open reading frame indicate the presence of degenerate NEL domain-encoding sequences (Hu et al. 2016).

The chromosomally encoded *Yersinia* LRR-NEL proteins represent candidate precursors of YopM and suggest a possible mechanism for explaining the provenance of the plasmid-borne *yopM* effector genes. LRR-NEL gene duplication would be followed by loss of the NEL domain and subsequent acquisition by the virulence plasmid (Hu et al. 2016). However, phylogenetic analysis of the LRR repeats sequences from chromosomal (LRR-NEL) and plasmid-borne (YopM) groups of *Yersinia* effectors demonstrated that these two groups represent distinct clades (Hu et al. 2016). These results imply that if the plasmid-borne YopMs evolved from chromosomal LRR-NELs, the acquisition event is old enough to have allowed significant divergence in repeat sequences and presumably, by extension, in substrate specificity and function.

Studies of *Yersinia* effectors have largely been limited to those encoded by pCD/pYV, and the translocation and functions of the chromosomally encoded LRR-NELs have not been tested in infection models. Nevertheless, given that bacteria do not produce E1 and E2 enzymes and use distinct, ubiquitin-independent proteolysis components, it seems likely that these genes have been retained over evolutionary time due to advantageous functions provided within host cells.

NEL-Only Effectors of Ralstonia. *Ralstonia solanacearum* is a phytopathogen that causes bacterial wilt in a range of plant hosts including tobacco, tomato, pepper, eggplant, and potato. *R. solanacearum* secretes more than 100 proteins into the extracellular milieu during culture in minimal media, and at least 60 of these are T3SS substrates known as Rips (*Ralstonia* injected proteins) (Poueymiro and Genin 2009). The *Ralstonia* NELs RipAW and RipAR (Rsp1475 and Rsp1236, respectively) lack LRR domains (Fig. 1a) but, as expected, demonstrate an in vitro poly-ubiquitination activity that is dependent on the presence of E1, E2, ubiquitin, and the catalytic cysteine (Nakano et al. 2017). These proteins also significantly suppress innate immune responses such as the production of reactive oxygen species and the expression of defense-related genes when expressed in leaves of *N. benthamiana* (Nakano et al. 2017). These studies characterizing the NEL-only effectors of *Ralstonia* provide an important insight, since they demonstrate that the NEL domains can modulate host physiology even in the absence of the LRR domains deemed important for substrate recognition and specificity in the canonical LRR-NELs of other organisms.

'Long-Form' LRR-NELs in Plant-Associated Pseudomonas spp. The genus *Pseudomonas* includes the human pathogen *P. aeruginosa*, plant pathogens from the *P. syringae* species complex, and environmental/commensal species such as

P. putida and *P. fluorescens*. Although *P. aeruginosa* does not encode NEL domain-containing proteins, *P. syringae* pv. *tomato* DC3000 (*Pto)* and *P. putida* KT2440 both encode multiple LRR-NEL proteins (two and four, respectively; Fig. 1a). Given that these six proteins are conspicuously longer than any of the other representative family members, we searched for similar long-form LRR-NELs among available databases by using NCBI BLASTP (Altschul et al. 1997) as well as by searching our own 'in-house' database derived from a de novo genome analysis pipeline (DeNoGAP; Thakur and Guttman 2016) that was applied to a collection of 367 *P. syringae* isolates.

In our *P. syringae* database, we identified a protein cluster that includes sequences of similar length and domain structure as the 'long-form' LRR-NELs encoded by *Pto* and *P. putida* (Fig. 1a). Specifically, proteins similar to the *Pto* LRR-NEL Pspto_4093 were identified in 66% (243 of 367) of the strains considered in our analysis and are broadly distributed among isolates from all of the primary *P. syringae* phylogroups (phylogroups 1, 2, 3, 4, 5, 6, and 10; Fig. 1b). These sequences were supplemented with additional sequences identified by sampling high-scoring BLASTP hits for each of the six long-form LRR-NELs shown in Fig. 1a and an alignment of this set of sequences was used for phylogenetic inference. Sequences represented in the resulting tree (Fig. 1b) are from genus *Pseudomonas*, including both the plant pathogen *P. syringae* and environmental/commensal strains like *P. putida* and *P. fluorescens*.

It is striking that the amino-terminal extensions ($\sim$ 1000 amino acids) of the long-form LRR-NELs lack similarity to any previously characterized protein domains (i.e., no hits are detected with an E-value less than the default threshold of 0.01 using the 50,369 PSSMs represented by CDD v3.16; Marchler-Bauer et al. 2011, 2015). If only the LRR-NEL portion of these atypical family members is functionally important, we would hypothesize that the amino-terminal extensions distinguishing these sequences would be under few evolutionary constraints and therefore be prone to degeneration. However, our amino acid alignment of *P. syringae* LRR-NELs with related sequences identified with BLASTP indicates sequence conservation along the entire length of the long-form LRR-NELs (not shown), suggesting that selection is acting to conserve these sequences, and robust in vitro ubiquitin ligase activity has been demonstrated for both Pspto_1492 and Pspto_4093 (Zhu et al. 2008). Furthermore, the Pspto_4093-related alleles have a genealogical structure that is congruent with the *P. syringae* core genome phylogeny (Fig. 1b), suggesting that this locus was present in the common ancestor of the *P. syringae* species complex and has been selectively maintained since. Notably, *P. putida* PP_2566 is present in the same subclade as *P. syringae* Pspto_4093 (clade 1A; Fig. 1b), while the other three *P. putida* sequences are members of a closely related subclade (clade 1B; Fig. 1b). Pspto_1492, on the other hand, is part of a distinct clade of *P. syringae* sequences (clade 2A; Fig. 1b). These findings together suggest that these unusual LRR-NELs provide important but adaptable functions whose evolution has been enabled by gene duplication and subsequent diversification into distinct lineages.

Finally, we wish to emphasize that neither Pspto_4093 nor Pspto_1492 are among the 29 previously confirmed effectors from *Pto* (Chang et al. 2005) and they both apparently lack promoter sequences characteristic of typical T3SS substrates in *P. syringae* (not shown). However, as for the chromosomally encoded LRR-NELs of *Yersinia* discussed above, we entertain the possibilities that these proteins are either (1) cryptic T3SS substrates, (2) delivered to host cells by a T3SS-independent mechanism, or (3) have evolved to contribute novel beneficial functions within the bacterial cell.

3 The YopJ Acetyltransferase Family

3.1 YopJ-like Acetyltransferases: General Features

The *Yersinia* YopJ proteins (known as YopP in *Y. enterocolitica*) were the first described representatives of an effector family (NCBI Conserved Domain Database entry cl07849; Marchler-Bauer et al. 2011, 2015) that includes members from diverse bacterial genera including pathogens of both animals (*Salmonella, Vibrio, Aeromonas*) and plants (*Xanthomonas*, *Ralstonia*, *P. syringae*), as well as plant symbionts (*Rhizobium*) (Orth 2000; Ma et al. 2006; Lewis et al. 2011; Table 1). YopJ and related proteins are acetyltransferases, with a conserved cysteine essential for acetyltransfer (Mittal et al. 2006; Mukherjee et al. 2006). This activity consumes the acetyl donor acetyl-coenzyme A (Ac-CoA) and requires inositol 6-*kis*phosphate (IP6) as a co-factor (Mittal et al. 2010). Most YopJ family members preferentially modify serine and threonine residues of their substrates, although certain members like PopP2 (*Ralstonia*) and VopA (*Vibrio*) are capable of acetylating lysines as well (Trosky et al. 2007; Tasset et al. 2010; Le Roux et al. 2015; Sarris et al. 2015).

The recently determined structure of HopZ1a, a YopJ family member from *P. syringae,* has provided a molecular basis for understanding the enzymatic activity characteristic of the YopJ family. The protein crystals obtained included two bound ligands—the co-factor IP6 and the post-acetyltransfer product of the spent acetyl donor, CoA (Zhang et al. 2016). This initial structure was later corroborated by structures of an additional YopJ family member, PopP2 from *R. solanacearum*, which additionally feature bound substrate and an acetylated catalytic cysteine intermediate (Zhang et al. 2017). Despite representing distinct sub-lineages of YopJ-related effectors (Fig. 2b), the structures of HopZ1a and PopP2 are quite similar overall (Fig. 2a). One conspicuous exception is presented by alpha-helix αD, which contributes to the substrate-binding interface in PopP2 (Fig. 2a). By mapping the conservation information derived from aligned *P. syringae* HopZ sequences (Sect. 3.5, below) onto the surface of the HopZ1a structure, we show that this αD helix contributes to a surface that is variable among closely related HopZ alleles. Although mostly variable, this 'front' face of the protein also includes a highly conserved pocket containing the active site and the acetyl donor Ac-CoA, while the 'back' face of the protein is more conserved overall and features the IP6 binding pocket (Fig. 2d). We speculate that the variability of the αD helix contributes to the differing substrate specificities among members of this effector family (Lewis et al. 2011; Ma and Ma 2016).

3.2 Evolutionary Relationships Among the YopJ Effectors of Animal Pathogens

To compare the sequence diversity among YopJ-like effectors, we aligned sequences obtained from the NCBI by BLASTP searches and from our in-house *P. syringae* database (see Sect. 3.5, below). A maximum likelihood phylogenetic tree describing this alignment is presented in Fig. 2b. Consistent with previous analyses (Ma and Ma 2016), our phylogeny indicates that *Yersinia* YopJ/YopP sequences are most similar to homologous effectors from other animal pathogens (*Aeromonas* AopP, *Salmonella* AvrA, and *Vibrio* VopA). Both the AvrA and VopA clades are composed of virtually identical sequences (Fig. 2b). The *Yersinia* YopJ/YopP sequences are also highly similar, although earlier reports have identified substitutions at three distinct positions that influence immune activation as measured by cytokine secretion (Ruckdeschel et al. 2001; Zheng et al. 2011). YopP from the O8 serotype of *Y. enterocolitica* is more proficient at induction of apoptosis in murine macrophages than the corresponding allele from serotypes O3 and O9 due to an R143S substitution in $\text{YopP}^{\text{O3/O9}}$ that results in reduced phosphorylation of IKK-β (Ruckdeschel et al. 2001). This position is located on the aforementioned helix αD, which is proximal to the substrate-binding surface of PopP2 (Fig. 2a). Therefore, it is likely that this mutation alters interactions between YopJ and one or more of its substrates. Similarly, two amino acid substitutions (L177F, E206K) in the YopJ allele from *Y. pestis* strain KIM are apparently responsible for induction of higher levels of cytotoxicity and caspase-1 cleavage compared to *Y. pestis* strain CO92 (Zheng et al. 2011). Of these two positions, L177 is found near the IP6 binding site and the phenotypic consequences of the substitution at this position may therefore reflect altered co-factor binding.

Finally, the most divergent group of YopJ-related sequences from animal pathogens are from *Bartonella* spp. (Fig. 2b). These organisms are facultative intracellular pathogens transmitted by blood-sucking arthropods, causing 'cat-scratch disease' and other zoonotic infections in humans (Chomel et al. 2004; Siamer and Dehio 2015). To our knowledge, these YopJ-like sequences have not previously been functionally characterized, but their striking diversity compared to their closest relatives may be a consequence of recent or ongoing adaptations to multiple distinct hosts. Intriguingly, although *Bartonella* spp. apparently lack a type III secretion system, a type IV secretion system has been described (Saenz et al. 2007). This raises the interesting possibility that these effectors may have adopted bacterial functions or alternatively that they are secreted by a distinct (T3SS-independent) mechanism.

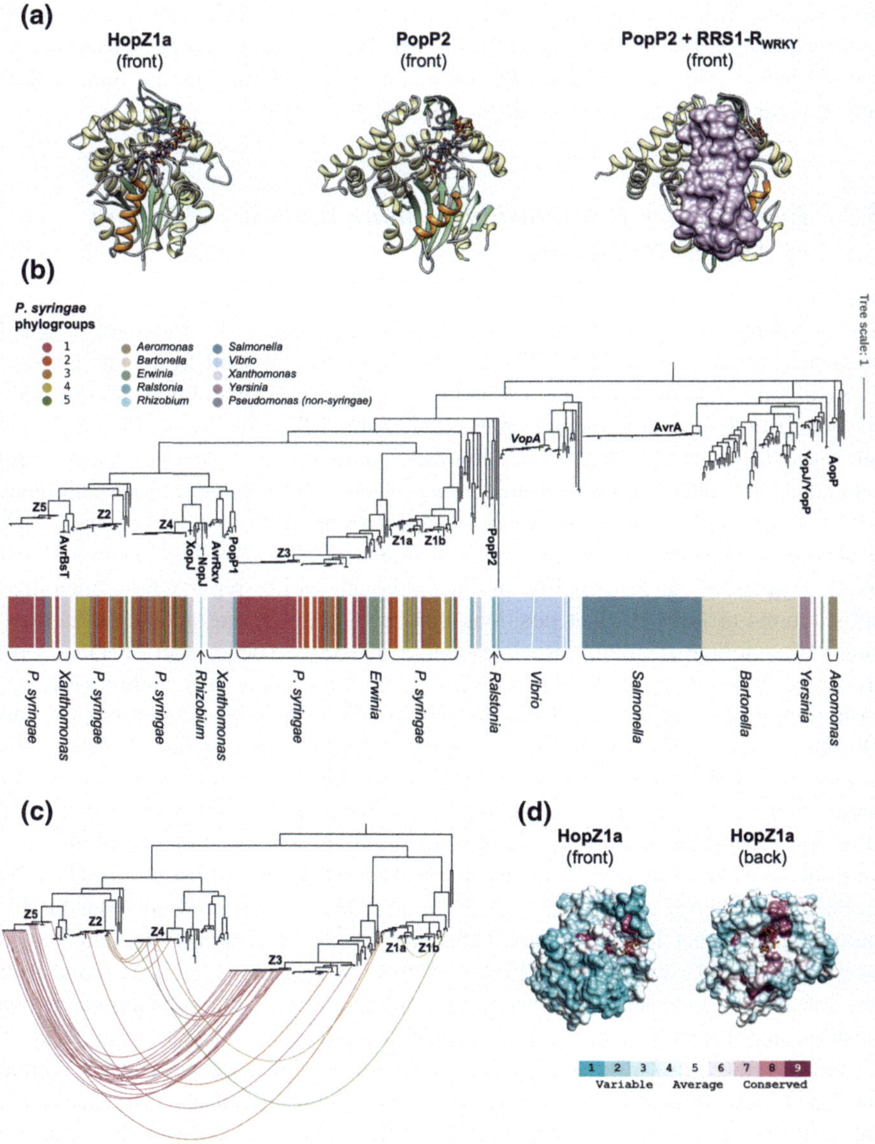
(a)
HopZ1a
(front)
PopP2
(front)
PopP2 + RRS1-R$_{WRKY}$
(front)
(b)
P. syringae
phylogroups
1
2
3
4
5
Aeromonas
Bartonella
Erwinia
Ralstonia
Rhizobium
Salmonella
Vibrio
Xanthomonas
Yersinia
Pseudomonas (non-syringae)
Tree scale: 1
VopA
AvrA
YopJ/YopP
AopP
Z5
Z2
Z4
Z3
Z1a
Z1b
AvrBsT
XopJ
NopJ
AvrRxv
PopP1
PopP2
P. syringae
Xanthomonas
P. syringae
P. syringae
Rhizobium
Xanthomonas
P. syringae
Erwinia
P. syringae
Ralstonia
Vibrio
Salmonella
Bartonella
Yersinia
Aeromonas
(c)
Z5
Z2
Z4
Z3
Z1a
Z1b
(d)
HopZ1a
(front)
HopZ1a
(back)
1 2 3 4 5 6 7 8 9
Variable
Average
Conserved

◂**Fig. 2** Diversity of the YopJ family of Acetyltransferases. **a** Experimentally determined protein structures of HopZ1a (PDB: 5KLQ), PopP2 (PDB: 5W40), and PopP2 with bound WRKY domain substrate (PDB: 5W3X). Beta strands are colored green and alpha helices are colored yellow, except for helix αD which is colored orange; the relative position of this structural element is notably distinct in the HopZ1a/PopP2 structures, and a polymorphism of the homologous sequences in *Yersinia* YopJ is implicated in altered host immune activation. **b** Phylogenetic relationships between YopJ/HopZ-related sequences from pathogens of both animals and plants. Database/BLASTP searches, alignment, and phylogenetic analysis of a representative subset of YopJ- and HopZ-related protein sequences was performed as described for Fig. 1b. Labels for *P. syringae* sequences identified with our in-house database are colored by phylogroup as in Fig. 1b, while in this case selected sequences identified by BLASTP search are also labeled by color to facilitate comparisons of sequence diversity between genera. **c** Distribution and co-occurrence of HopZ alleles found among *P. syringae* strains. HopZ sequences co-occurring within a single isolate are linked by arcs colored according to the phylogroup designation of that isolate. **d** Molecular surface of HopZ1a colored according to sequence conservation. The orientation of this surface representation of HopZ1a is the same as for the ribbon representations in (**a**) (front) or rotated about the *y*-axis by 180° (back). Per-residue conservation scores were calculated based on aligned *P. syringae* HopZ sequences using the ConSurf web server (Celniker et al. 2013)

3.3 YopJ and Related Effectors Target Immunity-Related Kinases

A role for *Yersinia* YopJ in manipulating host immunity was suggested by the observation of suppressed signaling through MAPK and NF-κB pathways in the presence of YopJ (Orth 1999, 2000). Subsequent investigations have demonstrated that YopJ can acetylate multiple kinases involved in MAP kinase signal cascades, as well as IKK, a positive regulator of the NF-κB pathway (Mittal et al. 2006; Mukherjee et al. 2006; Meinzer et al. 2012; Paquette et al. 2012). Collectively, these reports indicate flexibility in substrate preference and suggest a promiscuous enzymatic activity directed at kinase activation loops in general.

Despite this apparent flexibility, YopJ homologues from animal pathogens outside of the genus *Yersinia* seem to have maintained similar targets (Table 1). AopP is one of just two known effectors in *Aeromonas salmonicida* (Fehr et al. 2006), a γ-proteobacterial pathogen of fish and humans. Like YopJ, AopP also inhibits the NF-kB pathway (although downstream of IKK activation), but in contrast, AopP does not seem to influence MAPK signaling in response to pathway stimulation by treatment with epidermal growth factor (Fehr et al. 2006). VopA from *Vibrio parahaemolyticus* acetylates and inhibits the MAPKs JNK and p38 when murine macrophages are infected with a Δ*yopJ* strain of *Y. pseudotuberculosis* complemented by *vopA* (Trosky et al. 2007). *S. enterica* AvrA can inhibit JNK-dependent MAPK signaling and NF-κB activation in both mouse and fruit fly models (Jones et al. 2008). More distantly related effectors from *P. syringae* (Sect. 3.5, below), HopZ1a and HopZ3 also acetylate immunity-related kinases in susceptible plant hosts (Lewis et al. 2013; Lee et al. 2015). HopZ1a triggers an immune response in the model plant *Arabidopsis thaliana* that is dependent on a pseudokinase called ZED1; this finding has suggested a model wherein ZED1 acts as a decoy for (an) as-yet uncharacterized kinase target(s) (Lewis et al. 2013).

3.4 Pseudomonas Syringae HopZ Proteins

An early survey identified HopZ-encoding genes in 43 of 96 *P. syringae* isolates (45%) and found that these sequences could be grouped into three distinct clades, denoted HopZ1, HopZ2, and HopZ3 (Ma et al. 2006). More recently, new distinct alleles denoted HopZ4 (Üstün et al. 2014) and HopZ5 (Jayaraman et al. 2017) have been identified. As described above for the LRR-NELs (Sect. 2.3, above), we also queried the *P. syringae* sequence database for the presence of HopZ sequences. HopZ effectors are not as prevalent as the LRR-NELs but are present in approximately one-third (133 of 367; 36%) of the isolates, representing phylogroups 1 through 5 (Fig. 2b). These sequences (and close relatives obtained from the NCBI by BLASTP search) are also represented in the maximum likelihood phylogenetic tree described above in Sect. 3.2 (Fig. 2b).

3.5 Evolutionary Relationships Among HopZ Alleles

HopZ alleles are as distinct from each other as they are from the YopJ effectors of animal pathogens, and they should therefore be viewed as distinct sub-lineages rather than polymorphic variants of a *P. syringae*-specific effector (Fig. 2b). Indeed, although HopZ2, HopZ4, and HopZ5 alleles share a common ancestor, they are interspersed with sequences from diverse genera (Fig. 2b). For example, HopZ5 alleles are most closely related to sequences from *Xanthomonas* (including the previously characterized AvrBsT; Kim et al. 2010) and *Acidovorax* spp., and along with HopZ2 sequences, they share a common ancestor with sequences from *Brennaria*, *Burkholderia*, *Erwinia,* and *Xanthomonas*. The HopZ4 effectors share an ancestor with the HopZ2/HopZ5 clade, but are more closely related to the *Xanthomonas* effector XopJ (Noel et al. 2003), a clade of sequences from rhizobial species (e.g., *Bradyrhizobium*, *Mesorhizobium,* and *Cupriavidus* spp.) that includes the previously described NopJ from *Rhizobium* (Kambara et al. 2009), *Ralstonia* PopP1 (Lavie et al. 2002), as well as distinct but related sequences from *Xanthomonas* (AvrRxv, AvrXv4; Whalen et al. 1993; Roden et al. 2004).

HopZ1 and HopZ3 sequences form a second, distinct clade, and as above, are closely related to sequences from diverse genera. HopZ3 is related to previously described sequences from *Erwinia* spp. (Oh et al. 2005), as well as to novel sequences from *Marinomonas mediterranea*, *Vibrio mangrovi*, and *Pantoea agglomerans*. A HopZ1-related sequence is also present in *Robbsia andropogonis*, a newly described relative of *Burkholderia* spp. (Lopes-Santos et al. 2017).

Initial characterization of the available *P. syringae hopZ* genes suggested a genealogy that is largely consistent with the genome phylogeny/phylogroup structure (Ma et al. 2006), yet the presence of these effectors in bacterial pathogens from diverse genera indicates past horizontal gene transfer events and suggests potential for further effector mobility and acquisition by other bacterial lineages. Genes encoding HopZ1a (Sundin et al. 2004) and HopZ2 (Arnold et al. 2001) were originally identified on plasmids, while the *hopZ3* gene is chromosomally encoded but located within the highly dynamic 'exchangeable effector locus' (Alfano et al. 2000). The nature of the replicons (i.e., plasmid vs. chromosome) encoding other characterized HopZ alleles is unclear due to limitations of next-generation, short read-based sequencing methods.

Strikingly, one-third of the isolates encoding HopZ family effectors possess multiple distinct alleles (Fig. 2c). Additional alleles from the same effector family may have been acquired to restore a function previously lost through pseudogenization. Alternatively, two alleles may act on distinct targets and provide an additive or synergistic combined benefit to the bacterium. Finally, a second allele may have been acquired to suppress an effector-triggered immune response triggered by the first. For example, HopZ3 suppresses the *N. benthamiana* immune responses induced by HopZ1b (Zhou et al. 2009) as well as by four other unrelated *P. syringae* effectors (Vinatzer et al. 2006). Notably, this phenomenon is similar to the interactions between the *Yersinia* effectors YopM and YopE/YopT/YopJ (Sect. 2.3, above).

Overall, both the YopJ and LRR-NEL effector families have diversified into several distinct lineages, while the LRR-NEL effector family appears to have further diversified in a modular fashion through the gain and/or loss of domains.

4 Transcription Activator-Like Effectors (TALEs)

4.1 TALEs: General Features

The *Xanthomonas* genus contains species that infect the vasculature of hundreds of host plants, including rice and banana (Ryan et al. 2011). *Xanthomonas* species carry approximately 52 effector families including the transcription activator-like effectors (TALEs) (White et al. 2009). TALEs are a unique class of effectors that were first discovered in the plant pathogen *Xanthomonas campestris* pv. *vesicatoria* (Bonas et al. 1989). Since their initial discovery, TALE-like proteins have been found in three other bacterial genera: *Ralstonia* (RipTALs), *Burkholderia* (Bats), and marine bacteria (MOrTLs) (Cunnac et al. 2004; Mukaihara et al. 2004; Juillerat et al. 2014; de Lange et al. 2015). Harboring a nuclear localization signal, a DNA-binding domain, and an activation domain, TALEs mimic eukaryotic transcription factors and modulate host gene expression to benefit the pathogen (Yang et al. 2006; Kay et al. 2007; Chen et al. 2010). The DNA-binding domain of TALEs consists of variable numbers of repeats of a ~34 amino acid motif that is highly conserved at all positions except amino acids 12 and 13. These two amino acids are termed the Repeat Variable Di-residue (RVD), and they determine the nucleotide-binding specificity of each repeat (Boch et al. 2009; Moscou and Bogdanove 2009). Because the RVD controls nucleotide specificity, RVDs have been a major focus for studies of TALE structure, function, and evolution (Wilkins et al. 2015; Grau et al. 2016; Ruh et al. 2017; Erkes et al. 2017).

4.2 TALEs: RVD and Repeat Evolution

The deciphering and subsequent application of the RVD nucleotide-binding 'code' of TALES has generated substantial translational interest (Kuhn et al. 2016; Khan et al. 2017). One unexpected finding is that the theoretical diversity of RVDs is greater than the observed natural diversity (Erkes et al. 2017). For example, although there are twelve ways to code for the 'NS' RVD, only two have been observed out of the 516 TALEs analyzed (Erkes et al. 2017). Codon bias, RNA structure, and the stability of codon pairs do not appear to explain the conservation of RVD codons, but instead may reflect the role of recombination in creating new TALEs and RVDs (Erkes et al. 2017).

RVD sequences can be used to group TALES into distinct classes (Grau et al. 2016). Single base substitutions account for 93% of RVD differences between TALE classes, although deletions and duplications of one or more RVDs are also quite common (Yang et al. 2005; Booher et al. 2015; Schandry et al. 2016; Erkes et al. 2017). For example, the TALE class TalDS1 appears to have formed from TalAC through a single repeat deletion (Erkes et al. 2017).

4.3 TALEs: Genomic Organization and Distribution Among Xanthomonads

The distribution and mobility of TALEs within the genome have also been studied (Salzberg et al. 2008; Ferreira et al. 2015; Grau et al. 2016; Erkes et al. 2017; Denancé et al. 2018). Within the *Xanthomonas oryzae* pv. *oryzae* (Xoo) strains, TALEs are clustered in well-conserved genomic regions, while in *X. oryzae* pv. *oryzicola* (Xoc) strains, TALEs are more uniformly distributed throughout the genome (Salzberg et al. 2008; Grau et al. 2016; Erkes et al. 2017). The majority of TALE clusters in Xoo strains are flanked by inverted repeats, suggesting that transposons play a role in the mobility of these clusters (Erkes et al. 2017). Analysis of 116 TALEs from various *Xanthomonads* found that almost half of the TALEs were associated with inverted repeats and form mobile insertion cassettes (Ferreira et al. 2015). Likewise, the TALE composition of *Xanthomonas citri* pv. *fuscans*

4834-R includes plasmid-borne TALEs from *X. oryzae*, *X. citri*, *Xanthomonas axonopodis*, and *X. campestris* (Darrasse et al. 2013; Ferreira et al. 2015). While these patterns suggest an important role for horizontal gene transfer, Xoc TALEs are rarely associated with inverted repeats and mobile elements (Ferreira et al. 2015). Instead, TALEs from this species are more likely to be found in clusters separated by highly conserved spacer regions that resemble integrons, suggesting that in Xoc strains, TALEs may be added in a cassette-wise fashion (Nivina et al. 2016; Erkes et al. 2017).

Comparative and evolutionary genomic analyses of *Xanthomonas* TALEs have revealed some striking patterns (Ochiai et al. 2005; Wilkins et al. 2015; Ruh et al. 2017; Erkes et al. 2017; Denancé et al. 2018; Schandry et al. 2018). TALEs are numerous within both Xoo and Xoc, where as many as 28 TALEs can be found, which contrasts with other *Xanthomonas* species (e.g., *X. citri* pv. *fuscans* and *X. phaseoli* pv. *phaseoli*) that have only a few TALEs or none at all (da Silva et al. 2002; Bogdanove et al. 2010; Ryan et al. 2011; Barak et al. 2016). It has been postulated that the contrasting sizes of TALE repertoires are due to different evolutionary forces acting on different *Xanthomonas* lineages. For Xoo and Xoc, the chromosomal location of the abundant TALEs (Ochiai et al. 2005; Salzberg et al. 2008; Wilkins et al. 2015; Erkes et al. 2017), the convergent evolution of TALE host targets (Li et al. 2013b; Streubel et al. 2013; Cox et al. 2017), and the evolution of immunity genes that exploit the DNA-binding specificity of TALEs (Blanvillain-Baufumé et al. 2017; Hummel et al. 2017) suggest a long coevolutionary host-pathogen arms race (Ruh et al. 2017). In contrast, *Xanthomonads* like *X. citri* pv. *fuscans* and *X. phaseoli* pv. *phaseoli* carry only a small number of plasmid-borne TALEs (da Silva et al. 2002; Barak et al. 2016; Ruh et al. 2017). The mobility of these TALEs may allow them to spread readily among strains and may indicate a relatively recent acquisition compared to the Xoo and Xoc strains (Ruh et al. 2017).

4.4 TALEs: Truncated TALEs (TruncTALEs) and Interfering TALEs (ITALES)

Previously thought to be pseudogenes, both truncTALEs and iTALEs lack at least one essential feature found in TALEs, such as their DNA-binding or activation domains. While lacking TALE features, truncTALEs and iTALEs are both expressed and display immunity-suppressing capabilities by blocking the immune response elicited by TALE effectors (Ji et al. 2016; Read et al. 2016). The two classes of iTALEs, Tal3a and Tal3b, are able to suppress immunity triggered by the rice NLR Xa1 in response to a wide range of TALEs tested (Ji et al. 2016). By swapping the N-terminal region, the DNA-binding domain, and the C-terminal region of iTALEs with canonical TALEs, it was found that the iTALE N- and C-terminal regions are essential for immune suppression, while modifying the number and RVD sequence of the DNA-binding domain is not (Ji et al. 2016).

truncTALEs similarly require either/or both of their unique N- and C-terminal regions for immune suppression of the not-yet characterized rice NLR locus, *Xo1* (Read et al. 2016; Triplett et al. 2016). The RVD sequence of truncTALEs plays a partial role in Xo1-immune suppression while their nuclear localization signal and unique 28 amino acid repeats are not required (Read et al. 2016). Interestingly, although the N- and C-terminal regions of both truncTALEs and iTALEs are highly conserved among their respective groups, they both contain highly divergent repeat regions with diverse predicted DNA target sequences (Ji et al. 2016; Read et al. 2016). By comparing truncTALE RVD sequences from 12 different Xoo and Xoc strains, SNP patterns along with the strain phylogeny provide evidence that truncTALEs descended from a common ancestor and then diversified among pathovars through repeat duplications and deletions (Read et al. 2016).

The apparent dispensability of the DNA-binding domain and corresponding RVD sequences of iTALEs and truncTALEs for immune suppression can be attributed to their proposed mechanism of action. Xa1 and Xo1 are both thought to bind directly to TALEs through their N-terminal region and the beginning of their repeat region, leading to broad recognition of TALEs and making the RVD sequence irrelevant (Read et al. 2016). iTALEs and truncTALEs are thought to bind in a similar manner to Xa1 and Xo1, respectively, though at a higher affinity or concentration so as to displace and outcompete the immune-eliciting TALEs from binding to the NLR (Read et al. 2016). As both iTALEs and truncTALEs suppress immunity through protein-protein interactions, there is no selective pressure to maintain conserved DNA-binding domains, as DNA binding is not required (Read et al. 2016). In an electrophoretic mobility shift assay, the Tal2h truncTALE was unable to bind to 11 DNA probes that were predicted to be bound by the *Tal2h* RVD sequence (Read et al. 2016). Swapping the *Tal2h* RVD sequence to a standard TALE restored DNA binding, suggesting that the N-terminal or C-terminal regions of Tal2h suppress DNA binding of its repeat region (Read et al. 2016). It is interesting to speculate that both the truncTALEs and iTALEs have evolved to inhibit their own DNA binding to increase their effectiveness in immune suppression.

4.5 TALE Homologues: The 'TALE-Like' Effectors

TALE-like proteins have been identified in *R. solanacearum* (RipTALs), *Burkholderia rhizoxinica* (Bats), and marine bacteria (MOrTLs) (de Lange et al. 2013, 2014; Li et al. 2013a; Juillerat et al. 2014). They all share a common domain organization with a central repeat region that confers the DNA-binding ability common to all TALEs (de Lange et al. 2013, 2015; Juillerat et al. 2014).

The RipTALs of *R. solanacearum* rely on their RVDs to determine nucleotide specificity, and the RipTAL RVD code is almost identical to that of the *Xanthomonas* TALEs (de Lange et al. 2013). Although no RipTAL plant target genes have been identified to date, in silico predictions of host targets suggest that RipTALs from different classes have converged to target similar plant genes

(Schandry et al. 2016). This finding, in conjunction with a broad distribution among *R. solanacearum* phylogroups, suggests that the common ancestor of *R. solanacearum* may have horizontally acquired RipTALs from a *Xanthomonas* source (Fall et al. 2007; Heuer et al. 2007; Schandry et al. 2016).

The Bat proteins from *B. rhizoxinica* have been shown to bind DNA in a similar fashion to TALEs, but they lack both the nuclear localization signal and the activation domain (Juillerat et al. 2014). Identified on the basis of structural similarity, the DNA-binding domains of Bats and TALEs display less than 40% amino acid identity with no identical repeats observed in the Bat sequences (de Lange et al. 2014). Similar to RipTALs, Bats display much greater residue diversity in their non-RVD repeat residues, suggesting that different evolutionary constraints act upon Bat proteins (de Lange et al. 2014; Schandry et al. 2016). Only three Bat proteins have been discovered to date, and they are all plasmid-localized (de Lange et al. 2014). Surprisingly, despite structural and sequence divergence, Bat proteins can bind with the same code as TALEs and RipTALs (de Lange et al. 2014).

The most recently discovered TALE-like effectors are the two MOrTL proteins identified in a marine microbial metagenomics dataset (Juillerat et al. 2014; de Lange et al. 2015). MOrTL1 and MOrTL2 share only 30–40% amino acid identity to other TALE-likes, though they bind nucleotide sequences with the same code as all other TALE-likes (de Lange et al. 2015). MOrTL repeats show over 60% divergence from each other and other TALE-like proteins (de Lange et al. 2015). To investigate whether these repeats are sufficient to confer TALE function, MOrTL1 and MOrTL2 repeats were inserted into TALE repeat arrays. While MOrTL1 was sufficient and compatible with TALE repeats and allowed binding to the Bat1 target sequence, MOrTL2 repeats were not compatible with TALE repeats and did not confer target sequence binding (de Lange et al. 2015).

5 Conclusions

The LRR-associated Novel E3 ubiquitin Ligases (LRR-NELs), YopJ family acetyltransferases, and the transcription activator-like effectors (TALEs) are fascinating cases of broadly distributed type III effector families. These three families include alleles from diverse bacterial genera, highlighting the importance of these effectors for a range of different bacterial lifestyles. The conserved functional features of these proteins (i.e., catalytic residues, domain structures, and inter-domain contacts) point to broadly successful virulence strategies that are presumably effective in highly divergent hosts. These virulence strategies may either target analogous host systems, or they may alternatively represent general biochemical functions that provide broadly effective means to disrupt eukaryotic cellular homeostasis. Further understanding the function and targets of these effectors will reveal how pathogenic and symbiotic bacteria hijack and disrupt the cellular functions of their hosts, as well as how effector families diversify in response to host selection pressures.

References

Alfano JR, Charkowski AO, Deng WL et al (2000) The *Pseudomonas syringae* Hrp pathogenicity island has a tripartite mosaic structure composed of a cluster of type III secretion genes bounded by exchangeable effector and conserved effector loci that contribute to parasitic fitness and pathogenicity in plants. Proc Natl Acad Sci U S A 97:4856–4861. https://doi.org/10.1073/pnas.97.9.4856

Altschul SF, Madden TL, Schäffer AA et al (1997) Gapped BLAST and PSI-BLAST: a new generation of protein database search programs. Nucleic Acids Res 25:3389–3402. https://doi.org/10.1093/nar/25.17.3389

Anisimova M, Gascuel O (2006) Approximate likelihood-ratio test for branches: a fast, accurate, and powerful alternative. Syst Biol 55:539–552. https://doi.org/10.1080/10635150600755453

Arnold DL, Jackson RW, Fillingham AJ et al (2001) Highly conserved sequences flank avirulence genes: isolation of novel avirulence genes from *Pseudomonas syringae* pv. *pisi*. Microbiology 147:1171–1182. https://doi.org/10.1099/00221287-147-5-1171

Ashida H, Sasakawa C (2016) *Shigella* IpaH family effectors as a versatile model for studying pathogenic bacteria. Front Cell Infect Microbiol 5:100. https://doi.org/10.3389/fcimb.2015.00100

Ashida H, Toyotome T, Nagai T, Sasakawa C (2007) *Shigella* chromosomal IpaH proteins are secreted via the type III secretion system and act as effectors. Mol Microbiol 63:680–693. https://doi.org/10.1111/j.1365-2958.2006.05547.x

Ashida H, Kim M, Schmidt-Supprian M et al (2010) A bacterial E3 ubiquitin ligase IpaH9.8 targets NEMO/IKKγ to dampen the host NF-κB-mediated inflammatory response. Nat Cell Biol 12:66–73. https://doi.org/10.1038/ncb2006

Barak JD, Vancheva T, Lefeuvre P et al (2016) Whole-genome sequences of *Xanthomonas euvesicatoria* strains clarify taxonomy and reveal a stepwise erosion of type 3 effectors. Front Plant Sci 7. https://doi.org/10.3389/fpls.2016.01805

Bernal-Bayard J, Ramos-Morales F (2009) *Salmonella* type III secretion effector SlrP is an E3 ubiquitin ligase for mammalian thioredoxin. J Biol Chem 284:27587–27595. https://doi.org/10.1074/jbc.M109.010363

Bernal-Bayard J, Cardenal-Muñoz E, Ramos-Morales F (2010) The *Salmonella* type III secretion effector, *Salmonella* leucine-rich repeat protein (SlrP), targets the human chaperone ERdj3. J Biol Chem 285:16360–16368. https://doi.org/10.1074/jbc.M110.100669

Bhavsar AP, Brown NF, Stoepel J et al (2013) The *Salmonella* type III effector SspH2 specifically exploits the NLR Co-chaperone activity of SGT1 to subvert immunity. PLoS Pathog 9. https://doi.org/10.1371/journal.ppat.1003518

Blanvillain-Baufumé S, Reschke M, Solé M et al (2017) Targeted promoter editing for rice resistance to *Xanthomonas oryzae* pv. *oryzae* reveals differential activities for SWEET14-inducing TAL effectors. Plant Biotechnol J 15:306–317. https://doi.org/10.1111/pbi.12613

Boch J, Scholze H, Schornack S et al (2009) Breaking the code of DNA binding. Science (80) 1509:1509–1512. https://doi.org/10.1126/science.1178811

Bogdanove AJ, Schornack S, Lahaye T (2010) TAL effectors: finding plant genes for disease and defense. Curr Opin Plant Biol 13:394–401. https://doi.org/10.1016/j.pbi.2010.04.010

Bonas U, Stall RE, Staskawicz B (1989) Genetic and structural characterization of the avirulence gene avrBs3 from *Xanthomonas campestris* pv. *vesicatoria*. Mol Gen Genet 218:127–136. https://doi.org/10.1007/BF00330575

Booher NJ, Carpenter SCD, Sebra RP et al (2015) Single molecule real-time sequencing of *Xanthomonas oryzae* genomes reveals a dynamic structure and complex TAL (transcription activator-like) effector gene relationships. Microb Genomics 1. https://doi.org/10.1099/mgen.0.000032

Buchrieser C, Glaser P, Rusniok C et al (2000) The virulence plasmid pWR100 and the repertoire of proteins secreted by the type III secretion apparatus of *Shigella flexneri*. Mol Microbiol 38:760–771. https://doi.org/10.1046/j.1365-2958.2000.02179.x

Celniker G, Nimrod G, Ashkenazy H et al (2013) ConSurf: using evolutionary data to raise testable hypotheses about protein function. Isr J Chem 53:199–206. https://doi.org/10.1002/ijch.201200096

Chang JH, Urbach JM, Law TF et al (2005) A high-throughput, near-saturating screen for type III effector genes from *Pseudomonas syringae*. Proc Natl Acad Sci U S A 102:2549–2554. https://doi.org/10.1073/pnas.0409660102

Chen L-Q, Hou B-H, Lalonde S et al (2010) Sugar transporters for intercellular exchange and nutrition of pathogens. Nature 468:527–532. https://doi.org/10.1038/nature09606

Chomel BB, Boulouis HJ, Breitschwerdt EB (2004) Cat scratch disease and other zoonotic *Bartonella* infections. J Am Vet Med Assoc 224:1270–1279. https://doi.org/10.2460/javma.2004.224.1270

Chou Y-C, Keszei AFA, Rohde JR et al (2012) Conserved structural mechanisms for autoinhibition in IpaH ubiquitin ligases. J Biol Chem 287:268–275. https://doi.org/10.1074/jbc.M111.316265

Chung LK, Park YH, Zheng Y et al (2016) The *Yersinia* virulence factor YopM hijacks host kinases to inhibit type III effector-triggered activation of the pyrin inflammasome. Cell Host Microbe 20:296–306. https://doi.org/10.1016/j.chom.2016.07.018

Cordero-Alba M, García-Gómez JJ, Aguilera-Herce J, Ramos-Morales F (2016) Proteomic insight into the effects of the *Salmonella* ubiquitin ligase SlrP on host cells. Biochem Biophys Res Commun 472:539–544. https://doi.org/10.1016/j.bbrc.2016.03.014

Cox KL, Meng F, Wilkins KE et al (2017) TAL effector driven induction of a SWEET gene confers susceptibility to bacterial blight of cotton. Nat Commun 8:1–14. https://doi.org/10.1038/ncomms15588

Cunnac S, Occhialini A, Barberis P et al (2004) Inventory and functional analysis of the large Hrp regulon in *Ralstonia solanacearum*: identification of novel effector proteins translocated to plant host cells through the type III secretion system. Mol Microbiol 53:115–128. https://doi.org/10.1111/j.1365-2958.2004.04118.x

da Silva ACR, Ferro JA, Reinach FC et al (2002) Comparison of the genomes of two *Xanthomonas* pathogens with differing host specificities. Nature 417:459–463. https://doi.org/10.1038/417459a

Darrasse A, Carrère S, Barbe V et al (2013) Genome sequence of *Xanthomonas fuscans* subsp. *fuscans* strain 4834-R reveals that flagellar motility is not a general feature of xanthomonads. BMC Genomics 14. https://doi.org/10.1186/1471-2164-14-761

de Jong MF, Liu Z, Chen D, Alto NM (2016) *Shigella flexneri* suppresses NF-κB activation by inhibiting linear ubiquitin chain ligation. Nat Microbiol 1:16084. https://doi.org/10.1038/nmicrobiol.2016.84

de Lange O, Schreiber T, Schandry N et al (2013) Breaking the DNA binding code of *Ralstonia solanacearum* TAL effectors provides new possibilities to generate plant resistance genes against bacterial wilt disease. New Phytol 199. https://doi.org/10.1111/nph.12324

de Lange O, Wolf C, Dietze J et al (2014) Programmable DNA-binding proteins from *Burkholderia* provide a fresh perspective on the TALE-like repeat domain. Nucleic Acids Res 42:7436–7449. https://doi.org/10.1093/nar/gku329

de Lange O, Wolf C, Thiel P et al (2015) DNA-binding proteins from marine bacteria expand the known sequence diversity of TALE-like repeats. Nucleic Acids Res 43:10065–10080. https://doi.org/10.1093/nar/gkv1053

Denancé N, Szurek B, Doyle EL et al (2018) Two ancestral genes shaped the *Xanthomonas campestris* TAL effector gene repertoire. New Phytol 219:391–407. https://doi.org/10.1111/nph.15148

Diao J, Zhang Y, Huibregtse JM et al (2008) Crystal structure of SopA, a *Salmonella* effector protein mimicking a eukaryotic ubiquitin ligase. Nat Struct Mol Biol 15:65–70. https://doi.org/10.1038/nsmb1346

Dodds PN, Rathjen JP (2010) Plant immunity: towards an integrated view of plant–pathogen interactions. Nat Rev Genet 11:539–548. https://doi.org/10.1038/nrg2812

Dong C, Davis RJ, Flavell RA (2002) MAP kinases in the immune response. Annu Rev Immunol 20:55–72. https://doi.org/10.1146/annurev.immunol.20.091301.131133

Dong J, Xiao F, Fan F et al (2009) Crystal structure of the complex between *Pseudomonas* effector AvrPtoB and the tomato Pto kinase reveals both a shared and a unique interface compared with AvrPto-Pto. Plant Cell 21:1846–1859. https://doi.org/10.1105/tpc.109.066878

Edgar RC (2004) MUSCLE: multiple sequence alignment with high accuracy and high throughput. Nucleic Acids Res 32:1792–1797. https://doi.org/10.1093/nar/gkh340

Erkes A, Reschke M, Boch J, Grau J (2017) Evolution of transcription activator-like effectors in *Xanthomonas oryzae*. Genome Biol Evol 2:1–15. https://doi.org/10.1093/gbe/evx108

Evdokimov AG, Anderson DE, Routzahn KM, Waugh DS (2001) Unusual molecular architecture of the *Yersinia pestis* cytotoxin YopM: a leucine-rich repeat protein with the shortest repeating unit. J Mol Biol 312:807–821. https://doi.org/10.1006/jmbi.2001.4973

Fall S, Mercier A, Bertolla F et al (2007) Horizontal gene transfer regulation in bacteria as a "Spandrel" of DNA repair mechanisms. PLoS One 2. https://doi.org/10.1371/journal.pone.0001055

Fehr D, Casanova C, Liverman A et al (2006) AopP, a type III effector protein of *Aeromonas salmonicida*, inhibits the NF-kappaB signalling pathway. Microbiology 152:2809–2818. https://doi.org/10.1099/mic.0.28889-0

Ferreira RM, de Oliverira CP, Moreira LM et al (2015) A TALE of transposition: Tn3-Like transposons play a major role in the spread of pathogenicity determinants of *Xanthomonas citri* and other xanthomonads. MBio 6:e02505-14. https://doi.org/10.1128/mbio.02505-14

Galán JE, Collmer A (1999) Type III secretion machines: bacterial devices for protein delivery into host cells. Science 284:1322–1328. https://doi.org/10.1126/SCIENCE.284.5418.1322

Grau J, Reschke M, Erkes A et al (2016) AnnoTALE: bioinformatics tools for identification, annotation, and nomenclature of TALEs from *Xanthomonas* genomic sequences. Sci Rep 6:1–12. https://doi.org/10.1038/srep21077

Guindon S, Gascuel O (2003) A simple, fast, and accurate algorithm to estimate large phylogenies by maximum likelihood. Syst Biol 52:696–704. https://doi.org/10.1080/10635150390235520

Guindon S, Lethiec F, Duroux P, Gascuel O (2005) PHYML online—a web server for fast maximum likelihood-based phylogenetic inference. Nucleic Acids Res 33. https://doi.org/10.1093/nar/gki352

Guindon S, Dufayard JF, Lefort V et al (2010) New algorithms and methods to estimate maximum-likelihood phylogenies: assessing the performance of PhyML 3.0. Syst Biol 59:307–321. https://doi.org/10.1093/sysbio/syq010

Haraga A, Miller SI (2006) A *Salmonella* type III secretion effector interacts with the mammalian serine/threonine protein kinase PKN1. Cell Microbiol 8:837–846. https://doi.org/10.1111/j.1462-5822.2005.00670.x

Hauenstein AV, Zhang L, Wu H (2015) The hierarchical structural architecture of inflammasomes, supramolecular inflammatory machines. Curr Opin Struct Biol 31:75–83. https://doi.org/10.1016/j.sbi.2015.03.014

Hentschke M, Berneking L, Campos CB et al (2010) *Yersinia* virulence factor YopM induces sustained RSK activation by interfering with dephosphorylation. PLoS One 5. https://doi.org/10.1371/journal.pone.0013165

Heuer H, Yin YN, Xue QY et al (2007) Repeat domain diversity of avrBs3-like genes in *Ralstonia solanacearum* strains and association with host preferences in the field. Appl Environ Microbiol 73:4379–4384. https://doi.org/10.1128/AEM.00367-07

Hu Z, Chai J (2016) Structural mechanisms in NLR inflammasome assembly and signaling. In: Current topics in microbiology and immunology. Elsevier Ltd., pp 23–42. https://doi.org/10.1007/978-3-319-41171-2_2

Hu Y, Huang H, Hui X et al (2016) Distribution and evolution of *Yersinia* leucine-rich repeat proteins. Infect Immun 84:2243–2254. https://doi.org/10.1128/IAI.00324-16.Editor

Hueck CJ (1998) Type III protein secretion systems in bacterial pathogens of animals and plants. Microbiol Mol Biol Rev 62:379–433. https://doi.org/10.1111/j.1365-2958.2006.05301.x
Hummel AW, Wilkins KE, Wang L et al (2017) A transcription activator-like effector from *Xanthomonas oryzae* pv. *oryzicola* elicits dose-dependent resistance in rice. Mol Plant Pathol 18:55–66. https://doi.org/10.1111/mpp.12377
Israël A (2010) The IKK complex, a central regulator of NF-κB activation. Cold Spring Harb Perspect Biol 2. https://doi.org/10.1101/cshperspect.a000158
Jayaraman J, Choi S, Prokchorchik M et al (2017) A bacterial acetyltransferase triggers immunity in *Arabidopsis thaliana* independent of hypersensitive response. Sci Rep 7:3557. https://doi.org/10.1038/s41598-017-03704-x
Ji Z, Ji C, Liu B et al (2016) Interfering TAL effectors of *Xanthomonas oryzae* neutralize R-gene-mediated plant disease resistance. Nat Commun 7:1–9. https://doi.org/10.1038/ncomms13435
Jones JDG, Dangl JL (2006) The plant immune system. Nature 444:323–329. https://doi.org/10.1038/nature05286
Jones RM, Wu H, Wentworth C et al (2008) *Salmonella* AvrA coordinates suppression of host immune and apoptotic defenses via JNK pathway blockade. Cell Host Microbe 3:233–244. https://doi.org/10.1016/j.chom.2008.02.016
Jones JDG, Vance RE, Dangl JL (2016) Intracellular innate immune surveillance devices in plants and animals. Science (80) 354:aaf6395-aaf6395. https://doi.org/10.1126/science.aaf6395
Juillerat A, Bertonati C, Dubois G et al (2014) BurrH: a new modular DNA binding protein for genome engineering. Sci Rep 4:1–6. https://doi.org/10.1038/srep03831
Kambara K, Ardissone S, Kobayashi H et al (2009) Rhizobia utilize pathogen-like effector proteins during symbiosis. Mol Microbiol 71:92–106. https://doi.org/10.1111/j.1365-2958.2008.06507.x
Kay S, Hahn S, Marois E et al (2007) A bacterial effector acts as a plant transcription factor and induces a cell size regulator. Science (80) 318:648–651. https://doi.org/10.1126/science.1144956
Keszei AFA, Tang X, McCormick C et al (2014) Structure of an SspH1-PKN1 complex reveals the basis for host substrate recognition and mechanism of activation for a bacterial E3 ubiquitin ligase. Mol Cell Biol 34:362–373. https://doi.org/10.1128/MCB.01360-13
Khan M, Subramaniam R, Desveaux D (2016) Of guards, decoys, baits and traps: pathogen perception in plants by type III effector sensors. Curr Opin Microbiol 29:49–55. https://doi.org/10.1016/j.mib.2015.10.006
Khan Z, Khan SH, Mubarik MS et al (2017) Use of TALEs and TALEN technology for genetic improvement of plants. Plant Mol Biol Report 35:1–19. https://doi.org/10.1007/s11105-016-0997-8
Khan M, Seto D, Subramaniam R, Desveaux D (2018) Oh, the places they'll go! A survey of phytopathogen effectors and their host targets. Plant J 93:651–663. https://doi.org/10.1111/tpj.13780
Kim NH, Choi HW, Hwang BK (2010) *Xanthomonas campestris* pv. *vesicatoria* effector AvrBsT induces cell death in pepper, but suppresses defense responses in tomato. Mol Plant Microbe Interact 23:1069–1082. https://doi.org/10.1094/MPMI-23-8-1069
Kuhn R, Wurst W, Wefers B (2016) TALENs methods and protocols. Humana Press, New York. https://doi.org/10.1007/978-1-4939-2932-0
Lamkanfi M, Dixit VM (2010) Manipulation of host cell death pathways during microbial infections. Cell Host Microbe 8:44–54. https://doi.org/10.1016/j.chom.2010.06.007
Larock CN, Cookson BT (2012) The *Yersinia* virulence effector YopM binds caspase-1 to arrest inflammasome assembly and processing. Cell Host Microbe 12:799–805. https://doi.org/10.1016/j.chom.2012.10.020
Lavie M, Shillington E, Eguiluz C et al (2002) PopP1, a new member of the YopJ/AvrRxv family of type III effector proteins, acts as a host-specificity factor and modulates aggressiveness of *Ralstonia solanacearum*. Mol Plant-Microbe Interact 15:1058–1068. https://doi.org/10.1094/MPMI.2002.15.10.1058

Le Roux C, Huet G, Jauneau A et al (2015) A receptor pair with an integrated decoy converts pathogen disabling of transcription factors to immunity. Cell 161:1074–1088. https://doi.org/10.1016/j.cell.2015.04.025

Lechtenberg BC, Mace PD, Riedl SJ (2014) Structural mechanisms in NLR inflammasome signaling. Curr Opin Struct Biol 29:17–25. https://doi.org/10.1016/j.sbi.2014.08.011

Lee J, Manning AJ, Wolfgeher D et al (2015) Acetylation of an NB-LRR plant immune-effector complex suppresses immunity. Cell Rep 13:1670–1682. https://doi.org/10.1016/j.celrep.2015.10.029

Lewis JD, Lee A, Ma W et al (2011) The YopJ superfamily in plant-associated bacteria. Mol Plant Pathol 12:928–937. https://doi.org/10.1111/j.1364-3703.2011.00719.x

Lewis JD, Lee AH-Y, Hassan Ja et al (2013) The *Arabidopsis* ZED1 pseudokinase is required for ZAR1-mediated immunity induced by the *Pseudomonas syringae* type III effector HopZ1a. Proc Natl Acad Sci U S A 110:18722–18727. https://doi.org/10.1073/pnas.1315520110

Li L, Atef A, Piatek A et al (2013a) Characterization and DNA-binding specificities of *Ralstonia* TAL-like effectors. Mol Plant 6:1318–1330. https://doi.org/10.1093/mp/sst006

Li T, Huang S, Zhou J, Yang B (2013b) Designer TAL effectors induce disease susceptibility and resistance to *Xanthomonas oryzae* pv. *oryzae* in rice. Mol Plant 6:781–789. https://doi.org/10.1093/mp/sst034

Lopes-Santos L, Castro DBA, Ferreira-Tonin M et al (2017) Reassessment of the taxonomic position of *Burkholderia andropogonis* and description of *Robbsia andropogonis* gen. nov., comb. nov. Antonie Van Leeuwenhoek 110:727–736. https://doi.org/10.1007/s10482-017-0842-6

Ma K, Ma W (2016) YopJ family effectors promote bacterial infection through a acetyltransferase activity. Microbiol Mol Biol Rev 80:1011–1027. https://doi.org/10.1128/MMBR.00032-16

Ma W, Dong FFT, Stavrinides J, Guttman DS (2006) Type III effector diversification via both pathoadaptation and horizontal transfer in response to a coevolutionary arms race. PLoS Genet 2:2131–2142. https://doi.org/10.1371/journal.pgen.0020209

Maekawa T, Kufer TA, Schulze-Lefert P (2011) NLR functions in plant and animal immune systems: so far and yet so close. Nat Immunol 12:818–826. https://doi.org/10.1038/ni.2083

Marchler-Bauer A, Lu S, Anderson JB et al (2011) CDD: a conserved domain database for the functional annotation of proteins. Nucleic Acids Res 39:225–229. https://doi.org/10.1093/nar/gkq1189

Marchler-Bauer A, Derbyshire MK, Gonzales NR et al (2015) CDD: NCBI's conserved domain database. Nucleic Acids Res 43:D222–D226. https://doi.org/10.1093/nar/gku1221

Meinzer U, Barreau F, Esmiol-Welterlin S et al (2012) *Yersinia pseudotuberculosis* effector YopJ subverts the Nod2/RICK/TAK1 pathway and activates caspase-1 to induce intestinal barrier dysfunction. Cell Host Microbe 11:337–351. https://doi.org/10.1016/j.chom.2012.02.009

Michiels T, Wattiau P, Brasseur R et al (1990) Secretion of Yop proteins by Yersiniae. Infect Immun 58:2840–2849

Michiels T, Vanooteghem JC, Lambert de Rouvroit C et al (1991) Analysis of virC, an operon involved in the secretion of Yop proteins by *Yersinia enterocolitica*. J Bacteriol 173:4994–5009. https://doi.org/10.1128/jb.173.16.4994-5009.1991

Mittal R, Peak-Chew S-Y, McMahon HT (2006) Acetylation of MEK2 and I kappa B kinase (IKK) activation loop residues by YopJ inhibits signaling. Proc Natl Acad Sci U S A 103:18574–18579. https://doi.org/10.1073/pnas.0608995103

Mittal R, Peak-Chew SY, Sade RS et al (2010) The acetyltransferase activity of the bacterial toxin YopJ of *Yersinia* is activated by eukaryotic host cell inositol hexakisphosphate. J Biol Chem 285:19927–19934. https://doi.org/10.1074/jbc.M110.126581

Moscou MJ, Bogdanove AJ (2009) A simple cipher governs DNA recognition by TAL effectors. Science (80) 326:1501. https://doi.org/10.1126/science.1178817

Mukaihara T, Tamura N, Murata Y, Iwabuchi M (2004) Genetic screening of Hrp type III-related pathogenicity genes controlled by the HrpB transcriptional activator in *Ralstonia solanacearum*. Mol Microbiol 54:863–875. https://doi.org/10.1111/j.1365-2958.2004.04328.x

Mukherjee S, Keitany G, Li Y et al (2006) *Yersinia* YopJ acetylates and inhibits kinase activation by blocking phosphorylation. Science (80) 312:1211–1214. https://doi.org/10.1126/science.1126867

Nakano M, Oda K, Mukaihara T (2017) *Ralstonia solanacearum* novel E3 ubiquitin ligase (NEL) effectors RipAW and RipAR suppress pattern-triggered immunity in plants. Microbiol (United Kingdom) 163:992–1002. https://doi.org/10.1099/mic.0.000495

Nivina A, Escudero JA, Vit C et al (2016) Efficiency of integron cassette insertion in correct orientation is ensured by the interplay of the three unpaired features of *attC* recombination sites. Nucleic Acids Res 44:7792–7803. https://doi.org/10.1093/nar/gkw646

Noel L, Thieme F, Gabler J et al (2003) XopC and XopJ, two novel type III effector proteins from *Xanthomonas campestris* pv. *vesicatoria*. J Bacteriol 185:7092–7102. https://doi.org/10.1128/JB.185.24.7092

Notti RQ, Stebbins CE (2016) The structure and function of type III secretion systems. Microbiol Spectr 4:1–30. https://doi.org/10.1128/microbiolspec.vmbf-0004-2015

O'Neill LAJ (2002) Signal transduction pathways activated by the IL-1 receptor/toll-like receptor superfamily. Curr Top Microbiol Immunol 270:47–61

Ochiai H, Inoue Y, Takeya M et al (2005) Genome sequence of *Xanthomonas oryzae* pv. *oryzae* suggests contribution of large numbers of effector genes and insertion sequences to its race diversity. Jpn Agric Res Q 39:275–287. https://doi.org/10.6090/jarq.39.275

Oh C-S, Kim JF, Beer SV (2005) The Hrp pathogenicity island of *Erwinia amylovora* and identification of three novel genes required for systemic infection. Mol Plant Pathol 6:125–138. https://doi.org/10.1111/j.1364-3703.2005.00269.x

Orth K (1999) Inhibition of the mitogen-activated protein kinase kinase superfamily by a *Yersinia* effector. Science (80) 285:1920–1923. https://doi.org/10.1126/science.285.5435.1920

Orth K (2000) Disruption of signaling by *Yersinia* effector YopJ, a ubiquitin-like protein protease. Science (80) 290:1594–1597. https://doi.org/10.1126/science.290.5496.1594

Paquette N, Conlon J, Sweet C et al (2012) Serine/threonine acetylation of TGF -activated kinase (TAK1) by *Yersinia pestis* YopJ inhibits innate immune signaling. Proc Natl Acad Sci 109:12710–12715. https://doi.org/10.1073/pnas.1008203109

Poueymiro M, Genin S (2009) Secreted proteins from *Ralstonia solanacearum*: a hundred tricks to kill a plant. Curr Opin Microbiol 12:44–52

Quezada CM, Hicks SW, Galan JE (2009) A family of *Salmonella* virulence factors functions as a distinct class of autoregulated E3 ubiquitin ligases. Proc Natl Acad Sci U S A 106(12):4864–4869. https://doi.org/10.1073/pnas.0811058106

Ratner D, Orning MPA, Proulx MK et al (2016a) The *Yersinia pestis* effector YopM inhibits pyrin inflammasome activation. PLoS Pathog 12. https://doi.org/10.1371/journal.ppat.1006035

Ratner D, Orning MPA, Starheim KK et al (2016b) Manipulation of interleukin-1β and interleukin-18 production by *Yersinia pestis* effectors YopJ and YopM and redundant impact on virulence. J Biol Chem 291:9894–9905. https://doi.org/10.1074/jbc.M115.697698

Raymond B, Young JC, Pallett M et al (2013) Subversion of trafficking, apoptosis, and innate immunity by type III secretion system effectors. Trends Microbiol 21:430–441. https://doi.org/10.1016/j.tim.2013.06.008

Read AC, Rinaldi FC, Hutin M et al (2016) Suppression of Xo1-mediated disease resistance in rice by a truncated, non-DNA-binding TAL effector of *Xanthomonas oryzae*. Front Plant Sci 7:1–14. https://doi.org/10.3389/fpls.2016.01516

Roden J, Eardley L, Hotson A et al (2004) Characterization of the *Xanthomonas* AvrXv4 effector, a SUMO protease translocated into plant cells. Mol Plant-Microbe Interact 17:633–643. https://doi.org/10.1094/MPMI.2004.17.6.633

Rodrigues JA, López-Baena FJ, Ollero FJ et al (2007) NopM and NopD are rhizobial nodulation outer proteins: identification using LC-MALDI and LC-ESI with a monolithic capillary column. J Proteome Res 6:1029–1037. https://doi.org/10.1021/PR060519F

Rohde JR, Breitkreutz A, Chenal A et al (2007) Type III secretion effectors of the IpaH family are E3 ubiquitin ligases. Cell Host Microbe 1:77–83. https://doi.org/10.1016/j.chom.2007.02.002

Ronald PC, Beutler B (2010) Plant and animal sensors of conserved microbial signatures. Science (80) 330:1061–1064. https://doi.org/10.1126/science.1189468

Rosadini CV, Kagan JC (2015) Microbial strategies for antagonizing Toll-like-receptor signal transduction. Curr Opin Immunol 32:61–70. https://doi.org/10.1016/j.coi.2014.12.011

Ruckdeschel K, Richter K, Mannel O, Heesemann J (2001) Arginine-143 of *Yersinia enterocolitica* YopP crucially determines isotype-related NF-κB suppression and apoptosis induction in macrophages. Infect Immun 69:7652–7662. https://doi.org/10.1128/IAI.69.12.7652-7662.2001

Ruh M, Briand M, Bonneau S et al (2017) *Xanthomonas* adaptation to common bean is associated with horizontal transfers of genes encoding TAL effectors. BMC Genom 18:670. https://doi.org/10.1186/s12864-017-4087-6

Ryan RP, Vorhölter F-J, Potnis N et al (2011) Pathogenomics of *Xanthomonas*: understanding bacterium–plant interactions. Nat Rev Microbiol 9:344–355. https://doi.org/10.1038/nrmicro2558

Saenz HL, Engel P, Stoeckli MC et al (2007) Genomic analysis of *Bartonella* identifies type IV secretion systems as host adaptability factors. Nat Genet 39:1469–1476. https://doi.org/10.1038/ng.2007.38

Salzberg SL, Sommer DD, Schatz MC et al (2008) Genome sequence and rapid evolution of the rice pathogen *Xanthomonas oryzae* pv. *oryzae* PXO99A. BMC Genom 9:1–16. https://doi.org/10.1186/1471-2164-9-204

Sarris PF, Duxbury Z, Huh SU et al (2015) A plant immune receptor detects pathogen effectors that target WRKY transcription factors. Cell 161:1089–1100. https://doi.org/10.1016/j.cell.2015.04.024

Schandry N, de Lange O, Prior P, Lahaye T (2016) TALE-like effectors are an ancestral feature of the *Ralstonia solanacearum* species complex and converge in DNA targeting specificity. Front Plant Sci 7:1–16. https://doi.org/10.3389/fpls.2016.01225

Schandry N, Jacobs JM, Szurek B, Perez-Quintero AL (2018) A cautionary TALE: how plant breeding may have favoured expanded TALE repertoires in *Xanthomonas*. Mol Plant Pathol 19:1297–1301. https://doi.org/10.1111/mpp.12670

Scott NE, Hartland EL (2017) Post-translational mechanisms of host subversion by bacterial effectors. Trends Mol Med 23:1088–1102. https://doi.org/10.1016/j.molmed.2017.10.003

Siamer S, Dehio C (2015) New insights into the role of *Bartonella* effector proteins in pathogenesis. Curr Opin Microbiol 23:80–85. https://doi.org/10.1016/j.mib.2014.11.007

Singer AU, Rohde JR, Lam R et al (2008) Structure of the *Shigella* T3SS effector IpaH defines a new class of E3 ubiquitin ligases. Nat Struct Mol Biol 15:1293–1301. https://doi.org/10.1038/nsmb.1511

Streubel J, Pesce C, Hutin M et al (2013) Five phylogenetically close rice SWEET genes confer TAL effector-mediated susceptibility to *Xanthomonas oryzae* pv. *oryzae*. New Phytol 200:808–819. https://doi.org/10.1111/nph.12411

Stuart LM, Paquette N, Boyer L (2013) Effector-triggered versus pattern-triggered immunity: how animals sense pathogens. Nat Rev Immunol 13:199–206. https://doi.org/10.1038/nri3398

Sundin GW, Mayfield CT, Zhao Y et al (2004) Complete nucleotide sequence and analysis of pPSR1 (72,601 bp), a pPT23A-family plasmid from *Pseudomonas syringae* pv. *syringae* A2. Mol Genet Genomics 270:462–476. https://doi.org/10.1007/s00438-003-0945-9

Suzuki S, Mimuro H, Kim M et al (2014) *Shigella* IpaH7.8 E3 ubiquitin ligase targets glomulin and activates inflammasomes to demolish macrophages. Proc Natl Acad Sci 111:E4254–E4263. https://doi.org/10.1073/pnas.1324021111

Tasset C, Bernoux M, Jauneau A et al (2010) Autoacetylation of the *Ralstonia solanacearum* effector PopP2 targets a lysine residue essential for RRS1-R-mediated immunity in *Arabidopsis*. PLoS Pathog 6. https://doi.org/10.1371/journal.ppat.1001202

Thakur S, Guttman DS (2016) A De-Novo Genome Analysis Pipeline (DeNoGAP) for large-scale comparative prokaryotic genomics studies. BMC Bioinformatics 17:260. https://doi.org/10.1186/s12859-016-1142-2

Triplett LR, Cohen SP, Heffelfinger C et al (2016) A resistance locus in the American heirloom rice variety Carolina Gold Select is triggered by TAL effectors with diverse predicted targets and is effective against African strains of *Xanthomonas oryzae* pv. *oryzicola*. Plant J 87:472–483. https://doi.org/10.1111/tpj.13212

Trosky JE, Li Y, Mukherjee S et al (2007) VopA inhibits ATP binding by acetylating the catalytic loop of MAPK kinases. J Biol Chem 282:34299–34305. https://doi.org/10.1074/jbc.M706970200

Trülzsch K, Sporleder T, Igwe EI et al (2004) Contribution of the major secreted Yops of *Yersinia enterocolitica* O: 8 to pathogenicity in the mouse infection model. Infect Immun 72:5227–5234. https://doi.org/10.1128/IAI.72.9.5227-5234.2004

Üstün S, König P, Guttman DS, Börnke F (2014) HopZ4 from *Pseudomonas syringae*, a member of the HopZ type III effector family from the YopJ superfamily, inhibits the proteasome in plants. Mol Plant Microbe Interact 27:611–623. https://doi.org/10.1094/MPMI-12-13-0363-R

Vinatzer BA, Teitzel GM, Lee MW et al (2006) The type III effector repertoire of *Pseudomonas syringae* pv. *syringae* B728a and its role in survival and disease on host and non-host plants. Mol Microbiol 62:26–44. https://doi.org/10.1111/j.1365-2958.2006.05350.x

Wang F, Jiang Z, Li Y et al (2013) *Shigella flexneri* T3SS effector IpaH4.5 modulates the host inflammatory response via interaction with NF-κB p65 protein. Cell Microbiol 15:474–485. https://doi.org/10.1111/cmi.12052

Wei C, Wang Y, Du Z et al (2016) The *Yersinia* Type III secretion effector YopM Is an E3 ubiquitin ligase that induced necrotic cell death by targeting NLRP3. Nat Cell Death Dis 7: e2519. https://doi.org/10.1038/cddis.2016.413

Whalen MC, Wang JF, Carland FM et al (1993) Avirulence gene avrRxv from *Xanthomonas campestris* pv. *vesicatoria* specifies resistance on tomato line Hawaii 7998. Mol Plant Microbe Interact 6:616–627

White FF, Potnis N, Jones JB, Koebnik R (2009) The type III effectors of *Xanthomonas*. Mol Plant Pathol 10:749–766. https://doi.org/10.1111/j.1364-3703.2009.00590.x

Wilkins KE, Booher NJ, Wang L, Bogdanove AJ (2015) TAL effectors and activation of predicted host targets distinguish Asian from African strains of the rice pathogen *Xanthomonas oryzae* pv. *oryzicola* while strict conservation suggests universal importance of five TAL effectors. Front Plant Sci 6. https://doi.org/10.3389/fpls.2015.00536

Xin DW, Liao S, Xie ZP et al (2012) Functional analysis of NopM, a novel E3 ubiquitin ligase (NEL) domain effector of *Rhizobium* sp. strain NGR234. PLoS Pathog 8. https://doi.org/10.1371/journal.ppat.1002707

Yang B, Sugio A, White FF (2005) Avoidance of host recognition by alterations in the repetitive and C-terminal regions of AvrXa7, a type III effector of *Xanthomonas oryzae* pv. *oryzae*. Mol Plant-Microbe Interact 18:142–149. https://doi.org/10.1094/MPMI-18-0142

Yang B, Sugio A, White FF (2006) *Os8N3* is a host disease-susceptibility gene for bacterial blight of rice. Proc Natl Acad Sci U S A 103:10503–10508. https://doi.org/10.1073/pnas.0604088103

Yau R, Rape M (2016) The increasing complexity of the ubiquitin code. Nat Cell Biol 18:579–586. https://doi.org/10.1038/ncb3358

Zhang Y, Higashide WM, McCormick BA et al (2006) The inflammation-associated *Salmonella* SopA is a HECT-like E3 ubiquitin ligase. Mol Microbiol 62:786–793. https://doi.org/10.1111/j.1365-2958.2006.05407.x

Zhang Z-M, Ma K-W, Yuan S et al (2016) Structure of a pathogen effector reveals the enzymatic mechanism of a novel acetyltransferase family. Nat Struct Mol Biol 23:847–852. https://doi.org/10.1038/nsmb.3279

Zhang Z, Ma K-W, Gao L et al (2017) Mechanism of host substrate acetylation by a YopJ family effector. Nat Plants 3:17115. https://doi.org/10.1038/nplants.2017.115

Zheng Y, Lilo S, Brodsky IE et al (2011) A *Yersinia* effector with enhanced inhibitory activity on the NF-κB pathway activates the NLRP3/ASC/Caspase-1 inflammasome in macrophages. PLoS Pathog 7. https://doi.org/10.1371/journal.ppat.1002026

Zheng Z, Wei C, Guan K et al (2016) Bacterial E3 ubiquitin ligase IpaH4.5 of *Shigella flexneri* targets TBK1 to dampen the host antibacterial response. J Immunol 196:1199–1208. https://doi.org/10.4049/jimmunol.1501045

Zhou H, Morgan RL, Guttman DS, Ma W (2009) Allelic variants of the *Pseudomonas syringae* type III effector HopZ1 are differentially recognized by plant resistance systems. Mol Plant Microbe Interact 22:176–189. https://doi.org/10.1094/MPMI-22-2-0176

Zhu Y, Li H, Hu L et al (2008) Structure of a *Shigella* effector reveals a new class of ubiquitin ligases. Nat Struct Mol Biol 15:1302–1308. https://doi.org/10.1038/nsmb.1517

Zouhir S, Cordero-Alba M, Cardenal-Muñoz E et al (2014) The structure of the Slrp–Trx1 complex sheds light on the autoinhibition mechanism of the type III secretion system effectors of the NEL family. Biochem J 144:135–144. https://doi.org/10.1042/BJ20140587

The manufacturer's authorised representative in the EU is Springer Nature Customer Service Centre GmbH, Europaplatz 3, 69115 Heidelberg, Germany. If you have any concerns regarding our products, please contact ProductSafety@springernature.com

Printed and bound by CPI Group (UK) Ltd, Croydon, CR0 4YY
30/09/2026
02225507-0008